MOLECULAR MARKERS OF BRAIN TUMOR CELLS

Molecular Markers of Brain Tumor Cells

Implications for Diagnosis, Prognosis and Anti-Neoplastic Biological Therapy

by

Bela Bodey

Department of Pathology and Laboratory Medicine,
Keck School of Medicine, University of Southern California, Los Angeles
& Childrens Center for Cancer and Blood Diseases,
Childrens Hospital Los Angeles, Los Angeles, CA, U.S.A.

Stuart E. Siegel

Department of Pediatrics,
Keck School of Medicine, University of Southern California, Los Angeles
& Childrens Center for Cancer and Blood Diseases,
Childrens Hospital Los Angeles, Los Angeles, CA, U.S.A.

and

Hans E. Kaiser

Department of Pathology,
School of Medicine, University of Maryland, Baltimore, MD, U.S.A.
& Department of Clinical Pathology,
University of Vienna, Vienna, Austria

KLUWER ACADEMIC PUBLISHERS
DORDRECHT / BOSTON / LONDON

A C.I.P. Catalogue record for this book is available from the Library of Congress

ISBN 978-90-481-6721-0 (PB)
ISBN 978-1-4020-2804-5 (e-book)

Published by Kluwer Academic Publishers,
P.O. Box 17, 3300 AA Dordrecht, The Netherlands.

Sold and distributed in North, Central and South America
by Kluwer Academic Publishers,
101 Philip Drive, Norwell, MA 02061, U.S.A.

In all other countries, sold and distributed
by Kluwer Academic Publishers,
P.O. Box 322, 3300 AH Dordrecht, The Netherlands.

Cover image:
Childhood pilocytic astrocytoma. Antigen detection employing anti-CD8 mouse anti-human monoclonal antibody. Tumor infiltrating cytotoxic lymphocytes are present throughout the astrocytic tumor tissue and are in close contact with the neoplastic cells. Frozen tissue section; magnification: 200x

Printed on acid-free paper

TABLE OF CONTENTS

CONTRIBUTORS

Professor Bela Bodey, M.D., D.Sc.
Department of Pathology and Laboratory Medicine, Keck School of Medicine, University of Southern California, Los Angeles, & Childrens Center for Cancer and Blood Diseases, Childrens Hospital Los Angeles, Los Angeles, CA, USA

Professor Stuart E. Siegel, M.D.
Department of Pediatrics, Keck School of Medicine, University of Southern California, Los Angeles, & Childrens Center for Cancer and Blood Diseases, Childrens Hospital Los Angeles, Los Angeles, CA, USA.

Professor Hans E. Kaiser, D.Sc.
Department of Pathology, School of Medicine, University of Maryland, Baltimore, MD, USA & Department of Clinical Pathology, University of Vienna, Vienna, Austria.

ACKNOWLEDGMENTS

We would like to dedicate this book to Dr. Bodey's wife, Dr. Victoria Psenko, who provided the ideal, conducive environment necessary for Dr. Bodey to pursue his research endeavors.

We would like to acknowledge both of Dr. Bodey's children, Vivian Bodey and Bela Bodey Jr., for their technical assistance through the years in the laboratory and for their assistance in preparing this book.

PREFACE

The last twenty years of brain tumor research has seen immunohistochemistry applied and develop from an experimental research technique to a nearly routine method of great importance in histopathology. The field of morphologic research in oncology has been revitalized and revolutionized by immunohistochemistry in that now functional aspects can be easily associated with morphological descriptions. It comes as no surprise that the scientific conferences of the past decade have generated great interest since immunohistochemistry has allowed researchers to development epoch-making discoveries in molecular oncology, practically delving into the molecular biologic aspects of cancerogenesis, cellular neoplastic transformation and the intimate mechanisms of neoplastic progression, and metastasis along with significant expansion of our knowledge concerning the processes that govern cell cycle, cell proliferation and differentiation, apoptosis, immune surveillance, angiogenesis and signal transduction control, without sacrificing the beauty of classical morphology.

The development of immunohistochemistry to its present place in research, diagnostics and therapy, of course, could not have been possible without the discovery of the methodology of monoclonal antibody production, one of the most important scientific discoveries of the twentieth century.

GROWTH FACTORS IN MAMMALIAN EMBRYOGENESIS AND NEOPLASTIC TRANSFORMATION

The elucidation of the molecular mechanisms underlying embryonic growth control is a key step in the attempt to understand embryonic development and the regulation of cell proliferation and its impairment in neoplastic transformation (1). The most extensive cellular proliferation and differentiation takes place during early ontogenesis, therefore it seems likely that growth factors have a major role at this time both in the regulation of cell proliferation and the process of immunophenotypic (IP) differentiation. The expression of growth factors and their receptors by neoplastically transformed cells is out of control; this can lead to unchecked and continuous cell division. Malignant cells may secrete some growth factors and simultaneously express their receptors (autocrine stimulation) (2). Neoplastic cells also express the receptors for the growth factors secreted by neighboring cells (paracrine stimulation). Murray and Kirschner (3) demonstrated the primitive nature of the embryonic division cycle (lack of G1 regulation), as compared to the more complex regulation in which growth factors act to modulate the growth and differentiation of somatic cells [proline directed protein phosphorylation- (4)].

An alternative to studying the embryo *in vivo* is to use *in vitro* experimental models, such as teratocarcinoma stem cells (EC-embryonal carcinoma cells) which share numerous biochemical, morphological and immunological properties with normal early pluripotent stem cells. Some growth factors are unique, appearing only during embryonic/fetal (ontogenetic) development, others especially when derived from adult tissues can be present more permanently (defined growth factors). The data accumulated from numerous studies has defined four growth factors involved in early embryogenesis: *1)* insulin-like growth factors (IGFs); *2)* epidermal growth factors (EGFs); *3)* transforming growth factors (TGFs) and *4)* platelet-derived growth factors (PDGFs) (5-8). PDGF also increases the production of oncogenes c-*myc* and c-*fos*. Epigenetic mechanisms appear to involve an interaction between mitogenic growth factors and factors which induce cell differentiation. Neuronal differentiation of PC12 cells was observed after microinjection of the *ras* oncogene protein (9) or after infection of these cells with *ras* containing retroviruses (10). The involvement of the *ras* protein in this process is further supported by the observation that microinjection of *ras*-specific antibodies inhibits the NGF-induced differentiation of PC12 cells (11). The mitogenic action of growth factors and the anti-proliferative effect of IFN may be independent of cell cycle events, as demonstrated in studies with vascular smooth muscle and endothelial cells. Growth factors such as TGF-β have also been implicated in allowing for the sustained growth of the neoplasm, as well as in inhibiting the anti-neoplastic immune response. It has recently been proposed that the cell membrane located receptors for peptide growth factor (PGF-R) can be regarded as specific targets for immunodetection and immunotherapy of human malignancies (12). PGF-Rs play a crucial role in the regulation of neoplastic cell proliferation and may behave as TAAs. PGF-Rs are often present in greater quantity on malignant cells and their cell surface expression is regulated by cytokines. Neoplastic cells can promote their own proliferation by secreting PGFs, which act in the paracrine and autocrine stimulation of the neoplasm mass (13, 14). PGF-R may, therefore, represent an ideal cellular target for at least two various immunotherapeutic approaches: 1) for conjugated or unconjugated MoABs and 2) for genetically engineered fusion proteins composed of PGF-R physiological ligands conjugated to genetically modified bacterial toxins. Other clinical studies have been performed describing the targeting of receptors of epidermal growth factor (EGF) and interleukin 2 (IL-2) on neoplastic cells.

The development of neoplastic cell-specific and targeted immunotherapies is of particular interest. It is great to see that clinical oncologists are now finally taking this approach, especially with all of its many unprobed possibilities. We predict that the next three decades will see the employment of individualized "cocktails" of conjugated antibody molecules, targeting multiple antigenic epitopes, as the main line of non-toxic and efficacious therapy of human neoplastic disease, especially in the treatment of residual and metastatic neoplasms.

REFERENCES

1. Jakobovits A: The expression of growth factors and growth factor receptors during mouse embryogenesis, in: *Oncogenes and Growth Control*, P. Kahn & T. Graf, eds., Springer Verlag, Berlin, pp. 9-17, 1986.
2. Sporn MB, Roberts AB: Autocrine growth factors and cancer. Nature (London) *313:* 745-747, 1985.
3. Murray AW, Kirschner MW: Cyclin synthesis drives the early embryonic cell cycle. Nature (London) *339:* 275-280, 1989.
4. Hall FL, Vulliet PR: Proline-directed protein phosphorylation. Current Opinion Cell Biol *3:* 176-184, 1991.
5. Gazit A, Igarishi H, Ciu IM: Expression of the normal human sis/PDGF-2 coding sequence induces cellular transformation. Cell *39:* 89-97, 1984.
6. Glick RP, Gettleman R, Patel K, Kirtikumar P, Lakshman R, Tsibris JCM: Insulin and insulin-like growth factor 1 in brain tumors. Binding and *in vitro* effects. Neurosurgery *24:* 791-797, 1989.
7. Fleming TP, Matsui T, Aaronson SA: Platelet-derived growth factor (PDGF) receptor activation in cell transformation and human malignancy. Exp Gerontol *27:* 523-532, 1992.
8. Henriksen R, Funa K, Wilander E, Backstrom T, Ridderheim M, Oberg K: Expression and prognostic significance of platelet-derived growth factor and its receptors in epithelial ovarian neoplasms. Cancer Res *53:* 4550-4554, 1993.
9. Bar-Sagi D, Feramisco J: Microinjection of the ras oncogene protein into PC12 cells induces morphological differentiation. Cell *42:* 841-848, 1985.
10. Noda M, Ko M, Ogura A, Liu DG, Amano T, Takano T, Ikawa Y: Sarcoma viruses carrying ras oncogenes induce differentiation-associated properties in a neural cell line. Nature (London) *318:* 73-75, 1985.
11. Hagag N, Halegoua S, Viola M: Inhibition of growth factor-induced differentiation of PC12 cells by microinjection of antibody to ras p21. Nature (London) *319:* 680-682, 1986.
12. Tagliaferri P, Caraglia M, Muraro R, Pinto A, Budillon A, Zagonel V, Bianco AR: Pharmacological modulation of peptide growth factor receptor expression on tumor cells as a basis for cancer therapy. Anti-Cancer Drugs *5:* 379-393, 1994.
13. Sporn MB, Todaro GJ: Autocrine secretion and malignant transformation of cells. N Engl J Med *308:* 878-880, 1980.
14. Sporn MB, Todaro GJ: Autocrine growth factors and cancer. Nature (London) *313:* 747-751, 1985.

I. MOLECULAR BIOLOGY OF TUMORS

Chapter 1

BRAIN TUMORS

1. INTRODUCTION

In the last three decades, major advances in molecular biology and subsequently in the biology of neoplastic diseases and fundamental genetic discoveries have improved our understanding of neoplastic cellular transformation and its full blown development into an advanced neoplastic progression (1). Cancer associated markers (CAMs) represent the biochemical or immunological counterparts of the morphology of tumors. The expression of immunocytochemically defined cancer associated markers is also related to the tissue of origin and is thus, not a random event.

During the past 25 years, the employment of antigenic epitope specific MoABs against oncofetal, neoplasm associated, cell lineage specific, endothelial, and cell proliferation related antigens in the diagnosis and biological assessment of prognosis in neoplastic disease gained increased importance. A sensitive direct correlation exists between the expression of certain molecules and the development of an invasive, highly malignant IP of neoplastic cells, allowing for the occurrence of neoplasm induced neoangiogenesis and metastasis.

2. MEDULLOBLASTOMA

Primary brain tumors remain the second most common type of solid neoplasms during childhood (younger than 15 years of age) and the posterior fossa is the most common region of the central nervous system (CNS) affected. Medulloblastomas are in fact the most common childhood brain tumor (2). The annual incidence of pediatric brain tumors appears to be on the rise caused partly by improvements of diagnostic neuroimaging and its increased availability (3). Despite significant increases in survival rate during the past decade, the great majority of pediatric patients with medulloblastomas (MEDs) or primitive neuroectodermal brain tumors (PNETs) still succumb to their disease. Advances have recently been made in the employment of chemotherapy for childhood brain tumors. Chemotherapy increases disease-free survival in high-risk MED/PNET patients and enables the reduction of radiation therapy in standard-risk patients (4). Radiation can be significantly delayed and neurotoxicity ameliorated in many infants using chemotherapy. Chemotherapy can cause reduction in size of low-grade glioma, optic glioma, and oligodendroglioma. High-grade glioma and ependymoma are relatively chemoresistant. A recent venture by scientists has been an attempt to assess the risk stratification in medulloblastomas. Gene expression profiling has been shown to predict medulloblastoma outcomes independent of clinical variables (5). In addition, Erb-B-2 expression and clinical risk factors haven been shown to together constitute a highly accurate disease risk stratification tool (6).

MEDs or PNETs represent embryonal tumors of ectodermal origin (7-12). Medulloblastomas may be derived from granule cells of the developing cerebellum. The cerebellar granule cell is the most numerous neuron in the nervous system and is the likely source of medulloblastomas (2). Leung and co-workers showed that Bmi1 is strongly expressed in proliferating cerebellar precursor cells in mice and humans (13). Using Bmi1-null mice, they demonstrated that Bmi1 plays a crucial role in clonal expansion of granule cell precursors both *in vivo* and *in vitro*. Deregulated proliferation of these progenitor cells, by activation of the sonic hedgehog (Shh) pathway, led to MED development. As such, they also linked overexpression of BMI1 and patched (PTCH), suggestive of SHH pathway activation, in a substantial fraction of primary human MEDs. BMI1 overexpression thus serves as an alternative or additive mechanism in the pathogenesis of MEDs. As we reported in one of our articles, this common primary, childhood, cerebellarly located malignancy was named MED by Bailey and Cushing (14) based on the brain developmental theory of Schaper (15), who described the presence of "apolar, indifferent cells in the external granular layer of cerebellum" and named them as "medulloblasts" or the common neural stem cells. Despite a number of morphological, histochemical, and ultrastructural (transmission electron microscopic-TEM and scanning electron microscopic-SEM), and *in vitro* observations, evidence for the real existence of the hypothetical "medulloblast" is still lacking (16). In the great majority of cases, three differentiated cell types are found in childhood MEDs: neurons, glia, and mesodermal structures (*i.e.* muscle cells). As we reported, because of the presence of multiple differentiated cell types, these tumors were named after a postulated cerebellar stem cell, the "medulloblast", which would give rise to the differentiated cells found in the tumors. A group of researchers at the Massachusetts Institute of Technology (MIT) described a cell line with the properties expected of the postulated medulloblast (17). The rat cerebellar cell line (named ST15A) expressed an intermediate filament, nestin, that is characteristic of neuroepithelial stem cells. ST15A cells can differentiate, gaining either neuronal or glial properties. At the same time several clonal cells can also differentiate into muscle cells. These *in vitro* results suggest that a single neuroectodermal cell can give rise to the different cell types found in MED. Immunocytochemical observations also demonstrated the expression of nestin in human MED tissue and in a MED-derived cell line. Both the properties of the ST15A cell line and the expression of nestin in MED support a neuroectodermal stem cell origin for this childhood tumor. Hart and Earle (18) introduced the concept of PNET, to characterize brain tumors containing 95% or more "small and undifferentiated" cells.

Children with tumors expressing high levels of the neurotrophin-3 receptor, TrkC, have a more favorable outcome (19). During development, TrkC is present in the most mature granule cells. Favorable MEDs may originate from more highly differentiated granule cells. Morphologically MEDs are hypercellular and their microenvironment can be heterogeneous, containing areas of mixed cell populations, neuronal, glial, and/or mesodermal structures. These tumors have the tendency to seed along the cerebrospinal axis and invade the cerebrospinal cavity.

Sixty-three patients with cerebellar MED, treated between 1963 and 1992, were observed by Nishio and co-workers (20). 10 out of 63 patients have survived beyond the Collins' risk period. These included 6 males and 4 females who ranged in age from 6 months to 12 years at the time of diagnosis. A total removal of the tumor was achieved in 4 patients, while there was subtotal removal in 3, and partial removal in 3. Histologically, 6 tumors were classified as classical type MED and 4 were diagnosed as being a desmoplastic type. Postoperatively, 9 patients received craniospinal radiation therapy, and one received local radiation to the primary site. During the follow-up period of 3.9-25.4 years, 5 patients have been in continuous remission for from 14.2 to 25.4 years and are leading normal lives, 2 have survived for 18.1 and 18.5 years with mild to moderate neurological deficits, while the remaining 3 died after the Collins' risk period. Two out of these last 3 patients were under the age of one year at the time of onset, while the remaining one died after a second recurrence. The observations led to the conclusion that that careful follow-up is needed for all long-term survivors even after the Collins' risk period, especially for those who were under the age of one year at onset and who failed in the initial treatments.

In another study, 27 primary MEDs were analyzed using comparative genomic hybridization and a novel statistical approach to evaluate chromosomal regions for significant gain or loss of genomic DNA (21). An array of nonrandom changes was found in most samples. Two discrete regions of high-level DNA amplification of chromosome bands 5p15.3 and 11q22.3 were observed in 3 of 27 tumors. Nonrandom genomic losses were most frequent in regions on chromosomes 10q (41% of samples), 11 (41%), 16q (37%), 17p (37%), and 8p (33%). Regions of DNA gain most often involved chromosomes 17q (48%) and 7 (44%). These findings suggest a greater degree of genomic imbalance in MED than has been recognized previously and highlight chromosomal loci likely to contain oncogenes or tumor suppressor genes that may contribute to the molecular pathogenesis of childhood MED.

Appropriate prognostic indicators for MEDs in children are of the utmost importance. Immunocytochemical study assessed the prognostic values of N-myc expression in MEDs (22). Nineteen cases of MED or supratentorial PNET (sPNET) were observed for N-myc expression. Sixteen of the observed cases were N-myc-positive, and only three did not express N-myc at detectable levels. N-myc-positive patients had a tendency towards poor disease outcome (p=0.1125). Extended immunohistochemical observations revealed that N-myc-negative tumors were more differentiated towards glial cell lineage than N-myc-positive ones. N-myc-negative and GFAP-positive patients (n=2) tended to survive longer than N-myc-positive and GFAP-negative patients (n=13). The authors concluded that in MED and sPNET patients, N-myc expression may be an appropriate indicator of poor prognosis and more embryonic cell differentiation.

At the beginning of the twenty-first century, no therapeutic regimen can reliably cure PNET. A young age at diagnosis and an advanced stage of the tumor based on the grading of Chang and co-workers should be associated with an unfavorable clinical outcome. The prognostic importance of cell differentiation was addressed with the use of Rorke's classification for PNETs (23, 24), separated into five groups: 1) glial, 2) neuronal, 3) ependymal, 4) multipotential, and 5) without cell differentiation. The cellular classification of brain tumors is based on both histopathological cell and tissue characteristics and location in the brain. Cellular undifferentiated neuroectodermal tumors of the cerebellum have historically been referred to as MEDs, while tumors of identical histology in the pineal region are diagnosed as pineoblastomas. Pineoblastoma and MED are very similar but not identical. The nomenclature of pediatric brain tumors is controversial and potentially confusing (23-35).

The current World Health Organization (WHO) classification groups together both infratentorial neoplasms (MEDs) and their supratentorial counterparts as primitive neuroectodermal tumors (PNETs), implying a common origin. A number of neuropathologists advocate abandoning the traditional morphologically based classifications such as MED in favor of a terminology that relies more extensively on the cell phenotypic characteristics of the tumor. In such a system, MED is referred to as primitive neuroectodermal tumor (PNET) and then subdivided on the basis of cellular differentiation. Nomenclature of neoplasms containing poorly differentiated cells or densely cellular neuroepithelial tumors was simplified to reflect the current state of knowledge of neuroembryology and neuro-oncology, although the authors recognized that such a proposal would likely perpetuate the long-standing and continuing controversy relative to the nature and origin of these neoplasms (23, 24).

The most recent World Health Organization classification of brain tumors still maintains the medical term MED for posterior fossa located, undifferentiated (or dedifferentiated) neoplasms. It also maintains separate categories for cerebral PNETs and for pineal small round cell tumors (pineoblastomas). The pathologic classification of pediatric brain tumors is a specialized area that is undergoing constant evolution (35).

3. GLIAL TUMORS

> "This peculiarity of the membrane, namely, that it becomes continuous with the interstitial matter, the real cement, which binds the nervous elements together and that in all its properties it constitutes a tissue different from the other forms of connective tissue, has induced me to give it a new name, that of neuro-glia (nerve cement)."
>
> \- Rudolf Virchow, April 3, 1858 (36)

Malignant childhood ASTRs represent tumors appearing within the neuro-glial or macroglial central nervous system (CNS) (37); they account for over 50% of all intracranial tumors (38-40). Astrocytomas can grow anywhere in the CNS, but in children they usually occur in the brain stem, the cerebrum, or the cerebellum. To be more accurate ASTRs account for about 68 percent of the primary brain tumors occurring in children younger than age 20 (41). The most common brain tumors develop from glial cell precursors (astrocytes, oligodendrocytes, ependymocytes). Glial tumors [mainly astrocytomas (ASTRs)], especially glioblastomas (GBM), are characterized by hypercellularity, pleomorphism, a high number of cell mitoses, CIP heterogeneity, various grades of necrosis, and multiple endothelial cell proliferations related to newly generated, tumor-related capillaries (see Table II). Furthermore, glial tumors are characterized by high grade local invasiveness and a relatively low metastatic tendency. Cairncross (42) interpreted the histogenesis of ASTRs in the light of parallel concepts emerging from investigations in myeloproliferative disorders (43, 44). According to the stem cell hypothesis, ASTRs originate from a common pluripotent, neuroectodermally derived precursor cell, whose progeny retain the ability to differentiate and do so along astrocytic lines (45). In the last decade it was reported that mutations of P53 gene are present in more than two-thirds of secondary GBMs, but rarely occurs in primary GBMs, suggesting the presence of divergent genetic pathways in their histogenesis (46). The majority of malignant glial tumors are incurable with the current classical therapeutic modalities, including surgical resection, radiotherapy, and chemotherapy (47). This may well be the direct result of the biological variability of these tumors, *e.g.* multiple stem cell lines, intrinsic and acquired multidrug resistance.

The molecular pathogenesis of human ASTRs has been intensely studied during the past few years. Genetic alterations of chromosome 17p are associated with pilocytic ASTRs (World Health Organization (WHO) grade I), mutations at 17p and 19q are common in AAs (WHO grade III) and abnormal chromosomes 17p, 19q and 10 are associated with the most malignant GBM (WHO grade IV) (48). It is well established that low-grade ASTRs have an intrinsic tendency for progressive IP dedifferentiation toward higher-grade, more malignant ASTRs.

The presence of gemistocytes in low-grade ASTRs is regarded as a sign of poor prognosis because the majority of gemistocytic ASTRs rapidly progress to AA or GBM (49). To elucidate the role of gemistocytes in ASTR progression, Watanabe and co-workers assessed the fraction of neoplastic gemistocytes, bcl-2 expression, p53 mutations, p53 immunoreactivity (PAb 1801), and proliferative activity (MIB-1) in 40 low-grade astrocytomas (grade II) with histologically proven progression to AA (grade III) or GBM (grade IV). Astrocytoma progression took significantly less time in patients with a low-grade astrocytoma containing more than 5% gemistocytes (35 months) than in those with lesions containing less than 5% gemistocytes (64 months; p=0.038). All 11 astrocytomas with more than 5% gemistocytes contained a p53 mutation, whereas the incidence of p53 mutations in ASTRs with less than 5% gemistocytes was 61% (p=0.017). In low-grade ASTRs the p53 labeling index of gemistocytes (7.4%) was significantly higher than in all neoplastic cells (3.2%, p=0.0014). Gemistocytes also showed a significantly higher bcl-2 expression than all neoplastic cells, with a mean bcl-2 labeling index of 15.6% vs. 2.7% in low-grade ASTRs (p=0.0004),

20.9% vs. 3.0% in AA (p=0.002), and 30.2% vs. 5.2% in GBMs (p=0.0002). In contrast, in gemistocytes a significantly lower proliferating activity was identified than the mean of all neoplastic cells, with a mean MIB-1 labeling index of 0.5% vs. 2.6% in low-grade ASTRs, 1.5% vs. 11.6% in AA, and 1.7% vs. 16.6% in GBMs (p<0.0001). These data show that low-grade ASTRs with a significant fraction of gemistocytes progress more rapidly and typically carry a p53 mutation. The vast majority of gemistocytes are, however, in a nonproliferative state (G0 phase), which suggests terminal differentiation. Their accumulation within ASTRs may be due to bcl-2 mediated escape from apoptosis.

The prognosis for children with high grade gliomas remains somewhat unpredictable because histologic features alone provide an imperfect assessment of the biologic behavior of a given lesion (50). Whereas some patients experience prolonged disease control after surgery and adjuvant therapy, others with lesions that appear comparable exhibit rapid disease progression and death.

Table 1-1. Histological Types of Childhood Brain Tumors and Their Relative Incidence (after Vats, 1997).

TUMOR	INCIDENCE (%)
Supratentorial	40
Gliomas	21.5
Ependymomas	2.7
Craniopharyngioma	12.4
Infratentorial	54.9
Medulloblastoma (MED)	30.4
Cerebellar astrocytoma (ASTR)	7.3
Ependymoma	53
Brain stem glioma	11.8
Others	4.4

REFERENCES

1. Mischel PS, and Cloughesy TF: Targeted molecular therapy of GBM. Brain Pathol *13:* 52-61, 2003.
2. Jensen P, Smeyne R, Goldwitz D: Analysis of cerebella development in math1 null embryos and chimeras. J Neurosci *24(9):* 2202-2211, 2004.
3. Kun LE: Brain tumors. Challenges and directions. Pediatr Clin North Am. *44:* 907-917, 1997.
4. Kedar A: Chemotherapy for pediatric brain tumors. Semin Pediatr Neurol. *4:* 320-332, 1997.

5. Fernandez-Teijeiro A, Betensky RA, Sturla LM, Kim JY, Tamayo P, Pomeroy SL: Combining gene expression profiles and clinical parameters for risk stratification in medulloblastomas. J Clin Oncol *22:* 994-998, 2004.
6. Gajjar A, Hernan R, Kocak M, Fuller C, Lee Y, McKinnon PJ, Wallace D, Lau C, Chintagumpala M, Ashley DM, Kellie SJ, Kun L, Gilbertson RJ: Clinical, histopathologic, and molecular markers of prognosis: toward a new disease risk stratification system for medulloblastoma. J Clin Oncol *22:* 984-993, 2004.
7. Becker LE, Hinton D: Primitive neuroectodermal tumors of the central nervous system. Human Pathol *14:* 538-550, 1983.
8. Becker LE, Hinton D: Primitive neuroepithelial tumors of the central nervous system. *In:* Feingold M, ed. Pathology of Neoplasia in Children and Adolescents, Philadelphia: WB Saunders, pp. 397-418, 1986.
9. van den Berg SR, Herman MM, Rubinstein LJ: Embryonal central neuroepithelial tumors: current concepts and future challenges. Cancer Metast Rev *5:* 343-364, 1987.
10. Triche T: Primitive neuroectodermal tumors. Arch Pathol Lab Med *111:* 311-312, 1987.
11. Packer RJ: Childhood tumors. Curr Opin Pediatr. *9:* 551-557, 1997.
12. Rickert CH, Probst-Cousin S, Gullotta F: Primary intracranial neoplasms of infancy and early childhood. Childs Nerv Syst. *13:* 507-513, 1997.
13. Leung C, Lingbeek M, Shakhova O, Liu J, Tanger E, Saremaslani P, Van Lohuizen M, Marino S: Bmi1 is essential for cerebellar development and is overexpressed in human medulloblastomas. Nature *428:* 337-341, 2004.
14. Bailey P, Cushing H: Medulloblastoma cerebelli, a common type of midcerebellar glioma of childhood. Arch Neurol Psychiatry *14:* 192-224, 1925.
15. Schaper A: Einige kritische Bemerkungen zu Lugaro's Aufsatz: "Ueber die Histogenese der Körner der Kleinhirnrinde." Anat Anz *10:* 422-426, 1895.
16. Zeltzer PM, Bodey B, Marlin A, Kemshead J: Immunophenotype profile of childhood medulloblastomas and supratentorial primitive neuroectodermal tumors using sixteen monoclonal antibodies. Cancer *66:* 273-283, 1990.
17. Valtz NL, Hayes TE, Norregaard T, Liu SM, McKay RD: An embryonic origin for medulloblastoma. New Biol *3:* 364-371, 1991.
18. Hart MN, Earle KM: Primitive neuroectodermal tumors of the brain in children. Cancer *32:* 890-897, 1973.
19. Pomeroy SL, Sutton ME: Goumnerova LC, Segal RA: Neurotrophins in cerebellar granule cell development and medulloblastoma. J Neurooncol *35:* 347-352, 1997.
20. Nishio S, Morioka T, Takeshita I, Fukui M: Medulloblastoma: survival and late recurrence after the Collins' risk period. Neurosurg Rev *20:* 245-249, 1997.
21. Reardon DA, Michalkiewicz E, Boyett JM, Sublett JE, Entrekin RE, Ragsdale ST, Valentine MB, Behm FG, Li H, Heideman RL, Kun LE, Shapiro DN, Look AT: Extensive genomic abnormalities in childhood medulloblastoma by comparative genomic hybridization. Cancer Res *57:* 4042-4047, 1997.
22. Moriuchi S, Shimizu K, Miyao Y, Hayakawa T: An immunohistochemical analysis of medulloblastoma and PNET with emphasis on N-myc protein expression. Anticancer Res *16:* 2687-2692, 1996.
23. Rorke LB: The cerebellar medulloblastoma and its relationship to primitive neuroectodermal tumors. J Neuropathol Exp Neurology *42:* 1-15, 1983.
24. Rorke LB, Gilles FH, Davis RL, Becker LE: Revision of the World Health Organization classification of brain tumors for childhood brain tumors. Cancer *56:* 1869-1886, 1985.
25. Zülch KJ: Histologic classification of tumours of the central nervous system. International Histological Classification of Tumours, No. 21, World Health Organization, Geneva, 1979.

26. Zülch KJ: Principles of the new World Health Organization (WHO) classification of brain tumors. Neuroradiology *19:* 59-66, 1980.
27. Russel DS, Rubinstein LJ: Tumors of central neuroepithelial origin. In: Pathology of tumors of the Nervous System. Edward Arnold, London, 1989, p 83-247.
28. Gilles FH: Classification of childhood brain tumors. Cancer *56:* 1850-1857, 1985.
29. Dehner LP: Peripheral and central primitive neuroectodermal tumors: a nosologic concept seeking a consensus. Arch Pathol Lab Med *110:* 997-1005, 1986.
30. Kernohan JW, Sayre GP: Tumors of Central Nervous System. *Fas*cicle 35, Atlas of Tumor Pathology. Armed Forces Institute of Pathology, Washington, 1952, pp 17-129.
31. Becker LE: An appraisal of the World Health Organization classification of tumors of the central nervous system. Cancer *56:* 1858-1864, 1985.
32. Kleihues P, Burger PC, Scheithauer BW: Histological typing of tumors of the central nervous system. In: World Health Organisation International Histological Classification of Tumours. 2nd Edition, SpringerVerlag, Berlin-Heidelberg-New York, 1993.
33. Burger PC: Tumors of the central nervous system. Washington DC, 1994.
34. Szymas J: Histologic classification of central nervous system tumors by the World Health Organization. Pol J Pathol *45:* 81-91, 1994.
35. Burger PC: Revising the World Health Organization (WHO) Blue Book--'Histological typing of tumours of the central nervous system'. J Neurooncol *24:* 3-7, 1995.
36. Virchow R: Cellular pathology. New York: RM de Witt, 1860.
37. Williams BP, Abney ER, Raff MC: Macroglial cell development in embryonic rat brain: studies using monoclonal antibodies, flourescence-activated cell sorting and cell culture. Dev Biol *112:* 126-134, 1985.
38. Katsura S, Suzuki J, Wada T: Statistical study of brain tumours in the neurosurgical clinics in Japan. J Neurosurg *16:* 570-580, 1959.
39. von Deimling A, Louis DN, Wiestler OD: Molecular pathways in the formation of gliomas. Glia *15:* 328-338, 1995.
40. Kleihues P, Soylemezoglu F, Schauble B, Scheithauer BW, Burger PC: Histopathology, classification, and grading of gliomas. Glia *15:* 211-221, 1995.
41. Children's Cancer Research Fund: http://www.childrenscancer.org/research_archive5.jhtml
42. Cairncross JG: The biology of astrocytoma: lessons learned from chronic myelogenous leukemia-hypothesis. J Neuro-Oncol *5:* 99-104, 1987.
43. Greaves M, Janossy G, Francis G, Minowada J: Membrane phenotypes of human leukemic cells and leukemic cell lines: Clinical correlates and biological implications. *In:* Differentiation of normal and neoplastic hemopoietic cells (Clarkson B, Marks PL, Till J, eds). Vol 5, pp 823-841, Cold Spring Harbor Conferences on Cell Proliferation, Cold Spring Harbor Laboratory, New York, 1977.
44. Burns BF: Molecular genetic markers in lymphoproliferative disorders. Clin Biochem *22:* 33-39, 1989.
45. Bodey B, Zeltzer PM, Saldivar V, Kemshead J: Immunophenotyping of childhood astrocytomas with a library of monoclonal antibodies. Int J Cancer *45:* 1079-1087, 1990.
46. Watanabe K, Tachibana O, Sata K, Yonekawa Y, Kleihues P, Ohgaki H: Overexpression of the EGF receptor and p53 mutations are mutually exclusive in the evolution of primary and secondary glioblastomas. Brain Pathol *6:* 217-223, 1996.
47. Bullard DE, Gillespie Y, Mahaley MS, Bigner DD: Immunobiology of human gliomas. Semin Oncol *13:* 94-109, 1986.
48. Ohgaki K, Schäuble B, zur Hausen A, von Ammon K, Kleihues P: Genetic alterations associated with the evolution and progression of astrocytic brain tumors. Virchows Arch *427:* 113-118, 1995.

49. Watanabe K, Tachibana O, Yonekawa Y, Kleihues P, Ohgaki H: Role of gemistocytes in astrocytoma progression. Lab Invest *76:* 277-284, 1997.
50. Pollack IF, Campbell JW, Hamilton RL, Martinez AJ, Bozik ME: Proliferation index as a predictor of prognosis in malignant gliomas of childhood. Cancer *79:* 849-856, 1997.

Chapter 2

IMMUNOPHENOTYPIC CHARACTERIZATION OF INFILTRATING POLY- AND MONONUCLEAR CELLS IN CHILDHOOD BRAIN TUMORS

1. INTRODUCTION

Tumors of the central nervous system (CNS) are poorly responsive to the three classic modality of conventional anti-neoplastic therapy, including surgery, radiation, and chemotherapy. The median survival time of patients treated with surgery alone is 17 weeks which may be extended to 37 weeks through the combination of surgical resection of the tumor mass, radiotherapy, and chemotherapy (1). Infiltration of various CNS tumors by lymphocytes has been observed (2-9).

A fourth, recently developed therapeutic modality in malignant tumor therapy called "adoptive cellular immunotherapy (ACIT)" has been observed to be useful in advanced metastatic, often terminal neoplasm cases. In the majority of cases, the primary tumor and its metastases are infiltrated by a heterogeneous population of poly- and mononuclear leukocytes, including the tumor-associated or tumor-specific antigen-directed cytotoxic T lymphocyte (CTL) clone of the tumor infiltrating lymphocytes (TIL). These "*killer*" cells represent the host's main immunological effector cells and are MHC class I restricted and specifically lyse tumor cell targets. However, the physiologic function of these TILs has yet to be completely understood since they may represent the host's tumor-targeted cellular immune response or simply a cell clone component of a nonspecific inflammatory infiltrate. Immunotherapy has already been employed in various human cancers and can also be employed in brain tumor cases because tumor infiltrating leukocytes have been observed within the lesions.

The scientific aim of this immunocytochemical study was to characterize the cell surface immunophenotype (IP) of these tumor infiltrating poly- and mononuclear leukocytes with a well characterized library of MoABs directed against cell membrane localized, leukocyte differentiation antigens.

1.1 Results

We observed the expression of various cell membrane located leukocyte cell line differentiation antigens in the leukocyte infiltrates of 76 primary childhood brain tumors, including 34 medulloblastomas/PNETs and 42 astrocytomas. Leu 2/a antigen expression was demonstrated on 58/76 childhood brain tumors establishing the presence of the $CD8^+$, MHC class I restricted cytotoxic T lymphocytes (CTL). These killer cells usually represented 1-10% of all the cells in the tumor frozen section (+), but in some instances constituted 30-44% of all cells (++). $CD4^+$, MHC class II restricted, helper T lymphocytes were present in 65/76 brain tumors and constituted 1-10% of all cells (+). 74/76 childhood brain tumors were infiltrated by macrophages (Leu $M5^+$ cells), and these effectors represented 1-10% of all cells (+) in the tumor frozen section. Of the 76 primary childhood brain tumors observed, 76/76 expressed leukocyte common antigen (LCA), establishing the presence of infiltrating leukocytes. 76/76 pediatric brain tumors also expressed HLA-A,-B,-C and HLA-DR thus demonstrating an MHC class I restriction of the neoplastically transformed cell population as well as further illustrating the presence of subsets and clones of immunological effector cells within the tumor. MoAB UJ 308 detected premyelocytes and mature granulocytes, with unknown functional significance, in 60/76 childhood brain tumors. Natural killer (NK) cells were not defined within any of the tumors we observed.

Solid human tumors are frequently characterized with a markedly heterogeneous poly- and mononuclear cell infiltrate containing macrophages, granulocytes, various subpopulations of T lymphocytes and in some cases such rare cell populations as antibody producing plasma cells and mast cells (10-13). This type of infiltration may vary from florid to none at all, and the phenomenon rarely follows a consistent or predictable pattern.

According to the literature, in our systematic study, we observed some kind of infiltrating leukocytes in 100% (76/76) of the childhood brain tumors cases we examined. We demonstrated the presence of various infiltrating leukocytes on quick-frozen tissue sections in an *in situ* and *ex vivo* manner, thus allowing for a completely accurate observation of the components of the infiltrates as they are in the realm of the tumor (unlike in flow cytometric analyses where these cells are observed in a culture system and are exposed to a different microenvironment of tissue culture media, bringing about IP alterations (14,15) and a misleading dominance of $CD4^+$ and $CD8^+$ T lymphocytes).

Although immunocytochemical techniques have been previously applied to the study of various primary intracranial tumors, ours is the first report as far as we can determine that not only targeted the two lymphocyte subclasses (cytotoxic and helper), NK cells and possibly macrophages, but also other leukocytes comprising the host's nonspecific immune response such as granulocytes and the MHC restriction of all cells within the brain tumors. Our observations of cell-surface markers present on the various cells comprising the leukocytic infiltrate further clarified our ideas of mechanisms of homing immunological effector cells to the site of the tumor and CTL immunization against various cells among the heterogeneous tumor cell population.

We observed an extremely heterogeneous population of infiltrating leukocytes ranging from the already well investigated $CD8^+$ cytotoxic T lymphocytes and $CD4^+$ helper T lymphocytes to the intriguing presence of premyelocytes and mature granulocytes. We did not observe a predominant presence of neither cytotoxic nor helper T cells, but rather these two types of cells simply represented a small cell clone component of the whole infiltrate. Our observation of no NK cell infiltration at the site of the brain tumors is consistent with observations of rare to slight presence of NK cells in various other brain neoplasms (9, 16). NK cell activity can thus be explained as a nonspecific initial wave of the complete anti-tumor response of the immune system which then gives way to the sustained activity of macrophages, T lymphocytes, and other leukocytic components of the inflammatory infiltrate.

Since 1986 when TILs were first identified as an "anti-tumor" host's cellular response in their functional role by antigen specific (TAA directed) lysis of neoplastic cells (17), many attempts at increasing their efficacy in tumor eradication have been conducted. Lysis of tumor cells is accomplished by a subpopulation of T lymphocytes in the TIL: the $CD8^+$, cytotoxic, MHC Class I restricted T lymphocytes (CTL). In our observation, we established the presence of these killer cells in 58/76 brain tumors. CTL are tumor-targeted, MHC class I restricted, cytolytic cells which also employ specific T cell receptors to mediate their specific anti-neoplastic activities (15, 18-21).

Tumors effectively evade this antigen-specific immune response by down-regulating or losing their cell-surface MHC class I molecules (22). Thus CTL are not reactive with these neoplastically transformed cells. Another very important molecule in tumor-T cell interaction and antigen directed cytolysis is the intercellular adhesion molecule-1. Recently the immuno-inhibitory shedding of ICAM-1 (soluble ICAM-1) has been identified and has been shown that it inhibits the interaction between T cells and tumor cells. This molecule has been found to be shed by various human melanoma cell lines and binds to the ICAM-1 receptors on T-cells and thus leaves no place for the T cells to bind to the tumor cells' ICAM-1 molecules (23). This further establishes the critical nature of adhesion molecules in mediating intimate cell-cell interactions between various cells at the site of the tumor. It may also present a tumor defense mechanism against antigen-specific lysis by activated T cells.

Another molecule involved in tumor defense mechanisms is the transforming growth factor beta (TGF-β). TGF-β is a tumor derived (autocrine regulation) molecule which has been shown to suppress the *in vitro* generation of CTL from TIL of peripheral blood lymphocytes (24), and thus its *in vivo* secretion at the tumor site could be responsible for the intensive $CD8^+$, cytotoxic T lymphocyte suppression (25,26). Brain tumors have been shown to express various, predominantly low levels of MHC class I molecules as well as ICAM-1 and to produce TGF-β and these observations explain the inability to effectively isolate and expand infiltrating immunological effectors (CTL) from these neoplasias (26). This combination of factors probably represents a common tumor biological phenomenon and apparently renders the infiltrating cells incapable of proliferation and considerably lowers their immunological efficacy. This probably explains the inability of the infiltrating effector cells to overcome tumor progression.

How these cells get to the site of the tumor has been a query long etched in the minds of researchers and several possibilities have been proposed. The most basic is a tumor-specific accumulation of leukocytes brought on by tumor associated or tumor specific antigens. But this explanation has been abandoned for a more general possibility. Site-specific rather than tumor-specific accumulation of these infiltrates has been proposed through observations that leukocytes infiltrating various cutaneous neoplasms express a variety of adhesion molecules such as the integrin, aEb7 (homing receptor), which appears to be involved in the binding of intraepithelial lymphocytes to epithelial cells and the cutaneous lymphocyte-associated antigen, the T cell ligand for E-selectin, located on the surface of endothelial cells, which may mediate the homing of lymphocytes to sites of chronic cutaneous inflammation (27-29).

Our observation of the presence of mature granulocytes among the infiltrating leukocytes in 60/76 tumor cases also substantiates the theory that infiltration occurs due to inflammatory “signals” that cause a nonspecific immune response to occur.

Our ideas concerning the homing of these infiltrates to the site of the tumor have great repercussions in observations concerning the specific immune response to the tumor. Various problems are faced during this response. For instance, the release of many as yet unidentified chemical "radicals" that may well have bearing upon the efficacy of the tumor infiltrating immune effector cell population. During the passage of the neutrophil through the endothelium, numerous cell adhesion molecules, such as intercellular adhesion molecule-1 (ICAM-1) are utilized, creating close physical contact between the cells. Proteases and oxygen products (mainly H_2O_2) are released by neutrophils following their activation. The neutrophil-derived H_2O_2 readily diffuses into endothelial cells, triggering a chain reaction that ends in the production of the hydroxyl radical (HO), the toxic oxygen product responsible for endothelial cell injury.

Various inflammatory mediators, including tumor necrosis factor-α (TNF-α), a lymphokine released during the immune response, have been shown to be actively involved in this chain reaction when present at the cite of neutrophils passing through the endothelium (30). Such findings, coupled with our hypothesis of an inflammatory "signal" eliciting a nonspecific immune response to the site of the tumor mass, lead to questions regarding the interaction between the heterogeneous, tumor infiltrating mononuclear cell mass and the endothelium through which they must pass in order to reach the tumor cell mass. What is the nature of the chemicals released during the interaction between immune effector cells and endothelial cells and do these as yet unidentified chemicals reduce the efficacy of these effector cells against the tumor?

In view of our results, we propose that the leukocytic infiltrate, comprised of various immune effector cells including TIL, is first attracted to the tumor site as part of a nonspecific immune response to an inflammatory "signal" or necrotic transformations in the tumor mass. Furthermore, following these necrotic changes in the neoplastic cell mass, monocytes/macrophages, acting in their antigen presenting cell role, consume the necrotic cells and present previously "hidden," tumor-associated and tumor-specific antigens (TAAs and TSAs) to the other effector cells in the infiltrate, thus establishing immunization against the tumor mass as a process which occurs *in situ.* Thus, the infiltrating leukocytes are in a "developmental" stage when they first arrive at the tumor site, and this development begins with the initial nonspecific reaction and evolves into a specific reaction following *in situ* immunization.

Neoplastically transformed cells undergo constant microevolution. Natural selection of the most advantageous surface IP involves constant modulation of previous IPs. Progressive dedifferentiation characterizes all neoplastically transformed cells. During this process, numerous "novel" cell surface antigens appear, are modified and thus do present the host's immune system with some immunogenic elements. The leukocytic inflammatory infiltrate contains cells with diverse capabilities including neutrophils, macrophages and DCs as professional antigen presenting cells (APCs), as well as T lymphocytes. *In situ* activation of TAA specific cytotoxic T lymphocyte (CTL) clones occurs and thousands of neoplastic cells are lysed. However, as we would expect from any population in danger of extinction, the cells of the neoplastically transformed mass proceed with their microevolution and numerous clones of tumor cells survive each repeated attack by the immune system through secretion of immuno-inhibitory cytokines, such as TGF-β which has both ECM modulatory and direct suppressive effects on CTL generation from peripheral blood lymphocytes (24,25,31,32), downregulation of MHC molecules (26,33), loss of adhesion (34) and costimulatory molecules and induction of clonal T cell anergy (23,35), among other as yet uncovered ways. This process continues until the "creation" (ironically as it may sound, by the host's immune system) of highly resistant, poorly immunogenic, and extremely aggressive clones of tumor cells. This is the reality of cancer progression: a back-and-forth struggle between host and tumor, with evolutionarily dynamic exchanges throughout the entire process.

The expression of apoptosis related cell surface molecules on the surface of both tumor cell and CTL surfaces (*FasR-FasL* system) raises a distinct possibility of active PCD induction in CTL by tumor cells. Juxtacrine interactions between CTL and neoplastically transformed cells, coupled with observations that tumor cells can modulate the intracellular, signaling domains of cell surface receptors to elicit responses quite often contrary to the expected, may even provide a way for CTL to enhance the proliferation and dedifferentiation of cancer cells. Adoptive therapies using CTL raised against autologous neoplastically transformed cells *in vitro* should be employed in the control of minimal residual disease following surgical resection of the primary malignant growth. Further studies should establish the clinical significance of PCD-related protein expression and assess the possibility of targeting such molecules in the therapy of human neoplasms.

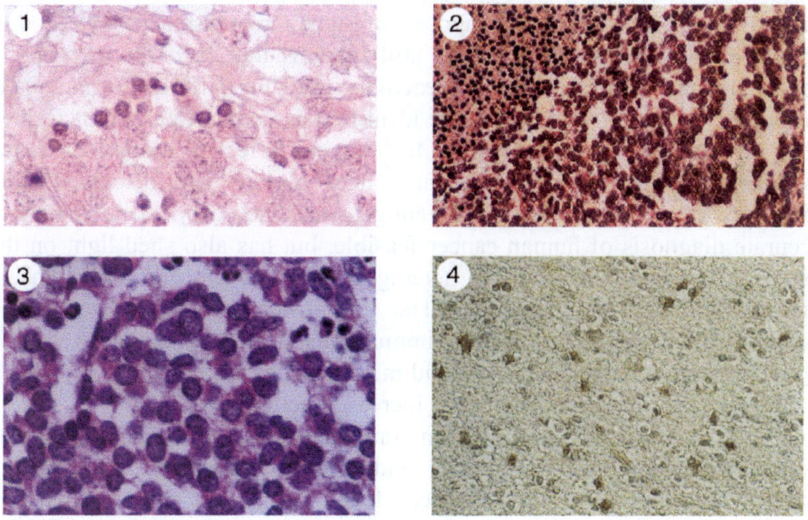

Figure 2-1.

1. Presence of tumor infiltrating leukocytes (TIL) in MED. An important cell clone is composed of CD8+, tumor associated antigen (TAA) directed, cytotoxic lymphocytes. Alkaline phosphatase conjugated, streptavidin-biotin enhanced immunohistochemical antigen detection. Magnification: 500x.
2. Childhood MED/PNET. The HOX-B4 positive tumor tissue contains a massive, mixed population of tumor infiltrating leukocytes (TIL). Alkaline phosphatase conjugated, streptavidin-biotin enhanced immunohistochemical antigen detection. Magnification: 200x.
3. Childhood MED/PNET. Expression of HOX-B3 embryonal antigen. Alkaline phosphatase conjugated, streptavidin-biotin enhanced immunohistochemical antigen detection technique. Magnification: 500x.
4. Childhood ASTR. Overexpression of p53. Peroxidase based antigen detection technique. Magnification: 500x.

2. THE SIGNIFICANCE OF IMMUNOHISTOCHEMISTRY

This chapter details the diagnostic significance of immunohistochemistry which has developed during the last quarter century. Certainly, the rapid advancement of monoclonal antibody producing technology has been of great significance in assuring the place of immunohistochemistry in the modern accurate microscopic diagnosis of human neoplasms, as a method of choice in histopathology. The fact still remains that in order to properly assess any immunohistochemical reactivity used for differential diagnostic purposes, the target cells have to be identified as neoplastically transformed cells by routine histopathological techniques. Selected groups of target molecules of great significance in cancer biology are discussed. The discovery of neoplasm associated antigens has not only made the more accurate diagnosis of human cancer feasible, but has also shed light on the extensive immunophenotypical heterogeneity of even the most closely linked human malignancies. The identification of disseminated neoplastically transformed cells by immunohistochemistry has allowed for a clearer picture of cancer invasion and metastasis, as well as the evolution of the tumor cell-associated IP towards increased malignancy.

During the last three decades immunohistochemistry has revolutionized both the research and diagnostic endeavors of pathologists, including neuropathologists (36). It is a great feeling to know that through our extensive research observations on brain tumors, which have been well documented in numerous articles, we have been able to contribute to these exciting developments. Advanced immunohistochemistry employs hundreds of well characterized, highly specific antibodies recognizing only the desired, target antigens. The highly sensitive methods of immunocytochemistry have expanded our understanding of the biology and immunobiology of brain tumors (as well as a wide array of other tumor types) and have led to the development and refinement of ever more reliable ways of diagnosis and differential diagnosis of CNS tumors, among many others. Immunoassaying (immunoreactivity) is also capable of detecting specific infectious agents, from ordinary bacteria to spirochetes, fungi, parasites and especially viruses, as well as identifying possible causative agents of CNS diseases. Certainly, the most important principle of the evaluation of an immunocytochemical result in mammalian neoplasms remains that positive immunoreactivity is significant only when it occurs in neoplastic cells previously recognized by standard (usually the routine hematoxylin and eosin stain) microscopic morphology.

The main goal of molecular oncology during the 1980s and 1990s has been the discovery of a marker which would be specific for tumor cells, as such, or for a tumor of specific tissue or organ differentiation line; and we agree that this goal has not been fulfilled (37). Presently, we consider a tumor marker as an increasing level of a substance which is not unique for neoplastically transformed cells, and is indeed expressed in normal cells. The marker could be localized in the cytoplasm of neoplastic cells or on their surface. The marker may also be produced by the tumor-surrounding tissue (paracrine secretion) under the direct influence of the neoplastic cells. Therefore, the immunocytochemical results of identifying a marker should be interpreted with great caution and should be always confronted with the results of routine histopathology, as well as other observations. This is valid even if the immunohistochemical results greatly improve the proper diagnosis and differential diagnosis.

Immunohistological procedures should also be based on strictly standardized antigen detection techniques (38). Histopathology and immunohistochemistry continue to be popular methods for predicting outcome in patients with malignant gliomas (39). Recently, traditional histopathological studies have stressed the importance of endothelial proliferation in the diagnosis of GBM. Immunohistochemical proliferation markers, in particular MIB-1, may be useful in assessing oligodendroglioma behavior, whereas their role in malignant ASTRs is less clear. Similarly, new studies on p53 and epidermal growth factor receptor (EGFR) immunohistochemistry in gliomas have demonstrated only limited predictive values. Nonetheless, the importance of immunohistochemistry in tumor biology is certain, and the rest of this monograph is devoted in great part to the important contributions of this technique.

In the last decade, there is increasing recognition of polyphenotypic high-grade malignancies in non-CNS tumor literature. Some of these tumors have been regarded as variants of PNET or as extrarenal malignant rhabdoid tumors (MRTs). Jay and co-authors have described two posterior fossa neoplasms, both of which displayed a "polyphenotypic" expression of neural, epithelial, myogenic, and glial markers, including synaptophysin, neurofilament, vimentin, glial fibrillary acidic protein, S-100, neuron-specific enolase, desmin, S antigen, MIC2, cytokeratin, epithelial membrane antigen, and carcinoembryonic antigen (CEA) (40). One tumor showed complex intercellular junctions, cytoplasmic intermediate filaments, well-developed rough and smooth endoplasmic reticulum and Golgi apparatus, cilia, and neurosecretory granules. The other neoplasm showed pools of glycogen, desmosomes, and tonofilaments. The histological and ultrastructural appearances were inconsistent with glioma, PNET, meningioma, ependymoma, choroid plexus carcinoma, sarcoma, germ cell tumor, and other tumors in the WHO classification. Although the polyphenotype raises the issue that these may represent variants of MRT or the atypical teratoid-rhabdoid tumor, the morphologic findings in the two cases were very dissimilar. The two cases presented by the authors underscore the problems in nosology and classification of polyphenotypic tumors of the CNS. This is particularly significant, as therapeutic protocols for PNET, MRT, and non-CNS polyphenotypic tumors are quite different.

Mixed gliomas have been difficult to define and subsequently diagnose due to the paucity of literature specifically examining this group of tumors (41, 42). Thirty mixed gliomas in which the minor glial component comprised at least 20% of the total tumor were observed: 20 oligoastrocytomas (OA) and ten malignant oligoastrocytomas (MOA). Nineteen patients were male (mean age 36 years) and 19 patients presented with seizures. The tumor was located in the frontal lobe in 17 patients and the temporal lobe in nine patients. The duration of preoperative symptoms in 25 patients ranged from five days to 14 years (mean 2.6 years). A mean follow-up of four years was available in 29 patients. Fourteen patients, seven with OA and seven with MOA, had recurrent tumor. One patient with MOA and four patients with OA (three with tumor progression and one with extensive leptomeningeal spread) died as a result of their tumor one to five years after diagnosis. Eighteen patients received chemotherapy and/or radiation therapy. Twenty-five tumors were immunostained with antibody to p53 protein. p53 nuclear staining was seen in 5/16 OA and 3/9 MOA. Immunoreactivity was observed only in neoplastically transformed astrocytes. One of the four patients who died with OA was p53 positive. Three recurrent MOAs were p53 positive. It was concluded that mixed gliomas most frequently occur in the frontal lobe and the majority of patients present with seizures; that there is no obvious association of p53 detection in mixed gliomas with tumor grade or behavior; and that similar to pure fibrillary astrocytomas, a subset of OA and MOA may be associated with p53 alterations.

Ganglioglioma is a rare, mixed neuronal-glial neoplasm of the central nervous system that occurs in young patients and has a benign clinical course (43). Twenty-seven specimens were studied by routine histochemistry, 21 specimens by immunochemistry, and 14 specimens were examined at the ultrastructural level by Hirose and co-workers in an attempt to define the immunophenotypic and morphologic features of ganglioglioma. The age of the 27 patients, 14 males and 13 females, ranged from 3 to 52 years (mean 22 years). The most commonly affected site was the temporal lobe (13 patients). Three patients experienced a local recurrence. Microscopically, the tumors were comprised of well differentiated, somewhat abnormal neurons as well as glial cells, the latter including astrocytes of fibrillary (59%) and pilocytic (41%) type. Scant mitotic activity was observed in 2 tumors (7%). Glial cells of all tumors were immunoreactive for glial fibrillary acidic protein (GFAP), S-100 protein, and vimentin. Ki-67 labeling indices ranged from 0.6 to 10.5% (mean 2.7%) and p53 labeling indices between 1.1 to 42.4% (mean 15.6%). Ki-67 and p53 labeling indices in recurrent tumors were significantly higher than those of nonrecurrent ones (p=0.036 and p=0.026, respectively). No examples of anaplastic transformation were encountered. Immunohistochemically, many neuronal cells were positive for synaptophysin (100%), Class 3 β-tubulin (100%), neurofilament protein (90%), and chromogranin A (86%), in addition to S-100 protein (71%) and, occasionally, vimentin (24%). Ultrastructural characteristics of neuronal cells included the presence of numerous, 100-230 nanometer dense core granules within both perikarya and cell processes, well developed rough endoplasmic reticulum, microtubules within cell processes, and synapses associated with clear vesicles. Astrocytic cells usually contained abundant intermediate filaments; their cell membranes, when abutting the stroma, were covered by basal lamina. It is certain that gangliogliomas are comprised of well differentiated neuronal cells and glial cells that are very often of pilocytic type. No cells with features intermediate between neurons and glia were observed. Neuronal cells are characterized by prominent neurosecretory features distinct from those of normal neurons in the central nervous system. Furthermore, higher Ki-67 and p53 labeling indices may indicate more aggressive behavior.

Nerve growth factor (NGF) is important to the survival, development, and differentiation of neurons (44). Its action is mediated by a specific cell surface transmembrane glycoprotein, nerve growth factor receptor (NGFR). NGFR expression was examined by immunohistochemistry in human fetal and adult adrenal medullary tissue, peripheral nervous system (PNS) neuroectodermal tumors (neuroblastoma, ganglioneuroblastoma, ganglioneuroma), pediatric primitive neuroectodermal tumors (PNETs) of the CNS, and CNS gliomas. Sixty-nine tumors in total were probed in this manner. Nerve growth factor receptor immunoreactivity was confined to nerve fibers and clusters of primitive-appearing cells in the fetal adrenal, and to nerve fibers and ganglion cells of the adult adrenal medulla; adrenal chromaffin cells were negative. In PNS neuroectodermal tumors, there was NGFR expression in tumor cells of 6 of 11 neuroblastomas and 6 of 6 ganglioneuroblastomas or ganglioneuromas. Thirteen of thirty-five CNS PNETs showed NGFR positivity. In most CNS PNETs, NGFR was restricted to scattered single or small groups of cells, but two tumors with astro-glial differentiation showed much more extensive immunoreactivity. Most ASTRs (11 of 14) and all ependymomas (3 of 3) were intensely NGFR positive.

In an earlier observation, MED-L cells with the 75kD low-affinity nerve growth factor receptor (p75NGFR) and MED-H cells with the proto-oncogene tropomyosin receptor kinase product (p140trk) were isolated selectively from a parent MED-3 cell line derived from cerebellar MED by panning, and the interaction of NGF with these cell lines was analyzed (45). NGF treatment induced neuronal differentiation, growth inhibition, and tyrosine phosphorylation of p140trk in MED-H cells, but not in MED-L cells. The cells of MEDs express the functional NGFR, p140trk, which regulates their differentiation and growth.

In another study, positron emission tomography (PET) with the amino acid tracer L-[1-C-11]-tyrosine was evaluated in 27 patients with primary and recurrent brain tumors (46). Patients underwent either static (n=14) or dynamic PET (n=13), with quantification of protein synthesis rate (PSR) and tumor-to-background ratio. Findings were compared with histopathological findings. Primary brain tumor was proved in 22 patients histologically, as well as metastatic cancer of unknown origin, primary non-Hodgkin lymphoma, meningioma, atypical infarction, and vasculitis in one patient each. At PET, 20 of 22 primary tumors, the metastasis, and non-Hodgkin lymphoma were correctly depicted. A false-positive finding was obtained with the infarction, and the meningioma and vasculitis were not depicted. The calculated sensitivity was 92%; specificity, 67%; and accuracy, 89%. There were no statistically significant relationships between histologic findings, PSR, and tumor-to-background ratio. The results strongly suggested that L-[1-C-11]-tyrosine is a valid tracer for early diagnosis of brain tumors and allowed quantification of PSR. PSR was further measured in human brain tumors using L-[1-(11)C]-tyrosine PET, and the PSR was compared to histopathological parameters of intratumoral cell proliferation and protein synthesis (47). The authors observed 20 patients who had a positive brain tumor biopsy and who also underwent PET analysis. Formalin fixed, paraffin tissue sections were stained employing the MoAB MIB-1, targeted against the core Ki-67 antigen, and nucleolar organizer regions (NORs) were measured as argyrophilic NORs (AgNORs). The PSR was determined using a kinetic model. PSR (in nmol/ml/min) ranged from 0.44 to 1.99 (mean 0.97), Ki-67 labeling indices ranged from 0.9% to 33.5% (mean 9.5%) and AgNOR area (in mm^2/cm^2) ranged from 0.13 to 0.85. No relationship was found between PSR and Ki-67 labeling neither index nor AgNOR area. These results suggest that the PSR and the intratumoral proliferation, as measured by Ki-67, are two independent processes.

Deregulation of telomerase, a ribonucleoprotein polymerase that compensates progressive loss of telomeric $(TTAGGG)_n$ repeats during DNA replication, has been suggested to facilitate tumorigenesis and cellular immortality by providing unlimited proliferation capacity for cancer cells

(48). The relationship between tumor proliferation activity and *in situ* expression of the telomerase RNA component was investigated in 46 human grades I to IV ASTRs. Heterogeneously distributed telomerase RNA expression was detected from all of the tumor samples as well as from normal human brain tissue. However, expression of telomerase RNA was significantly increased in highly malignant tumors (p=0.024) and in tumors that showed increased proliferation activity determined by MIB-1 immunohistochemistry (p=0.014). Interestingly, increased telomerase RNA levels were observed in a subgroup of grade II ASTRs that showed significant increase in proliferation activity (p=0.047), indicating that the telomerase RNA component is already up-regulated in the early stages of ASTR pathogenesis. Telomeric repeat amplification assays revealed telomerase activity in 4/6 GBMs and in 1 rapidly proliferating grade II ASTR. These results suggest that increased tumor proliferation activity triggers telomerase activation *via* mechanisms that involve increased production of the RNA component of telomerase.

Abnormal p53 detected by immunocytochemistry has been identified to be a predictor of poor disease outcome in a variety of human malignant neoplasms, including brain tumors (49-51). The role of p53 alteration in the pathogenesis of intracranial neuroectodermal tumors was observed by Ng and co-workers (52). Formalin fixed paraffin sections of 196 brain tumors were employed in the study. The results demonstrated up to 40% immunoreactivity (presence of p53 alteration) in high grade ASTRs (AA and GBM). p53 alterations were also detected in 11% of well differentiated ASTRs. The extent of tumor cell immunolabeling also increased from the low to high grade ASTRs. Only rare cases of oligodendrogliomas and MEDs showed positive immunoreactivity, whereas ependymomas and choroid plexus tumors were uniformly negative. It seems that p53 alterations play a role in the immunophenotype progression of ASTR cells from low to high grade, as reported by us in a previous study (54). Small, anaplastic ASTR cells were the predominant cell type expressing aberrant p53 protein.

Glioblastoma multiforme (GBM), the most malignant neoplasm of the human CNS, develops rapidly *de novo* (primary glioblastoma) or through malignant cellular IP progression from low-grade or anaplastic ASTR (so-called secondary glioblastoma) (54). It was recently reported that mutations of the p53 gene are present in more than two-thirds of secondary glioblastomas, but rarely occur in primary GBMs, suggesting the presence of divergent genetic pathways (55). Primary and secondary GBMs were screened by immunohistochemistry for murine double minute 2 (MDM2) overexpression and by differential PCR for gene amplification. Tumor cells immunoreactive to MDM2 were found in 15/29 primary GBMs, but in only 3/27 secondary glioblastomas (p=0.0015). MDM2 amplification occurred in

2 primary GBMs, but in none of the secondary GBMs. Only 1/15 primary GBMs overexpressing MDM2 contained a p53 mutation. These results suggest that MDM2 overexpression with or without gene amplification constitutes a molecular mechanism of escape from p53-regulated growth control, operative in the evolution of primary GBMs that typically lack p53 mutations.

In another article, p53 and MDM2 oncoprotein expression was evaluated in paraffin-embedded tissues from 61 patients with CNS gliomas (53 ASTRs and 8 oligodendrogliomas) and related to proliferation-associated markers [*i.e.* proliferating cell nuclear antigen (PCNA), Ki-67, and nuclear organizer regions (NORs)], as well as epidermal growth factor receptor (EGFR) (56). The authors employed a library of MoABs including PC-10, MIB-1, DO-1, 1B1O and EGFR 113 and the colloid silver nitrate (AgNOR) technique. MDM2 and p53 were co-expressed in 28% of the glioma cases. A p53-positive/MDM2-negative cellular IP was observed in 15% and a p53-negative/MDM2-positive IP in 20% of cases. There was a positive correlation of p53 and MDM2 expression with grade and proliferation indices. Univariate analysis in the group of diffuse astrocytomas showed that older age, high histological grade, high PCNA labeling index and high AgNOR score were associated with reduced overall survival ($p<0.05$). p53 labeling index, Ki-67 labeling index, AgNOR score, as well as tumor location and grade influenced disease-free survival ($p<0.05$), whereas the only parameters affecting post-relapse survival were histologic grade and Ki-67 labeling index ($p<0.1$). Multivariate analysis revealed that age, radiotherapy, PCNA labeling index and p53 labeling index were the independent predictors of overall survival. p53, Ki-67, MDM2, and EGFR labeling indices, as well as grade and type of therapy were independent predictors of disease-free survival, and grade was the only independent predictor of post-relapse survival. The authors concluded that p53 and MDM2 labeling indices, as well as EGFR and proliferation marker (PCNA and Ki-67) expression represent useful indicators of overall and disease-free survival in diffuse ASTR patients.

Formalin fixed, paraffin embedded sections of 84 oligodendrogliomas (63 primary tumors, 21 recurrences), 21 GBMs with oligodendroglial growth pattern (15 primaries, 6 recurrences) and 17 mixed gliomas was observed for the presence of mutations in exons 5-9 by means of single stranded conformation polymorphism (SCCP), temperature gradient gel electrophoresis (TGGE) and direct DNA sequencing. In parallel, p53 protein accumulation was determined by means of immunocytochemistry (57). The percentage of mutations was found to be higher than previously reported (6/44 grade II oligodendrogliomas, 4/19 grade III oligodendrogliomas, 4/15 glioblastomas). In four cases, the mutations lead to distinct changes in the primary or secondary structure of the protein (cysteine ➔ tyrosine, proline ➔ leucine) and were associated with marked accumulation of p53 protein. A significant correlation between p53 protein accumulation and TP53 gene aberrations was found ($p<0.001$), although p53 protein accumulation was detected more often than TP53 gene anomalies, indicating that factors other than TP53 gene mutation may also lead to p53 oncoprotein accumulation in brain tumor cells. A significant correlation was found for p53 protein accumulation and tumor grade, but not TP53 gene mutations. In conclusion, evaluation of p53 protein accumulation reflected the clinical course of oligodendrogliomas better than the mere presence of TP53 gene mutations.

The *c-myc* protein, which has been reported as a transcription factor having an unclear but apparent dual role in promoting cell proliferation and PCD, may be related to the relative absence of necessary growth factors for producing growth arrest (58). The gene encoding p53 has been described as the most commonly disrupted gene in human neoplasms, suggesting a tumor-suppressor function for wild-type p53 (53,59-64). High levels of p53 are associated with arrest of the cell cycle in G1, which allows for necessary repairs to be carried out following DNA damage and hypoxia (65). If the damage cannot be repaired, cell death by apoptosis is triggered. Mutations of the p53 gene lead to a loss of this critical DNA integrity and cell cycle monitoring function, which allows cells with damaged DNA to proliferate, obviously key in the production of neoplastically transformed cells. The bcl-2 family of proteins can inhibit (bcl-XL and Mcl-1) or induce (bax, bcl-XS, bcl-XL and bad) PCD in several cell systems (66-68). Bcl-2, a 25-26 kD protein, has been described in mitochondria, the nuclear envelope, and in the endoplastic reticulum. Recent data suggests that the activity of bcl-2 is regulated by a 21 kD protein bax, with extensive amino acid homology to bcl-2 (69). Bax is a homodimer capable of forming heterodimers with bcl-2. Bad is another protein that can interact with bcl-2. It has homology to bcl-2 within the homology domains BH1 and BH2. In mammalian cells, the bad protein is able to selectively heterodimerize with bcl-2 and bcl-XL, but not with other bcl-2 family related proteins. When bad is in a heterodimer with bcl-XL, it displaces bax from bcl-XL and promotes apoptosis.

Recently published evidence has emphasized the importance of PCD or apoptosis in the maintenance of tissue homeostasis and pathogenesis in the growth and progression of human neoplasms. This study, analyzed in breast cancer (BC), the significance of apoptosis in relation to the expression of p53 and bcl-2 proteins, tissue proliferation (defined by Ki-67 expression), hormone receptors, and tumor grade (70). Immunocytochemistry was performed for p53, bcl-2, estrogen receptor, progesterone receptor, and Ki-67 expression. Mutant p53 protein was detected using a mutant specific ELISA. Immunoreactivity of p53 significantly correlated with the presence of mutant p53 protein detected by ELISA (r=0.654, p=0.00001). An inverse correlation was observed between *bcl-2* expression and the extent of apoptosis (r=-0.33369, p=0.01912). The extent of apoptosis directly correlated with p53 protein accumulation (r=0.485, p=0.00041), Ki-67 immunoreactivity (r=0.435, p=0.001), histopathological grade (r=0.492, p=0.0003), BC size (r=0.326, p=0.023) and lymph node status (r=0.287, p=0.047). A direct correlation was also observed between p53 expression and Ki-67 immunoreactivity (r=0.623, p=0.0002). There was no statistically significant association between estrogen and progesterone receptor status and apoptosis. In addition, the TNM stage of the disease correlated with immunoreactivity of p53 (r=0.572, p=0.00012) and Ki-67 (r=0.3744, p=0.00818). Bcl-2, by inhibiting apoptosis, may cause a shift in tissue kinetics towards the preservation of genetically aberrant cells, thereby facilitating tumor progression. These results imply that rapidly proliferating tumors appear to have a high "cell turnover state" in which there may be an increased chance of apoptosis among the proliferating cells. The ability of apoptosis to also occur in the presence of mutant p53 protein suggests the existence of at least two p53-dependent apoptotic pathways, one requiring activation of specific target genes and the other independent of it.

Genetic observations have provided new insights into the regulation of developmental organization of PCD. During the development of the nematode *Caenorhabditis elegans,* two genes related to PCD were identified: ced3 and ced4 (two potential tumor suppressor genes), as well as, an inhibitor, a cell protector gene named ced9 (71-73). The ced9 gene is homologous to the bcl-2 proto-oncogene, discovered as a gene associated with a frequent translocation breakpoint in some B cell leukemias (74, 75). The ced3 gene encodes a cysteine protease that is homologous to the interleukin-1β-converting enzyme (ICE) (76). During the last few years, a number of novel genes involved in apoptosis, belonging to the ICE/ced-3 family and others, have been discovered: mouse Nedd 2 (77, 78), Ich-1, an ICE/ced-3 related gene that encodes both positive and negative regulators (79), TX/Ich-2/ICE_{rel}-II (80), the gene encoding CPP32, a novel human apoptotic protein with homology to the mammalian Il-1β-converting enzyme (81), Mch2 and Mch3 (82), ICE_{rel} II and ICE_{rel} III (83), ICE-LAP3 and ICE-LAP6 (84, 85), and MACH, a novel MORT1/FADD-interacting protease involved in CD95/*FasR*/APO-1, and TNF receptor-induced PCD (86). There is evidence that ICE family members are also auto-proteolytic, and are able to process one another's substrates, such as poly-ADP ribose polymerase (PARP), lamins, etc. (87-89).

Cadherins are Ca^{2+}-dependent cell adhesion molecules that play an important role in tissue formation and morphogenesis in multicellular organisms. In recent years, there have been reports of cadherin involvement in tumor invasion and metastasis (90). Twenty-two surgical specimens and some cultured cells were studied by immunohistochemical methods. No significant difference was observed in patients with AA, whereas decreased expression of N-cadherin was detected at the time of recurrence in those with GBM. In these groups, cerebrospinal fluid dissemination was found, and contralateral cerebral metastases and extracranial metastases were observed. It seems that decreased N-cadherin expression identified time of recurrence correlates with tumor invasion and dissemination of cerebrospinal fluid.

Desmosomes represent major intercellular adhesive junctions at basolateral membranes of epithelial cells and in other tissues (91). They mediate direct cell-cell contacts and provide anchorage sites for intermediate filaments important for the maintenance of tissue architecture. There is increasing evidence now that desmosomes in addition to a simple structural function have new roles in tissue morphogenesis and differentiation. Transmembrane glycoproteins of the cadherin superfamily of Ca^{2+}-dependent cell-cell adhesion molecules which mediate direct intercellular interactions in desmosomes appear to be of central importance in this respect. The complex network of proteins forming the desmosomal plaque associated with the cytoplasmic domain of the desmosomal cadherins, however, is also involved in junction assembly and regulation of adhesive strength.

Together with microtubules and actin microfilaments, approximately 11 nm wide, intermediate filaments (IFs) constitute the integrated, dynamic filament network present in the cytoplasm of metazoan cells. This network is critically involved in division, motility and other cellular processes (92). While the structures of microtubules and microfilaments are known in atomic detail, IF architecture is presently much less understood. The elementary 'building block' of IFs is a highly elongated, rod-like dimer based on an α-helical coiled-coil structure. Assembly of cytoplasmic IF proteins, such as vimentin, begins with a lateral association of dimers into tetramers and gradually into the so-called unit-length filaments (ULFs). Subsequently ULFs start to anneal longitudinally, ultimately yielding mature IFs after a compaction step. The assembly of nuclear lamins, however, starts with a head-to-tail association of dimers. Recently, X-ray crystallographic data were obtained for several fragments of the vimentin dimer. Based on the dimer structure, molecular models of the tetramer and the entire filament are now a possibility.

Tenascins (TNs) are a family of extracellular matrix glycoproteins. The first member of this family to be recognized, tenascin-C (TN-C), is known to be expressed in various tumors, including human ASTRs. Tenascin-X (TN-X) is the latest member of the TN family to be reported, and its expression in tumor tissues has not yet been examined (93). Expression of TN-X in glioma cell lines and human ASTRs was reported using immunoblot analysis employing anti-mouse TN-X antibodies. The expression pattern of TN-C and TN-X was also observed immunohistochemically in a series of 32 human ASTRs and tissue sections from five normal brains. Expression of TN-X was upregulated to a higher degree in low-grade ASTRs than in high-grade ASTRs. TN-X was mainly localized in the perivascular stroma around tumor vessels, and weakly expressed in the intercellular spaces among tumor cells. In contrast, TN-C was more strongly expressed in the intercellular spaces and in tumor vessels in high-grade ASTRs (AAs and GBMs) than in low-grade ASTRs. In the tissues expressing both TNs, the distribution of TN-X was often reciprocal to that of TN-C. These findings indicate that the expression of TN-C and TN-X in ASTRs is different, and that these glycoproteins could be involved in neovascularization in different manners.

Recent *in vitro* studies of the EGFR family have revealed complex signaling interactions involving the production of ligand-mediated heterodimers synergistic for the transformation of cells *in vitro* (94) and growth factors have been well established as necessary during the process of neoplastic transformation (95-98). In a series of 70 patients with childhood MED, immunocytochemistry and Western blotting analysis were conducted to characterize the expression patterns of all four EGFR family members (EGFR, HER-2, HER-3, and HER-4), as well as heregulin-α, a ligand for the HER-3 and HER-4 receptors. The majority of MED cases expressed two or more receptor proteins; coexpression of the HER2 and HER4 receptors occurred in 54%. Expression of heregulin-α was detected in 31% of the tumor cases observed by the authors. To investigate whether co-expression results in receptor heterodimerization, immunoprecipitation of protein extracts from primary tumors was conducted and various patterns of receptor interaction were observed, including that between HER-2 and HER-4. In multivariate 25-year survival analysis with clinicopathological disease features, no individual receptor or heregulin-α achieved significance. In contrast, when considered together in the multivariate model, co-expression of HER-2 and HER-4 demonstrated independent prognostic significance (p=0.006). These data suggest the hypothesis that HER2-HER4 receptor heterodimerization is of particular biological significance in MED, demonstrating the potential clinical significance of EGFR family heterodimerization in human neoplastic cells. The authors also analyzed expression of the AP-2 transcription factor implicated in the positive regulation of HER-2 and HER-3 gene transcription in malignant cells and found an association between AP-2 expression and not only HER-2 and HER-3, but also HER-4 levels in primary MEDs/PNETs.

The neurofibromatosis 2 (NF2) gene-encoded protein (merlin), may function as a molecular link between the cytoskeleton and the plasma membrane (99). Merlin is thought to play a crucial role as a tumor suppressor not only in hereditary NF2-related tumors, but also in sporadic tumors such as Schwannomas, meningiomas and gliomas. Using a merlin-expression vector system, specific antiserum against merlin was produced. Subsequently, the intracellular distribution of merlin in cultured glioma cells was observed, and merlin expression was further investigated in 116 human brain tumors. Immunofluorescence microscopy revealed that merlin was localized beneath the cell membrane and concentrated at cell-to-cell adhesion sites, where actin filaments are densely associated with plasma membrane. By immunohistochemistry, none of the Schwannomas from neither NF2 patients nor sporadic cases showed any immunoreactivity, while normal Schwann cells of cranial nerves were immunopositive. In meningiomas, merlin expression was frequently detected in the meningothelial subtype (8/10), but no expression could be detected in either the fibrous or the transitional variant. Most normal astrocytes were negative; however, reactive astrocytes often expressed merlin. Glioblastomas and AAs were found to be strongly positive, and focal positive staining was observed in fibrillary and pilocytic ASTRs. Thus, the loss of merlin appears to be integral to Schwannoma formation and the differential pathogenesis of meningioma subtypes. However, merlin alterations do not appear to play a critical role in neither the tumorigenesis nor the malignant transformation of neoplastic astrocytes.

Chang and co-workers have chemically characterized the gangliosides of the Daoy cell line in order to establish a model system for the study of ganglioside metabolism of human MEDs/PNETs (100). Cells comprising MEDs/PNETs contain a high concentration of gangliosides (143±13nmol LBSA/10^8 cells). The major species have been structurally confirmed to be GM2 (65.9%), GM3 (13.0%), and GD1a (10.3%). Isolation of individual gangliosides homogeneous in both carbohydrate and ceramide moieties by reversed-phase HPLC and analysis by negative-ion *Fas*t atom bombardment collisionally activated dissociation tandem mass spectrometry have allowed the authors to unequivocally characterize ceramide structures. In the case of GM2, 10 major ceramide subspecies were identified: d18:1-hC16:0, d18:1-C16:0, d18:0-C16:0, d18:1-C18:0, d18:1-C20:0, d18:1-C22:0, d18:2-C24:1, d18:1-C23:1, d18:1-C24:1, and d18:1-C24:0. Taken together with previous studies, these findings in human MED cells support the view that high expression and marked heterogeneity of ceramide structure are general characteristics of tumor gangliosides, molecules which are shed by tumor cells and are biologically active *in vivo*.

While the number of reports on macrophage infiltration of gliomas is increasing, the extent and mechanisms of macrophage recruitment remain unclear (101). To investigate whether monocyte chemoattractant protein-1 (MCP-1) plays a role in this process, *in situ* hybridization was performed for 22 GBMs, one AA and 4 grade II fibrillary ASTRs and reverse transcription-polymerase chain reaction was performed in 13 GBMs, one AA and three grade II ASTRs. High levels of MCP-1 mRNA were detectable in most GBMs, while a lower level was detected in grade II ASTRs. Many tumor-associated macrophages were observed by immunohistochemistry within most GBM cases, while the grade II ASTRs contained a lower number of infiltrating macrophages. The positive correlation between MCP-1 level and abundance of macrophagic infiltration suggests that MCP-1 has a role in the recruitment of macrophages to the site of the neoplasm. By combining *in situ* hybridization and immunohistochemistry, high levels of MCP-1 mRNA were shown both in tumor cells and within the tumor-associated macrophages. Reactive astrocytes and microglia along the boundary of the tumors also expressed MCP-1. In areas with T lymphocyte infiltration, larger numbers of MCP-1-positive cells with an enhanced level of expression were identified. The authors proposed that the mechanism of macrophage recruitment is, at least partly, affected by constitutive expression and T cell-mediated up-regulation of MCP-1 in tumor cells, as well as the tumor infiltrating macrophages. The production of MCP-1 by these macrophages establishes a positive amplification circuit for macrophage recruitment in gliomas.

In the last two decades, significant advances have been made in the identification of the soluble angiogenic factors, insoluble extracellular matrix (ECM) molecules and receptor signaling pathways that mediate control of angiogenesis. The question remains how the endothelial cells are capable of integrating these chemical signals with mechanical cues from their *in situ* tissue microenvironment so as to produce functional capillary networks that exhibit specialized form as well as function (102). These observations have revealed that ECM governs whether an endothelial cell will switch between growth, differentiation, motility, or apoptosis programs in response to a soluble stimulus based on its ability to mechanically resist cell tractional forces and thereby produce cell and cytoskeletal distortion. Transmembrane integrin receptors play a pivotal role in this mechanochemical transduction process because they both organize a cytoskeletal signaling complex within the focal adhesion and preferentially focus mechanical forces on this site. All three molecular filaments within the internal cytoskeleton (*i.e.* microfilaments, microtubules, and intermediate filaments) also contribute to the cell's structural and functional response to mechanical stress through their role as discrete support elements. A similar form of mechanical control also has been shown to be involved in the regulation of contractility in vascular smooth muscle cells and cardiac myocytes. The mechanism by which cells perform mechanochemical transduction and the implications of these findings for morphogenetic control are still not clear yet.

Experimental tumors of the CNS were observed with antibodies to quinolinate to assess the cellular distribution of this endogenous neurotoxin (103). In advanced F98 and RG-2 GBMs and E367 neuroblastomas in the striatum of rats, variable numbers of quinolinate immunoreactive cells were observed in and around the tumors, with the majority being present within tumors, rather than the brain parenchyma. The stained cells were morphologically variable, including round, complex, rod-shaped, and sparsely dendritic cells. Neuroblastoma and glioma cells were unstained, as were neurons, astrocytes, oligodendrocytes, ependymal cells, endothelial cells, and cells of the choroid plexus and leptomeninges. GFAP immunoreactivity was strongly elevated in astrocytes surrounding the tumors. Dual labeling immunohistochemistry with antibodies to quinolinate and GFAP demonstrated that astrocytes and the cells containing quinolinate immunoreactivity were morphologically disparate and preferentially distributed external and internal to the tumors, respectively, and no dual labeled cells were observed. Lectin histochemistry with Griffonia simplicifolia B4 isolectin and Lycopersicon esculentum lectin demonstrated numerous phagocytic macrophages and reactive microglia in and around the tumors, whose distribution was similar to that of quinolinate immunoreactive cells, although they were much more numerous. Dual labeling studies with antibodies to quinolinate and the lectins demonstrated partial co-distribution of these markers, with most double-labeled cells having the morphology of phagocytes. The present findings suggest the possibility that quinolinate may serve a functional role in a select population of inflammatory cell infiltrates during the host's immune response to primary brain neoplasms.

3. ORIGINAL IMMUNOHISTOCHEMICAL OBSERVATIONS

Our systematic and detailed cellular IP analyses of 82 childhood brain tumors [34 medulloblastomas (MEDs)/primitive neuroectodermal tumors (PNETs), 42 astrocytomas (ASTRs), 5 choroid plexus papillomas (CPPs) and 1 choroid plexus carcinoma (CPCs)], was conducted using over 55 MoABs. An indirect, four-step, enzyme linked [alkaline phosphatase (AP) or peroxidase], biotin-streptavidin based, antigen detection technique was employed. The following are the immunocytochemically identified, most characteristic IPs of brain tumor cells with the employed extensive library of more than 90 MoABs for:

<u>MED/PNETs:</u> Synaptophysin and Chromogranin A; HLA-A,B,C, HLA-DR; Vimentin, TE4, TE7, TE8, TE15; AE1 and AE3; TE3; CLA and UJ 308; Leu 2/a and Leu 3a/3b, NF-H and MAP-1, MAP-2, MAP-5; GFAP; EGFR; p53; HOX-B3, -B4, -C6; c-erb-2 (HER-2), c-erb-3 (HER-3), c-erb-4 (HER-4); Caspase-3; *FasR*; Survivin; CD105 (Endoglin); MMP-2, -3, -9, -10, -13.

<u>ASTRs:</u> GFAP, Vimentin, Tenascin, HLA-A,B,C, HLA-DR; CLA and UJ 308; UJ 167.11, A_2B_5, and Thy-1; Chromogranin A; NF-H, TE7; Leu-2/a and Leu 3a/3b; c-erb-2 (HER-2), c-erb-4 (HER-4); p53; Caspase-3, -6, -8, -9; *FasR*; CD105 (Endoglin), MMP-3, -10; MAGE-1.

<u>Choroid plexus papillomas (CPP):</u> GFAP, NF-H, NF-M, Vimentin and Cytokeratins (4/5).

The great majority of childhood brain tumors are infiltrated by host's immune effector cells directed against tumor cells. Leu-2/a$^+$ cells comprise the significant CD8$^+$ CTL population of tumor infiltrating leukocytes. CTL are MHC class I restricted, tumor-associated antigen (TAA) specific, cytotoxic cells and were identified in 65/76 (ASTRs and MEDs only) brain tumors. CTL usually represented 1-10% of all cells, but in some cases 30-44% of the cells were CD8$^+$. CD4$^+$, MHC class II restricted helper lymphocytes, detected using MoAB anti-Leu-3/a, were present in 65/76 brain tumors. Usually 1-10% of the observed cells reacted with MoAB anti-Leu 3/a. Macrophages (Leu-M5 antigen positive cells) were expressed in 74/76 brain tumors. Their number also represented 1-10% of all observed cells in the brain tumors. All 76 brain tumors contained cells that reacted positively with MoABs anti-HLA-A,-B,-C and anti-HLA-DR, demonstrating strong MHC antigen and overall leukocyte antigen expression. Leukocyte common antigen (LCA) expression was demonstrated, by the positive reaction of the cellular antigens with MoAB anti-HLe-1, in all 76 brain tumors studied. MoAB UJ 308 detected the presence of premyelocytes and mature granulocytes in 60/76 brain tumors. They were localized perivascularly, within the tumor tissue, or close to necrotic regions. Natural killer (NK) cells were not defined in these childhood brain tumors.

Our results allowed us to draw the following conclusions:

1) PNETs, ASTRs, CPPs, and CPCs display a heterogeneous IP;
2) both differentiated and immature neurofilament proteins are present in the great majority of childhood primary brain tumors;

3) gliomas, MEDs, and PNETs co-express GFAP, NF-H, NF-M, vimentin, and at least one cytokeratin;
4) neuronal differentiation is always present within childhood brain tumors;
5) neoplastically transformed astrocytes may be able to present antigens to infiltrating T lymphocytes since MHC molecules are expressed on their surfaces;
6) MHC class I and II molecules are not present on normal astrocytes *in vivo*, whereas *in vitro*, in culture astrocytes express MHC class I and II molecules, especially after interferon-γ preincubation;
7) 65/76 PNETs and ASTRs contained tumor infiltrating leukocytes (TIL), particularly the cytotoxic, MHC class I restricted and tumor associated antigen (TAA)-specific and directed, $CD8^+$ lymphocyte clone, representing the functionally compromised effector cells of the host's cellular immunological response;
8) granulocytes and premyelocytes participate among the TIL, probably as a response to intratumoral inflammation and necrotic changes;
9) the growth of solid neoplasms, including childhood brain tumors, even at clinically undetectable sizes (a few mm^3), as well as the generation of an invasive cellular immunophenotype (CIP) in neoplastic cells and distant metastases depends upon the continuous formation of new blood capillaries;
10) the newly organized tumor blood capillaries' most striking feature is the presence of markedly enlarged and disorganized perivascular spaces, measuring at least 2-5 mm, detected in all observed childhood primary brain tumors and was independent of the tumor's size, as well as its morphological and cellular differentiation features;
11) endothelial cells undergo rapid proliferation during neoplasm related and induced neo-vascularization which can be demonstrated immunomorphologically by the detection of the cell surface localized endoglin (EDG/CD105), a transforming growth factor-β (TGF-β) receptor and a proliferation-associated antigen (PAA) expressed on endothelial cell surfaces;
12) inhibition of angiogenesis as a form of anti-neoplastic therapy has been and should be extensively studied;
13) brain tumor cells often express an apoptosis related transmembrane glycoprotein *Fas*/APO-1 (CD95), member of the nerve growth factor/tumor necrosis receptor superfamily which is not present on normal cells in CNS, apoptosis being triggered by the binding of *Fas*/APO-1 to its natural ligand (*FasL*/APO-1L);
14) further studies should substantiate the importance of CD105 or EDG in the earliest possible detection, diagnosis and neoplasm related

angiogenesis inhibition-based treatment of mammalian solid neoplasms, especially childhood brain tumors; and

15) future observations of immunobiology, cell cycle regulation and endothelial cell proliferation should aid in the development of an individualized, fourth modality, immunotherapeutical regimen for primary childhood brain tumors.

Neoplasm associated antigens (NAAs) represent the biochemical or immunological counterparts of the morphological characteristics of neoplastically transformed cells. The expression of an immunocytochemically defined NAA can also be related to the tissue of origin and the particular developmental stage.

Understanding of the tumor biology and immunobiology of brain tumors was also made possible by the highly specific and sensitive methods of immunocytochemistry, and reliable diagnosis and differential diagnosis of CNS tumors has become a matter of IP search and characterization employing well chosen libraries of highly specific, antibodies.

Detection of cell lineage specific cytoskeletal proteins and markers of cell proliferation are significant not only for histopathological diagnosis and differential diagnosis, but also in expanding our understanding of the histogenesis, cell differentiation properties, and neoplastic growth characteristics during the development of the intratumoral cellular microenvironment in childhood brain malignancies.

A study identified the incidence of *c-myc* protein expression in MED/PNET and partially identified the mechanisms (in addition to c-myc gene amplification) that lead to increased protein expression. c-myc gene copy number, mRNA level, and protein expression was analyzed in a panel of MED/PNET cell lines. c-myc protein levels were assessed in *ex vivo* brain tumor specimens and cell lines employing immunohistochemical antigen detection with a c-myc-specific MoAB. In addition, Southern analysis confirmed c-myc gene amplification in the D425 MED cell line and rearrangement of one allele in D283 MED, which was analyzed further and appeared to represent a small deletion of the 3' of exon 3. *c-myc* transcript levels were dramatically elevated in both lines. Using a c-myc probe, fluorescence *in situ* hybridization (FISH) determined the presence of c-myc in 3 tandem copies at 8q24 in D283 MED and multiple copies as double minutes in D425 MED. c-myc protein expression was demonstrated by immunocytochemistry in the cells of 9/10 tumors and all cell lines, regardless of gene amplification status or level of mRNA expression. The results revealed that c-myc protein expression is common in MED/PNET brain tumor specimens and established cell lines. Elevated protein levels are observed in the absence of amplification, suggesting that multiple mechanisms of c-myc dysregulation may be involved in the histogenesis of MED/PNET. These results support a possible initiating role for c-myc protein in the development of childhood neuroectodermal malignancies.

Synaptophysin, a vesicular integral membrane protein, is specifically expressed in neuroendocrine tissues. According to cDNA cloning studies, it has a molecular weight of 33,300 Daltons, one potential N-glycosylation site at the vesicle inside, four major hydrophobic domains, as well as a C-terminus containing approximately 90 amino acids. The C- and N-termini of synaptophysin are located on the cytoplasmic side of the vesicular membrane. No signal sequence has been identified. Transfection of non-neuroendocrine cells with synaptophysin cDNA leads to the synthesis of synaptophysin-containing vesicles, which contain this protein in highly enriched form and have biophysical properties similar to the presynaptic vesicles of neurons. So far, the vesicular content has only been determined in rat neurons, where classical neurotransmitters such as biogenic amines and transmitter-active amino acids were found. Reconstitution of the purified protein in liposomes suggests a possible channel function of synaptophysin. Employing mono- and polyvalent antibodies against synaptophysin, a considerable number of studies in several laboratories have shown that this protein is a reliable marker molecule for neuroendocrine tumors of various degrees of differentiation.

4. INTERMEDIATE FILAMENTS (IFS)

> "Microfilaments, intermediate filaments, and microtubules are three major cytoskeletal systems providing cells with stability to maintain proper shape. Although the word "cytoskeleton" implicates rigidity, it is quite dynamic exhibiting constant changes within cells. In addition to providing cell stability, it participates in a variety of essential and dynamic cellular processes including cell migration, cell division, intracellular transport, vesicular trafficking, and organelle morphogenesis."
>
> Yoon and co-workers (104)

One of the basic concepts of tumor immunocytochemistry was that neoplastic cells typically retain the IF characteristic of their cell of origin (105-108). Anomalous simultaneous expression of intermediate filament proteins may reflect a stop in physiological cellular differentiation or dedifferentiation of neoplastic cells (106). In higher mammals, the typical cytoskeleton of astroglial cells contains a principal, 8-9 nm intermediate filament IF with the hallmark protein of glial differentiation GFAP. GFAP is the major intermediate filament of mature astrocytes, and its relatively specific expression in these cells suggests an important function (110). As a member of the cytoskeletal protein family, GFAP is thought to be important in modulating astrocyte motility and shape by providing structural stability to astrocytic processes (111). In the CNS of higher vertebrates, following injury, either as a result of trauma, disease, genetic disorders, or chemical insult, astrocytes become reactive and respond in a characteristic manner, the so-called process of astrogliosis. Astrogliosis is characterized by *Fas*ter production of extreme quantity of GFAP and it is well documented by employing immunohistochemistry with highly specific anti-GFAP MoABs. Recently, to study the role of the GFAP gene, mice have been genetically engineered carrying null alleles (no protein), modified alleles (altered protein), or added wild type alleles (elevated protein). Surprisingly, the absence of GFAP has relatively subtle effects on development. On the other hand, overexpression can be lethal, and led to the discovery that GFAP coding mutations are responsible for most cases of Alexander disease, a devastating neurodegenerative disorder. Future experiments with various GFAP mouse models will reveal much more about GFAP and the astrocyte function.

Cytokeratin filaments and IF-anchoring membrane domains of desmosomal type (112) are typical of glial cells. Qualitative IF analysis has been used to aid the neuropathological subclassification of nonglial human tumors (105, 113). However, Rungger-Brandle and co-workers (112) reported an epithelium-like cytoskeleton in normal astroglial cells of the optic nerves of certain amphibians (*Xenopus laevis*, *Rana ridibunda*, and *Pleurodeles waltlii*). The keratin nature of these IF proteins was determined with gel electrophoresis combined with immunoblotting and immunoflourescence. Vimentin expression, traditionally a marker of cells of mesoderm origin (114), has been detected in ASTRs (114-117). Co-expression of vimentin, GFAP, and several keratins has been demonstrated in glial tumors, in formalin-fixed and paraffin-embedded tissues (113, 118). Co-expression of glial fibrillary acidic protein (GFAP) and neurofilament type of IFs have been found in childhood ASTRs and MED/PNETs (119). Keratin has been reported in some ASTRs, but only in epithelial-like areas of very anaplastic tumors (120, 121).

The keratins are water-insoluble proteins, with molecular weights varying between 40 and 70 kDa, that form 10 nM dermatome-associated IF in cells of epithelial origin and in certain specialized epithelia origin and in certain specialized epithelia such as the epidermis (122-129). Two-dimensional electrophoresis has shown expression of about 20 different keratin subunits in mammalian species (128, 129). Usually only a few subunits are present in the form of coordinated pairs containing at least one acidic and one neutral-basic keratin. The subunit composition of other IFs (glial, vimentin, desmin, and neurofilament) is relatively simple (130). The protein structure of keratin is highly heterogeneous and depends on cell type, stage of ontogenesis, stage of epithelial differentiation, and the modulation effects of the external microenvironment upon growth and proliferation (128,131-135).

Keratin IFs are composed of dimers, consisting of one acidic and one basic subunit, forming an α-helical structure. All known epidermal keratins contain a central α-helical domain of highly conserved size and secondary structure. Keratins of four molecular weights (50, 56.5, 58, and 65 to 67 kDa) have been identified in normal human epidermis. The basal cells express four: two neutral-basic units of 56 and 58 kDa and two acidic keratin units of 48 and 50 kDa. Terminally differentiated suprabasal epidermal cells express two keratin units: an acidic of 56.5 kDa and a basic of 67 kDa. Comparisons of the α-helical sequences demonstrate, however, the presence of two distinct keratin sequence types: acidic units are Type 1 and neutral-basic units are Type 2 (134).

The immunocytochemical detection of cell lineage specific and nonspecific IF proteins in childhood brain tumor cells of glial origin has been detected (135, 136). Our data indicate that there are at least four types of IF proteins in neoplastically transformed glial cells.

4.1 Results

Our library of MoABs was assembled to allow catechetical demonstration of four different protein antigens, typical for the well known human IF subtypes.

4.1.1 Glial Fibrillary Acidic Protein (GFAP)

All of the tumors contained cells that were highly reactive with anti-GFAP MoAB as expected. One AA expressed GFAP in 10 to 50% of cells. In all remaining glial tumors (15 out of 16), we detected positivity in >50% of cells and in 13 out of 16 in >90% of cells. The intensity of the immunoreactivity (staining) was always high (A,B) and the staining pattern varied between fibrillary cytoplasm (normal) and a totally uncharacteristic cytoskeleton organization in the form of connected fibrils and small and large aggregates.

4.1.2 Neurofilament Triplet Proteins

The expression of normal neuronal cell lineage-specific proteins was demonstrated employing three MoABs reacting with neurofilament triplet proteins of cytoskeleton IFs (137). The majority (13 out of 16) of ASTRs expressed NF (H) in >1% of cells. Two pilocytic ASTRs were extremely rich in NF containing cells (>90%). In AAs the results were variable; five out of seven tumors demonstrated positive reaction in >1% of the cells, but in one specimen with oligodendrocytic differentiation none of the three neurofilament proteins were present. NF (L) was detected in six out of 16 tumors in >1% cells. NF (M) was also present in >1% cells of six out of 16 glial tumors.

4.1.3 Vimentin Expression

All 16 glial tumors demonstrated intense qualitative and quantitative vimentin expression in >10% of cells, and in nine out of 16 cases in >90% of cells.

4.1.4 Epithelial Cell Lineage Specific Antigens

Expression of different molecular weight keratins was observed with the employment of five keratin-antigen specific commercially available anti-keratin MoABs, well characterized by the original authors and ICN Biomedicals (138).

Type 1, acidic keratin, was detected with MoAB AE1 in six out of 16 glial tumors in >1% of cells. A basic and acidic keratin pair was demonstrated with MoAB AE3 in 15 out of 16 tumors in >1% of cells. This Type 2, 56 to 58 kDa or 65 to 67 kDa keratin, was present in >10% of the cells in 13 out of 16 glial tumors and in three out of 16 (two AAs and one pilocytic ASTR) in >90% of cells.

MoAB AE2 defined presence of another "keratin pair," containing both acidic (56.6 kDa) and basic (65 to 67 kDa) keratins. We were able to demonstrate its presence in five out of 16 tumors in >1% of cells. MoAB AE2 detected the whole α-helix only in 1 to 10% of astrocytomas (ASTR) cells. One pilocytic ASTR simultaneously contained Type 1 and 2 keratins in >10 to 50% of cells and a different keratin pair, detected with MoAB AE2, in only 1 to 10% of cells.

The expression of a basic, 64 kDa keratin divided ASTR tumor cells in two subpopulations: $AE5^+$ and $AE5^-$. This keratin was present in 7/16 glial tumors of different levels of anaplasia in >90% of cells. Three out of 16 tumors contained smaller amounts of this 64kDa basic keratin, and in six out of 16 glial tumors, it was totally absent.

The majority of glial tumors (14/16) also expressed a 51 kDa acidic keratin (detected with MoAB AE8) in >1% of cells; in five out of 16 astrocytomas in 1 to 10% of cells, in 10 out of 16 cases in >10%, and in six out of 16 tumors in >50% of the cells. The expression of this acidic keratin was found in >90% of cells in only three out of 16 tumors.

The three most significant results: (*a*) low molecular weight (68 kDa) neurofilament protein was detected in six out of 16 ASTRs; (*b*) partial co-expression of at least four major IF protein subunits GFAP, vimentin, neurofilament (usually H and M molecular weight) and some keratin subclass was detected in all ASTRs; and (*c*) NF (H) expression was demonstrated in 15 out of 16 ASTRs.

Similar expression pattern of IFs was not found in MED/PNETs in frozen sections, and after formalin fixation.

Our single labeling immunocytochemical technique (119) combined with the use of strictly serial sections, allowed us to determine the simultaneous co-expression of two or more different IF protein subtypes sometimes in single, neoplastic glial, and juxtaposed cells. Our method is adequate for simultaneous demonstration only of two IF proteins. The astrocytomas are not so heterogeneous in their cell composition that two adjacent cells will be totally different. In our opinion, the same cell line malignant astrocytes contained four different IF proteins in their neoplastically transformed cytoskeletons. The polyclonal, less specific character of antibodies produced in animal species other than that of the mouse did not allow a correct simultaneous double labeling technique of one section, with two or more antibodies.

4.2 Discussion

Brain tumor histogenesis is a very complex process, with multiple variations. We have identified the co-expression of multiple IF proteins and abnormally organized cytoskeleton structures in neoplastically transformed astroglial cells. Their co-expression in primary childhood brain tumor cells of glial origin to date was never reported in the literature. The multiple IF proteins detected by our 10 MoABs make it clear that none of them identify a true tumor-specific antigen (TSA) (139, 140).

Systematic immunocytochemical typing of IF proteins and its use for tumor diagnosis and possible subclassification are based on the hypothesis that neoplastic cells retain the IF type characteristic of their cell of origin (105-108, 114, 140). However, tumors that contain additional IF subclass-related proteins have been reported in the last four years (108, 114, 116, 119,

121, 139, 141). In our opinion, neoplastic glial cells retain only the ontogenetic, developmental sequence in their IF protein expression and the neoplastically transformed cells undergo a clonal progression.

Adult ASTR tumors contain IF proteins reacting positively with at least two keratin-directed MoABs regardless of neuropathological subtype of tumors (113). However, these tumors were formalin-fixed and paraffin-embedded tissues (113). According to Zipser and Schley (142), immunocytochemistry of fresh, frozen, and fixed and paraffin-embedded tissues cannot produce the same results. GFAP and vimentin are co-expressed in normal, reactive, and neoplastically transformed ASTR in formalin-fixed, paraffin-embedded adult glial tumors (106, 113, 115, 116, 143). Indeed, Wang and co-workers (117) reported the co-polymerization of GFAP and vimentin to form heteropolymers within the cytoskeleton organization of neoplastic astrocytes. GFAP and neuron-specific enolase (NSE) are co-expressed by the same reactive astrocytes, using a double labeling immunofluorescent detection technique (144). In frozen sections, the IF NF (H) is present within the majority of glial tumor cells, in all of our 16 cases, including the giant cells of GBM (119). Our results are an addition to the immunocytochemical results in wax-embedded cells of 78 previously reported glial tumors (145). In addition, frozen section technique allowed detection of two other neurofilament sub proteins (M and L), in both anaplastic and pilocytic ASTRs in some children. In some tumor cells, the NF is distributed in the form of clumps, balls, or cords, consistent with a structural breakdown of the fine architecture of the cytoskeleton organization during neoplastic transformation (109, 124, 140, 146, 147).

We are suspicious of immunochemical results obtained only from formalin-fixed paraffin-embedded tissues, whether or not enzyme digestion (*e.g.*, protease, trypsin, ficin, etc.) is used. Some MoABs (*e.g.*, Lal-130) bind to different antigenic epitopes after different fixatives (142), and we think it is time to separate immunocytochemical results on fixed and paraffin-embedded tissues from those using fresh frozen tissue sections.

This study was performed exclusively on fresh frozen sections. We have demonstrated, for the first time, co-expression of four IF protein subclasses: vimentin, GFAP, NF (H and sometimes M and L), and a keratin subclass. Previous reports about the molecular markers of primitive neuroectodermal tumors (PNETs) (147-149) did not detect the simultaneous expression of these four IFs. The study of Molenaar and co-workers (148) employed only formalin-fixed and paraffin-embedded tissues, with emphasis on PNETs. The other study of Gould and co-workers (149) was carried out on frozen sections, but with emphasis on the neuroendocrine differentiation of PNETs, with 11 gliomas being used as control tissues. In both studies, the lack of a detailed description of the ABC immunohistochemical method employed resulted in the report failing to identify the kind of secondary antibody that was used, $F(ab)_2$ fragment or the whole molecule of goat-anti-mouse IgG (119).

Neurofilament (L, 68 kDa) protein is the first to appear from the NF triplet proteins during the normal human brain ontogenesis (150), certainly not before nestin, and probably not before α-internexin. The percentage of NF (L) positive tumor cells was low (under 10%) related to retinoblastoma (151). Moreover, the strong NF (L) positivity of retinoblastoma cells (151) contrasts with weak expression in these astrocytomas ("B" and sometimes "C" grades of intensity). However, controversy existed concerning the similarity of low molecular weight NF and GFAP proteins (152). GFAP isolated from glial cells and the 50 kDa or less molecular weight of NF protein detected in myelin-free axons were biochemically similar (150, 152), but after cloning and sequencing of all proteins it became clear that GFAP and NF-L are different IF proteins. NF-L is also a Type IV intermediate filament, while GFAP belongs to Type III. A vimentin-GFAP, both Type III, protein transition occurs at the time of hyalinization (152).

For glial cells, we propose that vimentin and keratin IFs are ontogenetic, differentiation-stage-dependent cytoskeleton proteins and only at a higher level of cell specialization do cell-type specific intermediate filaments (GFAP and neurofilament triplet proteins) appear.

The molecular weight of keratin IF proteins present in pediatric ASTRs can be estimated from the well-characterized immunoreactivity pattern of the five anti-epithelial (keratin) MoABs. In our study, we used MoABs as separate reagents, detecting Type I (AE1) and Type 2 (AE3) keratins. The keratin IF profile of glial tumor cells is dominated by a keratin pair (α-helix) (acidic, 55 kDa, and basic 64 kDa) detected with MoAB AE3, Type 2 keratin that has been designed as a marker of "corneal type of epithelial differentiation" (130, 133), suggesting that this type of epithelial differentiation occurs in glial tumors. Unfortunately, MoAB AE2 detected another keratin pair (acidic 56.6 kDa and basic 65 to 67 kDa) in the majority of glial tumors that cross reacts with filaggrin, and its significance remains uncertain. MoAB AE5, directed against an independent basic keratin (64 kDa), was inconsistently positive in childhood glial tumors.

Co-expression of glial, neuronal, mesenchymal, and epithelial IF subclass proteins supports both the "theory of differentiation" and the "theory of stop of ontogenetic differentiation" proposed by Gould (109) as a possible mechanism of the origin and histogenesis of primary brain tumors. We propose that acquisition of intense expression of vimentin by glial tumor cells may indicate a significant level of cellular dedifferentiation or undifferentiation or a significant change from higher to lower levels of cell metabolism, possibly accounting for more aggressive tumor behavior and progression. Further observations, employing immunoblot and Western blot analyses, will determine the ontogenetic significance of the cytoskeleton and the cytoskeleton organization of neoplastic neuronal and glial cells.

Recently, it is increasingly evident that the cytoskeleton of living cells plays important roles in mechanical and biological functions of the cells. IFs contribute to the mechanical behaviors of living cells. Vimentin, a major structural component of IFs in many cell types, is shown to play an important role in vital mechanical and biological functions such as cell contractility, migration, stiffness, stiffening, and proliferation (153).

Two decades ago, cytoplasmic IFs were considered to be stable cytoskeletal elements contributing primarily to the maintenance of the structural and mechanical integrity of cells. However, recent studies of living cells have revealed that IFs and their precursors possess a remarkably wide array of dynamic and motile properties (154). These properties are in large part due to interactions with molecular motors such as conventional kinesin, cytoplasmic dynein, and myosin. The association between IFs and motors appears to account for much of the well-documented molecular cross talk between IFs and the other major cytoskeletal elements, microtubules, and actin-containing microfilaments. Furthermore, the associations with molecular motors are also responsible for the high-speed, targeted delivery of nonfilamentous IF protein cargo to specific regions of the cytoplasm where they polymerize into IFs.

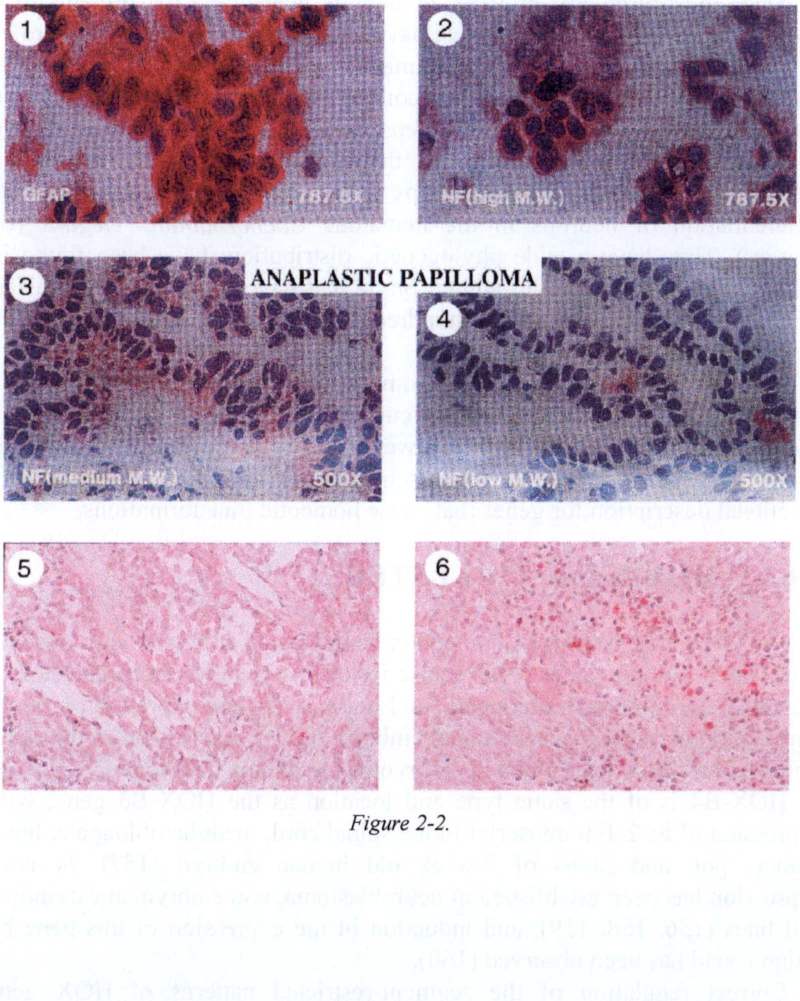

Figure 2-2.

1-4. Anaplastic papilloma of the choroid plexus. Intermediate filaments, GFAP and neurofilament subclasses detected employing an alkaline phosphatase conjugated immunohistochemical antigen detection. Magnification marked on the pictures.

5. MED/PNET. Pan-cytokeratin intermediate filament presence within the cytoplasms of tumor cells. Alkaline phosphatase conjugated, streptavidin-biotin enhanced immunohistochemical antigen detection technique. Magnification: 200x.

6. GBM. Expression of pan-cytokeratin among the intermediate filaments of this very malignant glial tumor. Alkaline phosphatase based immunohistochemistry. Magnification: 200x.

5. EXPRESSION OF HOMEOBOX B3, B4, AND C6 GENE PRODUCTS

Since their discovery in 1983, homeobox genes, and the proteins they encode, the homeodomain proteins, have turned out to play important roles in the developmental processes of many multicellular organisms. While certainly not the only developmental control genes, they have been shown to play crucial roles from the earliest steps in embryogenesis - such as setting up an anterior-posterior gradient in the egg of the fruit fly Drosophila melanogaster - to the very latest steps in cell differentiation - such as the differentiation of neurons in the nematode *Caenorhabditis elegans* (*C. elegans*). They have a wide phylogenetic distribution: have been found in baker's yeast, plants, and all animal phyla that have been examined so far. Since their original discovery, hundreds of homeobox genes have been described (155).

The first genes found to encode homeodomain proteins were Drosophila developmental control genes, in particular homeotic genes, from which the name "homeo"-box was derived. However, many homeobox genes are not homeotic genes; the homeobox is a sequence motif, while "homeotic" is a functional description for genes that cause homeotic transformations.

5.1 HOX-B and –C CLUSTERS

HOX-B3 is an Antp-type homeobox gene located at human chromosome 17q21-q22. The prevalent 3.9kb transcript of this gene has been detected in the spinal cord, medulla oblongata, and brain, with faint expression in lung, limbs, and gut of the 7-week human embryo. *In vitro* expression of the same fragment has been determined in embryonal carcinoma cells (156).

HOX-B4 is of the same type and location as the HOX-B3 gene, with expression of its 2.4kb transcript in the spinal cord, medulla oblongata, lung, kidney, gut, and limbs of 7-week old human embryo (157). *In vitro* expression has been established in neuroblastoma, and embryonal carcinoma cell lines (156, 158, 159), and induction of the expression of this gene by retinoic acid has been observed (160).

Correct regulation of the segment-restricted patterns of HOX gene expression is essential for proper patterning of the vertebrate hindbrain. The molecular basis of restricted expression of HOX-B2 in rhombomere 4 (r4) was examined by Maconochie and co-workers (161) using deletion analysis in transgenic mice to identify an r4 enhancer from the mouse gene. A bipartite HOX/PBX binding motif is located within this enhancer, and *in vitro* DNA binding experiments showed that the vertebrate labial-related protein HOX-B1 will cooperatively bind to this site in a PBX/EXD-dependent manner. The HOX-B2 r4 enhancer can be transactivated *in vivo* by the ectopic expression of HOX-B1, HOX-A1, and Drosophila labial in transgenic mice. In contrast, ectopic HOX-B2 and HOX-B4 are unable to induce expression, indicating that *in vivo* this enhancer preferentially responds to labial family members. Mutational analysis demonstrated that the bipartite HOX/PBX motif is required for r4 enhancer activity and the responses to retinoids and ectopic HOX expression. Furthermore, three copies of the HOX-B2 motif are sufficient to mediate r4 expression in transgenic mouse embryos and a labial pattern in Drosophila embryos. This reporter expression in Drosophila embryos is dependent upon endogenous *labial* and *exd*, suggesting that the ability of this HOX/PBX site to interact with labial-related proteins has been evolutionarily conserved. The endogenous HOX-B2 gene is no longer upregulated in r4 in HOX-B1 homozygous mutant embryos. On the basis of these experiments it has been concluded that the r4-restricted domain of HOX-B2 in the hindbrain is the result of a direct cross-regulatory interaction by HOX-B1 involving vertebrate PBX proteins as cofactors. This suggests that part of the functional role of HOX-B1 in maintaining r4 identity may be mediated by the HOX-B2 gene.

During mouse hindbrain development, HOX-B3 and HOX-B4 share an expression domain caudal to the boundary between rhombomeres 6 and 7. Transgenic analysis reveals that an enhancer (CR3) is shared between both genes and specifies this domain of overlap. Both the position of CR3 within the complex and its sequence are conserved from fish to mammals, suggesting it has a common role in regulating the vertebrate HOX-B complex. CR3 mediates transcriptional activation by multiple HOX genes, including HOX-B4, HOX-D4, and HOX-B5, but not HOX-B1. It also functions as a selective HOX response element in Drosophila, where activation depends on Deformed, Sex combs reduced, and Antennapedia but not labial. Taken together, these data show that a Deformed/HOX-B4 autoregulatory loop has been conserved between mouse and Drosophila. In addition, these studies reveal the existence of positive cross-regulation and enhancer sharing as two mechanisms for reinforcing the overlapping expression domains of vertebrate HOX genes. In contrast, Drosophila HOX genes do not appear to share enhancers and where they overlap in expression, negative cross-regulatory interactions are observed. Therefore, despite many well documented aspects of HOX structural and functional conservation, there are mechanistic differences in HOX complex regulation between arthropods and vertebrates (162).

HOX genes may also play important lineage-specific functions in a variety of somatic tissues including the hematopoietic system. It has been shown that certain members of the HOX A and B clusters, such as HOX-B3 and HOX-B4, are preferentially expressed in subpopulations of human bone marrow that are highly enriched for the most primitive hematopoietic cell types. To assess the role these genes may play in regulating the proliferation and/or differentiation of such cells, the overexpression of HOX-B4 in murine bone marrow cells was accomplished by retroviral gene transfer and analyzed subsequent effects on the behavior of various hematopoietic stem and progenitor cell populations both *in vitro* and *in vivo*. Serial transplantation studies revealed a greatly enhanced ability of HOX-B4-transduced bone marrow cells to regenerate the most primitive hematopoietic stem cell compartment resulting in 50-fold higher numbers of transplantable totipotent hematopoietic stem cells in primary and secondary recipients, compared with serially passaged neo-infected control cells. This heightened expansion *in vivo* of HOX-B4-transduced hematopoietic stem cells was not accompanied by identifiable anomalies in the peripheral blood of these mice. Enhanced proliferation *in vitro* of day-12 CFU-S and clonogenic progenitors was also documented. These results indicate HOX-B4 to be an important regulator of very early but not late hematopoietic cell proliferation and suggest a new approach to the controlled amplification of genetically modified hematopoietic stem cell populations (163).

HOX-B5 and HOX-B6 are adjacent genes in the mouse HOX-B locus and are members of the homeotic transcription factor complex that governs establishment of the mammalian body plan. To determine the roles of these genes during development, mice with a targeted disruption in each gene were generated by Rancourt and co-workers (164). Three phenotypes affecting brachiocervicothoracic structures were found in the mutant mice. First, HOX-B5$^-$ homozygotes have a rostral shift of the shoulder girdle, analogous to what is seen in the human Sprengel anomaly. This suggests a role for HOX-B5 in specifying the position of limbs along the anteroposterior axis of the vertebrate body. Second, HOX-B6$^-$ homozygotes frequently have a missing first rib and a bifid second rib. The third phenotype, an anteriorizing homeotic transformation of the cervicothoracic vertebrae from C6 through T1, is common to both HOX-B5$^-$ and HOX-B6$^-$ homozygotes. Quite unexpectedly, HOX-B5, HOX-B6 transheterozygotes (HOX-B5$^-$HOX-B6$^+$/HOX-B5$^+$HOX-B6$^-$) also show the third phenotype. By this classical genetic complementation test, these two mutations appear as alleles of the same gene. This phenomenon is termed nonallelic noncomplementation and suggests that these two genes function together to specify this region of the mammalian vertebral column.

An epitope-specific antibody against the protein product of the murine HOX-B5 gene was used to select an enriched library of HOX target sequences. Nucleotide sequencing identified a 910 bp DNA fragment, containing a consensus Antennapedia-like binding site, and identical in 640 bps at 3' end of the clone to the promoter of the SPI3 gene, which encodes a serine protease inhibitor protein (165). Using a 120 bp cDNA fragment as probe, SPI3 expression was detected mainly in the CNS of 15 day mouse embryos, a pattern which is similar to that of the HOX-B5 gene at this stage (166). Thus, the SPI3 gene is a candidate target of the HOX-B5 gene in vertebrate embryos (167).

The HOX-C cluster genes represent Antp-type homeobox genes, involved in transcriptional activation during embryonic development. The HOX-C6 homeobox is present in at least three different mRNAs, two of which have been somewhat characterized (C8S and C8L). The entire C8S (S for short) fragment is contained within the C8L fragment, with C8L containing an extra domain at the amino terminus. The transcripts coexist in the various cells (fibroblasts) and tissues (spinal cord, backbone, limbs, and skin) examined so far, but their ratio is variable. The expression of the HOX-C6 gene and other HOX-C cluster genes has been observed in various human malignancies.

5.1.1 RESULTS

The expression pattern of the HOX-B3, HOX-B4, and HOX-C6 homeobox gene products was examined immunocytochemically in childhood MEDs/PNETs employing an indirect, alkaline phosphatase conjugated antigen detection technique on formalin-fixed, paraffin-embedded tissue sections. Strong immunoreactivity (staining intensity) (A, B) of HOX-B3 and HOX-B4 was identified in the neuroectodermally derived, undifferentiated neoplastic cells of all MEDs/PNETs. The immunoreactivity was present in between 50% and 90% (+++), but usually over 90% (++++) of the neoplastic cells. HOX-C6 was detected at medium intensity (mostly B) in 50% to 90% (+++) of the MED/PNET cells.

5.2 HOX Genes and Cancer

Homeobox genes are broadly classified into two subclasses: HOX and non-HOX homeobox genes. A number of genes in both classes are expressed in a variety of hematopoietic cells. Two major categories of evidence implying the involvement of these genes in normal hematopoiesis have been demonstrated. First, the expression pattern of the homeobox genes in hematopoietic cells is lineage- and differentiation-stage specific. Second, enforced and suppressed expression of various homeobox genes cause defects in the hematopoietic cells of specific lineages. The reduction in myeloid, erythroid and B cell progenitors is found in mice with a disrupted HOXA9 gene. The thymuses of HOXB3-overexpressed marrow recipients contain a markedly decreased number of CD4/CD8 double-positive T cells. These examples suggest that the proper level of expression and timely down-regulation of some homeobox genes are necessary for normal hematopoiesis. Homeobox genes are also implicated in human and mouse leukemias. In human leukemias, a HOX gene (HOXA9) and two non-HOX homeobox genes (PBX1 and HOX11) are involved in chromosomal translocations. In pre-B acute lymphoblastic leukemia (ALL) with t(1;19) translocation, a fusion protein is created between E2A and a homeobox gene PBX. In T-cell ALL with t(10;14) translocation, the HOX 11 gene is deregulated. In acute myeloid leukemia (AML) with t(7;11) translocation, the HOX A9 gene is rearranged (168-170). In mouse leukemias, provirus integrations cause aberrant expression of several HOX and non-HOX genes (171).

Retroviruses induce myeloid leukemia in BXH-2 mice by the insertional mutation of cellular proto-oncogenes or tumor suppressor genes. Disease genes can thus be identified by proviral tagging through the identification of common viral integration sites in BXH-2 leukemia. Three genes whose expression is activated by proviral integration in BXH-2 leukemias have been described: HOX-A7, HOX-A9, and a PBX1-related homeobox gene, Meis1. Proviral activation of HOX-A7 or HOX-A9 is strongly correlated with proviral activation of Meis1 implying that HOX-A7 and HOX-A9 cooperate with Meis1 in leukemia formation (172, 173).

The relative levels of expression of HOX and related genes in phenotypically and functionally defined subpopulations of AML blasts and normal hematopoietic cells have been observed. Initially, a semi-quantitative RT-PCR technique was used to amplify total cDNA from total leukemic blast cell populations from 20 AML patients and light density cells from four normal bone marrows. Expression of HOX genes (A9, A10, B3 and B4), MEIS1 and MLL was easily detected in the majority of AML samples with the exception of two samples from patients with AML subtype M3 (which expressed only MLL). Low levels of HOXA9 and A10 but not B3 or B4 were seen in normal marrow while MLL was easily detected. PBX1a was difficult to detect in any AML sample but was seen in three of four normal marrows. Cells from nine AML patients and five normal bone marrows were FACS-sorted into $CD34^+CD38^-$, $CD34^+CD38^+$ and $CD34^-$subpopulations, analyzed for their functional properties in long-term culture (LTC) and colony assays, and for gene expression using RT-PCR. 93 ± 14% of AML LTC-initiating cells, 92 ± 14% AML colony-forming cells, and >99% of normal LTC-IC and CFC were $CD34^+$. The relative level of expression of the four HOX genes in amplified cDNA from $CD34^-$ as compared to $CD34^+CD38^-$ normal cells was reduced >10-fold. However, in AML samples this down-regulation in HOX expression in $CD34^-$ as compared to $CD34^+CD38^-$ cells was not seen ($p < 0.05$ for comparison between AML and normal). A similar difference between normal and AML subpopulations was seen when the relative levels of expression of MEIS1, and to a lesser extent MLL, were compared in $CD34^+$ and $CD34^-$ cells ($p < 0.05$). In contrast, while some evidence of down-regulation of PBX1a was found in comparing $CD34^-$ to $CD34^+$ normal cells it was difficult to detect expression of this gene in any subpopulation from most AML samples. Thus, the down-regulation of HOX, MEIS1 and to some extent MLL which occurs with normal hematopoietic differentiation is not seen in AML cells with similar functional and phenotypic properties (174, 175).

HOX-A9, Meis1 and PBX1 encode home domain containing proteins implicated in leukemic transformation in both mice and humans. HOX-A9, Meis1 and PBX1 proteins have been shown to physically interact with each other, as HOX-A9 cooperatively binds consensus DNA sequences with Meis1 and with PBX1, while Meis1 and PBX1 form heterodimers in both the presence and absence of DNA. Whether HOX-A9 could transform hemopoietic cells in collaboration with either PBX1 or Meis1 was investigated by Kroon and co-workers (176). Primary bone marrow cells, retrovirally engineered to overexpress HOX-A9 and Meis1a simultaneously, induced growth factor-dependent oligoclonal acute myeloid leukemia in <3 months when transplanted into syngenic mice. In contrast, overexpression of HOX-A9, Meis1a or PBX1b alone, or the combination of HOX-A9 and PBX1b failed to transform these cells acutely within 6 months post-transplantation. Similar results were obtained when FDC-P1 cells, engineered to overexpress these genes, were transplanted to syngenic recipients. These studies demonstrate a selective collaboration between a member of the HOX family and one of its DNA-binding partners in transformation of hemopoietic cells.

Retrovirally engineered overexpression of a 5'-located HOX gene, HOXA10, in murine bone marrow cells demonstrated effects strikingly different from those induced by overexpression of a 3'-located gene, HOXB4. In contrast to HOXB4, which causes selective expansion of primitive hematopoietic cells without altering their differentiation, overexpression of HOXA10 profoundly perturbed myeloid and B-lymphoid differentiation. The bone marrow of mice reconstituted with HOXA10-transduced bone marrow cells contained in high frequency a unique progenitor cell with megakaryocytic colony-forming ability and was virtually devoid of unilineage macrophage and pre-B-lymphoid progenitor cells derived from the transduced cells. Moreover, and again in contrast to HOXB4, a significant proportion of HOXA10 mice developed a transplantable acute myeloid leukemia with a latency of 19 to 50 weeks. These results thus add to recognition of HOX genes as important regulators of hematopoiesis and provide important new evidence of HOX gene-specific functions that may correlate with their normal expression pattern (177).

Alterations in HOX B6 and HOX B9 mRNA expression were observed during the *in vitro* differentiation of four myeloid leukemia cell lines because HOX B6 may be involved closely in myeloid differentiation. HL-60, NB4, NKM-1 and NOMO-1 were established from acute leukemia of M2, M3, M2 and M5 subtype of the French-American-British classification, respectively. All-trans retinoic acid (ATRA), TPA, and G-CSF were used as differentiation inducers. Each cell line was cultured with each inducer and total RNA was isolated on day 1, 2, 3, or 5. HOX B mRNA was detected by Northern blotting and RT-PCR methods. HOX B6 and HOX B9 mRNAs were constitutively expressed in NB4, NKM-1 and NOMO-1, but were expressed at very low levels in HL-60. HOX B6 and HOX B9 mRNAs were also expressed in fresh acute myelocytic leukemia blasts. HOX B6 mRNA expression in HL-60, NB4, and NKM-1 cultured with ATRA increased on day 3 and decreased on day 5. HOX B6 mRNA expression in NB4 and NKM-1 cultured with TPA decreased on day 3. HOX B9 mRNA expression displayed changes similar to those of HOX B6 mRNA in NB4 and NKM-1. These results indicate that myeloid leukemia cell lines express HOX B6 and HOX B9, and that their respective mRNA expressions in NB4 and HL-60 increase at a mid stage of myeloid differentiation by ATRA induction and then decrease during a late stage. HOX B6 mRNA expression decreased in monocytoid differentiation by TPA induction in NB4, HL-60 and NKM-1. HOX B6 antisense-oligonucleotide inhibited the proliferation of NB4 and NKM-1. These results suggest that HOX B gene expression is related to simultaneous activation of cellular proliferation and differentiation in leukemic cells (178).

HOX genes have shown a lineage-specific expression in hematopoiesis and are suggested as being involved in the expression of certain adhesion molecules in normal and neoplastic tissues (179). It has been demonstrated that HOXC4 and HOXC6, but not HOXC5, are expressed during lymphoid differentiation (180, 181), and HOXC4 has also been detected on activated lymphocytes (182). The expression of HOXC4, HOXC5 and HOXC6 was investigated in purified subpopulations of bone marrow in addition to 36 specimens of acute myeloid leukemias (AMLs), eight chronic myeloid leukemias (CMLs), several myeloid cell lines and cutaneous localizations of three myelomonocytic leukemias and one granulocytic sarcoma by RT-PCR and partly by RNA *in situ* hybridization (RISH). HOXC4 and HOXC6 transcripts were both detected by RT-PCR in 22/36 and 24/36 AMLs, respectively. The distribution of HOXC4 and HOXC6 gene expression over the different types of AML was largely similar and covered all types of AML. In contrast, HOXC5 gene expression was found in only 6/32 AMLs.

Expression of HOXC5 was restricted to AMLs of the granulocytic (FAB M1-M3), early monocytic (FAB M4) and early erythroid (FAB M6) lineage. In general, except in one FAB M5b case, no expression of HOXC5 was found in AMLs derived from late stages of monocytic (FAB M5) and megakaryocytic (FAB M7) lineages. As for HOXC4 and HOXC6, expression of HOXC5 was absent in CMLs. Using RISH significant HOXC4, HOXC5 and HOXC6 expression was found in a number of additionally studied AML samples of different FAB classification (M2, M4, M5b and M5b), (M2 and M5b) (M2, M4, M5b), respectively. In tissue localizations of leukemias a different expression pattern of HOXC4, HOXC5 and HOXC6 was found. In contrast to mature leukemic stages of myeloid differentiation, these skin localizations of leukemias expressed HOXC5 and HOXC6. HOXC4 expression was found both in leukemic cells derived from peripheral blood and from cutaneous localizations. Besides HOXC4 expression in monocytes, no expression of HOXC4, HOXC5 and HOXC6 was found in granulocytes and monocytes, colonies of growth factor-induced $CD34^{+}$ bone marrow cells. In earliest $CD34^{+}/CD38^{low/high}$ cell fractions of bone marrow only HOXC4 and in megakaryocytic cells both HOXC4 and HOXC6 were found. Thus, the expression patterns of these HOXC genes found in the limited number of cell fractions of normal bone marrow suggest that the expression patterns found in AMLs and CMLs might reflect the normal situation. Furthermore, the presence of HOXC5 and HOXC6 expression specifically in skin infiltrates of late differentiation stages of myeloid leukemias, suggests an additional role for these genes in the positioning of these myeloid cells in skin tissue (183).

HOXC4 and HOXC6 gene expression levels have been found to increase with differentiation of lymphoid cells. In contrast, HOXC5 is not expressed in the lymphoid lineage, but was found in lymphoid cell lines, representing the neoplastic equivalents of various differentiation stages of T and B lymphocytes. The expression pattern of HOXC4, HOXC5, and HOXC6 was investigated in 89 non-Hodgkin's lymphomas (NHLs) of different histologic subtypes and originating from different sites. Using RNA *in situ* hybridization and semiquantitative reverse transcription-polymerase chain reaction, expression of HOXC4 was observed in 83 of 88 and HOXC6 in 77 of 88 NHLs and leukemias investigated. In contrast, HOXC5 expression was found in only 26 of 87 NHLs and appeared to be preferentially expressed by two specific subsets of lymphomas, *i.e.* primary cutaneous anaplastic T-cell lymphomas (9 of 9) and extranodal marginal zone B-cell lymphomas (maltomas; 7 of 9). These results indicate that, in contrast to HOXC4 and HOXC6, HOXC5 shows a type- and site-restricted expression pattern in both T- and B-cell NHLs (184). Strong expression of HOXC4 and HOXC6 was also detected in all, except one, primary cutaneous lymphomas and all reactive cutaneous lymphoid infiltrates. Interestingly, a strong expression of HOXC5 in primary anaplastic $CD30^+$ large T-cell lymphomas was found. RISH was consistently negative for HOXC5 in all other types of primary cutaneous B- and T-cell lymphomas. However, by semiquantitative RT-PCR these lymphomas showed a weak expression of HOXC5 mRNA. Therefore, it was concluded that these lymphomas express low constitutive levels of HOXC5 mRNA. Furthermore, HOXC5 expression was consistently absent in reactive cutaneous lymphoid infiltrates, hyperplastic tonsils and lymph nodes, and peripheral blood lymphocytes either unstimulated or stimulated by a cocktail of CD3 and CD28 antibodies. As a strong expression of HOXC5 in primary cutaneous lymphomas was observed only in anaplastic large T-cell lymphomas and reactive control tissues lacked HOXC5 expression, these data strongly support a role for HOXC5 in the genesis of anaplastic large-T-cell lymphomas (185).

The various genes of the multiple HOX clusters have also been detected in a variety of epithelial tumors. An exhaustive survey of the expression of the 39 class I HOX genes was performed on normal and malignant human cervical keratinocytes by Alami and co-workers (186). The vast majority (34/39) of HOX genes are expressed in normal keratinocytes. Only HOXA2, HOXA7, HOXC5, HOXC8 and HOXD12 were found to be silent. Interestingly, this pattern is conserved in transformed keratinocytes (SiHa cells) except for the appearance of HOXC5 and HOXC8 mRNA. HOXC5 and HOXC8 expression was also observed in two other transformed keratinocytes cell lines of independent origins, Eil-8 and 18-11S3, and confirmed by *in situ* hybridization.

The profiling of differentially expressed genes from primary tumor samples using cDNA expression array can reveal new tumor markers as well as target genes for therapeutic intervention. Using cDNA expression array technology, an expression profile of genes that are associated with human cervical cancer was produced. Hybridization of the cDNA blotting membrane (588 genes on a single membrane) was performed with ^{32}P-labeled cDNA probes synthesized from RNA isolated from either normal cervix or cervical cancer. Parallel analyses of the hybridized signals enabled us to profile genes that were differentially expressed in cervical cancer. In each experiment, the extent of hybridization of each gene was evaluated by comparison with the most abundant mRNAs in the human cervix. These include myc proto-oncogene, 40S ribosomal protein S19, heat shock proteins, leukosialin S (CD43), integrin αL (CD11A), calgranulin (A), and CDK4 inhibitor (p16ink4). No detectable changes were observed in the expression levels of these genes. Several mRNAs, such as those encoding guanine nucleotide-binding protein Gs (α subunit), leukocyte adhesion protein (LFA1β), nuclear factor NF45, homeobox protein HOX-A1, and β-catenin were detected in increased levels in cervical cancer. Genes that showed decreased expression in cervical cancer tissue were a group of apoptosis-related proteins, cell adhesion molecules, nuclear transcription factors, and a homeobox protein (HOX7). For example, the expression levels of Smad1 and HOX7 were consistently decreased in all tumor tissues tested. Northern analysis of Smad1 and HOX7 RNA in primary cervical tumor tissues and cervical carcinoma cell lines indicated that, in general, the mRNA levels of these genes were decreased in human cervical cancer. The precise relationship between the altered expression of these genes and cervical tumorigenesis is a matter of further investigation (187). The relative expression of homeobox genes has also been studied in endometrial carcinomas. HOXD10 expression in human endometrial adenocarcinomas was assessed by densitometric comparison of co-amplified ^{32}P-labeled gene products separated by agarose gel electrophoresis. MRNA expression of HOXD10 relative to β-tubulin is significantly lower in endometrial carcinomas than in normal endometrium. Furthermore, the ratio of HOXD10 to β-tubulin expression was found to vary inversely with the histologic grade of the tumor (188).

Retinoic acid (RA) modulated the expression of genes encoding the HOX family of transcription factors, whose differential expression orchestrates developmental programs specifying anterior-posterior structures during embryogenesis, possibly inducing various effects on HOX gene expression based on the subunit organization of its receptors (189-191). HOX proteins bind DNA as monomers and heterodimers with PBX proteins. RA was shown to upregulate PBX protein abundance coincident with transcriptional activation of HOX genes in P19 embryonal carcinoma cells undergoing neuronal differentiation. However, in contrast to HOX induction, PBX upregulation is predominantly a result of post-transcriptional mechanisms. Interestingly, PBX1, PBX2, and PBX3 exhibit different profiles of upregulation, suggesting possible functional divergence. The parallel upregulation of PBX and HOX proteins in this model suggests an important role for transcriptional control by PBX-HOX heterodimers during neurogenesis, and argues for precise control by RA (192). At both high and low retinoic acid concentrations, HOX2 genes are sequentially activated in embryonal carcinoma cells in the 3' to 5' direction (193).

The products of PBX homeobox genes, which were initially discovered in reciprocal translocations occurring in human leukemias, have been shown to cooperate in the *in vitro* DNA binding with HOX proteins. Despite the growing body of data implicating HOX genes in the development of various cancers, little is known about the role of HOX-PBX interactions in the regulation of proliferation and induction of transformation of mammalian cells. It has been shown that both cellular transformation and proliferation induced by HOX-B4 and HOX-B3 are greatly modulated by the levels of available PBX1 present in these cells. Furthermore, the transforming capacity of these two HOX proteins depends on their conserved tetrapeptide and homeodomain regions which mediate binding to PBX and DNA, respectively. Thus, cooperation between HOX and PBX proteins modulates cellular proliferation and cooperative DNA binding by these two groups of proteins may represent the basis for HOX-induced cellular transformation (194-195).

The expression of HOX genes in primary and metastatic human small-cell lung cancer (SCLC) xenografted in nude mice was observed, and correlations with the histology and stage of SCLC progression were investigated by Tiberio and co-workers (196). The results show that different SCLCs display differential patterns of HOX gene expression. Furthermore, in SCLC, the number of actively expressed HOX genes might be substantially lower in metastatic cancers than in primary tumors. The alteration in HOX gene expression in SCLCs mainly concerns the HOX-B and -C loci. This finding suggests that downregulation of HOX genes may play a role in small-cell lung cancer progression, possibly through their implication in tumor suppression.

The expression patterns of the class I homeogenes HOX-B and HOX-C clusters in the presence of retinoic acid (RA) were studied in two human small-cell lung cancer (SCLC) cell lines and compared to that of NT2/D1 embryonal carcinoma cells. Contrasting with the sequential 3'-5' induction of the HOX genes observed after RA treatment of embryonic NT2/D1 cells, in the SCLC cells the responding genes (induced or down-regulated) were interspersed with insensitive genes (expressed or unexpressed), while no genomic alteration affected the corresponding clusters. These findings imply that HOX gene regulatory mechanisms are altered in non-embryonic SCLC cells, perhaps reflecting their ability to respond to more diversified stimuli, in relation with their origin from adult tissues (197). The expression pattern of the HOX-B gene cluster in four xenografted small-cell lung cancers was also compared to the methylation of the DNA in the corresponding genomic regions. In 90% (17/19) of the studied cases, the expressed genes were in methylated regions whereas 70% (12/17) of the unexpressed genes were in unmethylated regions. This specific behavior could correspond to a particular gene expression regulation mechanism of the HOX gene network. Since some genes (HOX-B2, HOX-B4, HOX-B7) were always inactive when unmethylated, this unexpected relationship might indicate their key function(s) in the HOX gene network (198).

Molecular characterization of three newly cloned HOXA1 transcripts from human breast cancer cells was reported by Chariot and co-workers (199). In addition, these alternatively spliced transcripts encode one homeodomain-containing protein and two products lacking the conserved DNA-binding domain. Moreover, the authors demonstrated that all three HOX-A1 transcripts are induced by retinoic acid, as well as progestin in MCF7 cells, suggesting that the HOX-A1 gene may be a key element in the establishment of the breast cancer cell phenotype (200).

Aberrant expression of HOX genes has also been reported in yolk sac tumors (primitive gonadal tumors) (201), primary and metastatic colorectal cancers (202), neuroblastomas (160, 203, 204), kidney cancer (205), and many articles discussing and reviewing their importance in mammalian embryogenesis and pathogenesis have appeared in the medical literature (206-211).

Angiogenesis is characterized by distinct phenotypic changes in vascular endothelial cells (EC). Evidence has emerged that the HOX-D3 homeobox gene mediates conversion of endothelium from the resting to the angiogenic/invasive state. Stimulation of EC with basic fibroblast growth factor (bFGF) resulted in increased expression of HOX-D3, integrin $\alpha v\beta 3$, and the urokinase plasminogen activator (uPA). HOX-D3 antisense blocked the ability of bFGF to induce uPA and integrin $\alpha v\beta 3$ expression, yet had no effect on EC cell proliferation or bFGF-mediated cyclin D1 expression. Expression of HOX-D3, in the absence of bFGF, resulted in enhanced expression of integrin $\alpha v\beta 3$ and uPA. In fact, sustained expression of HOX-D3 *in vivo* on the chick chorioallantoic membrane retained EC in this invasive state and prevented vessel maturation leading to vascular malformations and endotheliomas. Thus, it is quite clear that HOX-D3 regulates EC gene expression associated with the invasive stage of angiogenesis (212). Similar upregulation of bFGF by another HOX gene, HOX-B7, has been reported in melanomas (213).

The functions of the HOX gene products may be modulated by a variety of secreted factors, such as growth factors, cytokines and hormones, as alluded to previously in this article. It has been shown that expression of HOX-C6 in osteosarcomas and neuroblastomas is differentially regulated by rhBMP-2, TGF-β and activin-A. This suggests that specific HOX genes may be target genes for TGF-β superfamily members, and may represent an additional way in which growth factors exert their immense effects on development and carcinogenesis (214).

Homeobox genes have been shown to be intimately involved in the normal development of the nervous system. Our report of the expression of the HOX-B3, HOX-B4, and HOX-C6 gene products in childhood MEDs/PNETs represents one of the first describing the immunohistochemical detection of homeobox gene expression in intracranial malignancies. The exact mechanisms which regulate the re-expression of this class of oncofetal antigens in childhood brain tumors, as well as other solid tumors should be elucidated through further basic research. In conclusion, the utility of the expression of the various homeobox gene products as detected by immunocytochemical approaches in the earlier diagnosis and improved prognostication of solid human neoplasms, especially in childhood MEDs/PNETs should also be addressed by future scientific endeavors.

6. CELL PROLIFERATION

Uncontrolled cell proliferation is one of the most important signs of cellular malignancy and reflects significant genetic changes (215, 216). In the course of the past two years, numerous studies have sought to determine the dynamic cell kinetics of a variety of human neoplasms, which has proven helpful in the understanding of tumor growth patterns and contributed to the development of more effective anti-neoplastic therapy improving the prognosis associated with human cancer (217-221).

Enzinger and co-workers (222) in the WHO's *International Classification of Tumours* state that the presence of cell mitoses is a basic morphologic and diagnostic parameter in malignant tumors. Proliferative characteristics of brain tumors have been the subject of numerous studies, with the specific aim to correlate chemotherapeutical clinical trials to cell proliferation and cell cycle events (223-226). Among the primary brain tumors, GBM and anaplastic gliomas have been found to have the greatest growth fraction (227). In direct correlation, the mean labeling index (MLI) was found to be low (1% or even lower) in the well differentiated and clinically non-aggressive forms of ASTRs (228). Unfortunately a clear, molecular biologic marker for the potential of progressive disease and recurrence risk remains unavailable, but there is increasing experimental evidence that some proliferation markers can provide significant information (229-232). Intratumoral proliferative activity is an important factor in the prediction of further tumor biologic behavior, and can be a serious guide for the oncologist in the choice of correct treatment modality (233).

Among others, proliferating cell nuclear antigen (PCNA) has demonstrated a reliable correlation between premalignant lesions and the progressive potential for fully developed neoplastic disease (234). PCNA is a 46 kDa acidic, non-histone, nuclear protein required for DNA synthesis, and acts as the auxiliary protein of DNA-polymerase delta (235). Several experimental observations employing anti-PCNA MoABs have identified that PCNA accumulates in the cell nuclei during the S phase and can be considered as a proliferation marker (236). Eighty out of 250 cases of ASTR glioma collected from a practice served by a single clinical team over a 15-

year period were studied employing a full complement of clinical, follow up, histopathological analysis and PCNA immunocytochemistry for the determination of the PCNA labeling index (237). Cruz-Sanchez and co-workers sought to clarify the importance of various parameters as predictors of tumor behavior. A significant correlation with survival was found with histological grouping and PCNA labeling index. Furthermore, the authors report that the utilization of objective values (mitosis, cellular density and necrosis) appears to be useful in grading ASTR tumors.

Immunocytochemistry has proved to be a powerful method in the microscopic diagnosis, stage evaluation and prognosis prediction of brain tumors (238). The leading role of various growth factors, growth factor receptors, and tyrosine kinases in the control of human cell proliferation is now well established (227, 228, 239). Until recently, the actual, intimate steps of signal transduction, signal amplification, and the biochemical pathways stimulated by the mitogenic growth factors remained unresolved (240, 241). Natural polypeptide growth factors are able to direct the cascade of protein phosphorylation, using the activity of intracellular second mediators (241, 242).

Howard and Pelc (243) introduced the concept of the cell cycle in its present form, subdivided into four consecutive, G1, S, G2 and M phases (244). Most biochemically specific events that are cell cycle related have been discovered. The two most important events common to all cell cycles are S-phase, when chromosomes are replicated, and M-phase, when the replicated chromosomes are segregated into two daughter cells (245). It is during G1, the restriction point of somatic cell cycle, that external serum factors such as platelet derived growth factor (PDGF), epidermal growth factor (EGF), and insulin-like growth factor 1 (IGF-1) act to stimulate the transition to S-phase (239, 241, 242).

The existence of human nuclear antigens associated with cell proliferation was suggested by demonstration of autoantibodies in the sera of lupus erythematosus patients (246-248). The kinetics of brain tumor cells can be estimated, especially in the more rapidly growing tumors such as AAs, GBMs, MEDs and meningeomas (224,225,249-253). Expression of a human nuclear antigen, Ki-67, was detected in proliferating and dividing cells and a monoclonal antibody raised against it was used to recognize the growth fraction of surgical specimens of brain tumors (233, 255-264). This marker was present after the G1/S boundary (240), during S-, G2- and M-phase of the mitotic cycle (253). The Ki-67 antigen is denaturated by formalin fixation; therefore, immunocytological detection requires the use of frozen sections. Using the AMeX-technique for fixation and paraffin embedding, a German group of investigators was able to demonstrate Ki-67 expression in fixed and paraffin embedded neoplasm specimens (227,265).

The established cell proliferation marker designated proline-directed protein kinase (PDPK) is present in both normal and neoplastic cells during their whole mitotic cycle (266, 267). $p34^{cdc2}$ is a protein kinase that plays an important role in the control of the cell cycle and regulation of activity of the tumor suppressor gene (268). The G2-to-M transition represents an especially important cell cycle checkpoint that is regulated by $p34^{cdc2}$ (269). The M-Phase promoting factor (MPF) is actually a complex of $p34^{cdc2}$ and cyclin B (270). The more nuclearly rather than cytoplastically located PDPK is a heterodimer comprised of the catalytic subunit $p34^{cdc2}$ and the regulatory subunit $p58^{cyclinA}$ and is activated by tyrosine phosphorylation of the regulatory subunit (267). The induction of the cyclin A occurs late in the G1 stage, provoking the activation of the PDPK heterodimer complex possibly at the G1/S boundary (244,271). The persistence of the enzyme activity throughout the S phase provides a unique biochemical indicator of cell status: revealing not only which tumor cells are capable of dividing, but the proportion of cells that are actually committed to and progressing through these particular stages of the cell cycle (272-281).

In the last decade, G1-phase specific protein, named "cyclin A" was isolated in mammals (239-242, 267, 282). The discovered proline directed serine/threonine protein kinase (PDPK), in fact, represents a cytosolic kinase system to the biochemical signaling cascades activated by growth factors (241).

Nishizaki and co-workers (283) used flow-cytometric DNA analysis in 48 brain tumors, together with immunohistochemical measurement using MoAB Ki-67 and compared the results with BudR indices. Three PNETs were investigated for their BudR index, but only two results were mentioned - 15.0% and 7.3%, and only one Ki-67 index, which was extremely high at 26.0%. Nineteen glial tumors were observed. Two GBMs had a high Ki-67 index 18.2% and 18.8%.

The group of Deckert (277) observed proliferative activity in 133 brain tumors (exclusively in frozen sections). The PNET group was divided into nine MEDs with Ki-67 labeling indices 5%-42% (x = 17.9%) and two PNETs with 13.4%-14.0% (x = 13.7%). The calculated mean Ki-67 labeling indices, ranging from less than 1% in pilocytic ASTRs to 4.2% in isomorphic oligodendrogliomas, underline a benign/semibenign nature of low-grade astrocytomas.

6.1 Results

Frozen and formalin fixed, paraffin embedded tissue sections of each of sixteen PNETs were screened and found to be composed of immunophenotypically distinct, primitive, mostly undifferentiated or possibly, already redifferentiated neuroepithelial cell groups. The cellular microenvironments of the tumors were also composed of cells that morphologically resemble neoplastic astrocytes, oligodendrocytes, ependymal cells, mature ganglion-like cells, neurons and/or melanocytes. Identification of the proliferating population of childhood PNET cells has become easy to demonstrate, with employment of two mouse anti-human MoABs: anti-Ki-67 and anti-PCNA.

The mean labeling indexes (MLI's) (tumor labeling index) in the 16 investigated cases are as follows (anti-Ki-67 resp. anti-PCNA): *1)* PNET with astrocytic differentiation: 7.8% resp. 10.6%; *2)* PNET with ependymal differentiation: 10.4% resp. 15.2%; *3)* pigmented PNET: 8.8% resp. 12.6%; *4)* PNET: 2.8% resp. 6.2%; *5)* PNET: 11.6% resp. 16.8%; *6)* PNET: 2% resp. 5.1%; *7)* PNET: 5.2% resp. 8,6%; *8)* PNET: 3% resp. 5.8%; *9)* PNET: 4.2% resp. 6.4%; *10)* PNET: 1.4% resp. 3.2%; *11)* PNET: 7.6% resp. 10.2%; *12)* PNET: 10.2% resp. 14.4%; *13)* PNET: 3.2% resp. 5.2%; *14)* PNET: 4.8% resp. 6.6% *15)* PNET: 8.8% resp. 11.8% and *16)* PNET: 7.4% resp. 10.4%. MLI's of PNETs are within the range of 1.4% and 11.6% for the Ki-67 antigen and 3.2% to 16.8% for the PCNA nuclear antigen. The mean MLI's are 6.2% resp. 9.74. The three positive tissue controls, anaplastic astrocytomas (ANA-ASTRs) demonstrated unusually high MLI's: 14.6% resp. 19.8%; 12.2% resp. 16.8%; and 13.1% resp. 18.2%. They were also characterized by the presence of an active proliferation stage and aggressive clinical behavior during tumor progression.

According to an internationally accepted demarcation, the MLI and the proliferation ratio is relatively high when >8%, intermediate between 4% and 8%, and low or modest under 4%. Two PNETs which we observed demonstrated differentiation features towards an ependymal and melanocytic cell lineage and were characterized by an active proliferation state: 10.4% resp. 15.2% and 8.8% resp. 12.6%, respectively. The MLI of the third PNET, with astrocytic differentiation was high intermediate (7.8% resp. 10.6%). Every PNET observed contained cell groups undergoing neuritogenesis from neoplastically transformed cells and others resembling mature ganglion-like cells which were characterized by an extremely low to negative MLI.

In conclusion: 1) the 13 PNETs without overt cell lineage differentiation features, demonstrated low overall MLI's in 5/13 cases; intermediate MLI's in 6/13 cases; and only 2/13 showed rapid proliferation with high MLI's; 2) signs of differentiation towards a particular cell lineage results in a relatively high MLI and proliferation ratio; 3) our study provides further support for the possibility of normal neuritogenesis within the neoplastic microenvironment of childhood PNETs; and 4) particular component cell groups within a pediatric neuroectodermally-derived brain tumor may differentiate towards a wide variety of lineages, which are constituents of the cytoarchitecture of the normal brain.

We observed, but still lying under experimental biochemical and immunocytochemical characterization, two newly developed antisera (against cyclin A and D) to be useful in both frozen and formalin fixed and paraffin embedded tissues.

The antigenic epitopic structures attacked by the antibodies were not destroyed by the routine histologic fixation with 10% neutral formalin and the usual paraffin-wax embedding technique. The optimal working dilution of the antibody for frozen sections was 1:500 and for paraffin embedded material 1:100.

The twenty primary brain tumors we investigated demonstrated the presence of proliferative activity in their heterogeneous cell populations. The less differentiated neuroectodermal PNETs and anaplastic glial forms contain more mitotic cells. Low mitotic activity was detected in the normal postnatal choroid plexus, in choroid plexus papilloma and in the two choroid plexus malignancies (anaplastic choroid plexus papilloma and choroid plexus carcinoma).

All seven PNETs (100%) demonstrated high expression of the antigen from the cyclin A group in >90% of cells. The intensity of staining was always the highest possible (A). The CHLA-1 antibody was highly positive in the pigmented PNET (in >90% of the cells), significantly lower antigen expression was demonstrated in PNET with astrocytic differentiation and PNETs without signs of cell line differentiation (in 60 to 70% of the cells).

The cells of normal choroid plexus demonstrated 10 to 50% positivity against the two antisera. The intensity of staining remained high. The choroid plexus papilloma contains more cells than the normal plexus with significantly higher mitotic activity. The antigen expression was detected in over 50 to 90% of cells with both antisera; the staining intensity was high. The antigen expression returned to the pattern seen in normal choroid plexus if the choroid plexus papilloma showed the morphological signs of anaplastic redifferentiation. Both antigens were detected in 10 to 50% of the cells, but the intensity of staining was not as high. Our results determined a very low expression (under 1% to 10%) of the antigens in the case of continued redifferentiation to choroid plexus carcinoma. The intensity of the staining was also significantly lower.

The highest proliferative activity was defined in the one case of glioblastoma multiforme. High staining intensity expression of cyclin A was detected in >90% of the cell populations. The regulatory subunit epitopes were detected in 50 to 90% of the cells, with A,B pattern of staining intensity. No difference was found between the staining patterns of four ASTRs and three anaplastic ASTRs. Both antigens were present in 50 to

90% of the cells and the staining intensity was high. In three pilocytic ASTRs, the cyclin A expression was high in 50 to 90% of the cells. The other antigen was detected only in 10 to 50% of the cells. Both antigens demonstrated the highest immunoreactivity (staining intensity).

6.2 Discussion

One of the most extensively reported methods for assessing the quantity of the proliferative cell population in brain tumors is that of bromodeoxyuridine (BrdU) labeling and its subsequent immunocytochemical localization (284-290). 5-bromo-2-deoxyuridine (BrdU) is a halopyrimidine, an analogue of thymidine and cells, undergoing mitotic division *in vitro* and *in situ*, and can incorporate BrdU into their replicating DNA (74). BrdU and tritiated thymidine are incorporated into the neoplastic cell's DNA through cells in the S-phase and can be shown to be in the nucleus by MoABs using either FITC or immunocytochemistry. The only disadvantage to employing BrdU is that patients should receive a 15-20 minutes infusion of 250 mg/m(2) diluted in 100 ml NaCl) 4-6 hours before removal of neoplastic samples for diagnostic and therapeutic purposes (284, 285). The developed anti-BrdU MoAb was used for *in vitro* cell kinetic studies (291). The method could also be a useful aid in the selection of new clinical treatments (285).

Hoshino and co-workers investigated three patients with PNETs. Using paraffin sections after formalin fixation, the percentage of S-Phase fractions was 13.2±1.2%, 20.7±0.9%, and 11.6±1.2%. Results were higher than those occurred in GBMs and AAs (284). These results were similar to those obtained by tritiated-thymidine labeling (12.0±1.3% for PNET). Ostertag and co-workers tested 52 stereotactic brain tumor biopsies with MoAB Ki-67 (254). The antibody reacted with a nuclear protein expressed in G1, S, G2 and M phases of the cell cycle, but absent in the G0 phases. This study reports investigation of only one cerebellar PNET with a Ki-67 labeling index of 11.2. Very high percentages of Ki-67 positive cells were observed in three PNETs (40-50%) by Giangaspero and co-workers in their study on the growth fraction in 22 primary brain tumors (259). The study used frozen, fresh cut brain tissue biopsies, with the results showing extremely high percentages (60% of Ki-67^{+} cells in choroid plexus carcinoma and 10%-40% in six GBMs). Schlote and co-workers (276) immunostained thirty-two PNETs with MoAB Ki-67. Cryostat sections were used to characterize the proliferating growing parts in six PNETs. The mean labeling index demonstrated low rates: between 0.45 to 0.74 in various PNETs. After 1985, only a few number of studies have been reported that demonstrate only a small number of PNETs have already been investigated immunocytochemically for anti-Ki-67 antigen expression. The results have, however, demonstrated serious differences in the proliferative capacity of the individual PNETs, with morphological and functional cellular heterogeneity.

Proliferative activity may provide a potential correlate of tumor biological aggressiveness, and thus Pollack and co-workers examined the relationship between MIB-1 labeling index and clinical outcome in a series of 29 archival pediatric malignant non-brainstem gliomas from patients treated consecutively between 1975 and 1992, in which clinical, histologic, diagnostic, and therapeutic parameters were previously defined (293). All tumors were re-reviewed by two neuropathologists and classified as grade III or IV lesions based on contemporary guidelines. Among the specimens from the 26 patients who survived the perioperative period, a striking difference in outcome was apparent between tumors with MIB-1 indices <12 (n=10) and those with indices >12 (n=16). The authors reported that median progression free survival was >48 months in the low MIB-1 group, compared with only 6 months in the high MIB-1 group (p=0.014). Furthermore, median overall survival was >48 months in the low MIB-1 group, compared with only 16 months in the high MIB-1 group (p=0.012). MIB-1 index remained associated with survival after taking into account the effect of resection extent, which also correlated strongly with outcome in the cases observed. Although MIB-1 index was associated with histopathologic grade (grade III: 11.9±9.7 vs. grade IV: 27.3±19.0; p=0.015), it proved to be a much stronger predictor of outcome than histology. According to the authors, MIB-1 index may supplement routine histologic classification as a means for improving the accuracy of predicting the biologic behavior of childhood malignant gliomas and may provide a basis for stratifying patients in future malignant glioma studies and refining therapeutic decision-making.

Tissue factor (TF) is a cell surface glycoprotein that initiates the extrinsic coagulation protease cascade and it is expressed in various tumor cell types. TF belongs to the interferon receptor family, and it is one of the early immediate genes, suggesting that TF has a biological function other than hemostasis (293). Hamada and co-workers observed the expression of TF in glial tumors. Immunocytochemistry demonstrated the presence of TF in three glioma cell lines. Immunocytochemical analyses of 44 surgical specimens revealed that all gliomas were positive for TF, and 19/20 (95%) GBMs, 12/14 (86%) AAs and only 1/10 (10%) benign gliomas were moderately or strongly positive for TF. These results revealed that TF is expressed in glioma cells, and that the level of TF expression is correlated with the grade of malignancy of the glioma, suggesting that TF may participate not only in cell growth, but also in IP progression.

The deleted in colorectal cancer (DCC) gene, a candidate tumor suppressor gene on chromosome 18q21, encodes a neural cell adhesion molecule family protein that is most highly expressed in the nervous system (NS) (294). Reyes-Mugica and co-workers examined DCC expression in 57 resected human astrocytic tumors in order to address the hypothesis that DCC may play a role in glioma development and/or progression. Overall, low-grade ASTR were predominantly DCC positive (15/16), whereas high-grade tumors expressed the DCC protein significantly less often (27/41; p=0.03). The authors were able to assess the relationship between DCC expression and tumor progression in 15 patients who initially presented with a low-grade ASTR and subsequently recurred with a GBM. Within this panel of paired lesions from the same patient, 14/15 low-grade tumors expressed the DCC protein, whereas only 7/15 corresponding glioblastomas were DCC positive. Furthermore, secondary GBM resulting from malignant progression of low-grade ASTRs were more often DCC negative (8/15) compared with primary or *de novo* GBM (6/26; p=0.05). The author concluded that that their findings implicate DCC inactivation in glioma progression and also demonstrate that DCC expression is preferentially, but not exclusively, lost in the genetic pathway to secondary GBM.

The β-catenin, glycogen synthase kinase 3b (GSK-3b), and adenomatous polyposis coli (APC) gene products interact to form a network that influences the rate of cell proliferation (295). Medulloblastoma occurs as part of Turcot's syndrome, and patients with Turcot's who develop MEDs have been shown to harbor germ-line APC mutations. Although APC mutations have been observed and not identified in sporadic MEDs, the status of the β-catenin and GSK-3b genes has not been evaluated in this neuroectodermal tumor. Observations have revealed that 3/67 MEDs harbor β-catenin mutations, each of which converts a GSK-3b phosphorylation site from serine to cysteine. The β-catenin mutation seen in the tumors was not present in matched constitutional DNA in the two cases where matched DNA was available. A loss of heterozygosity analysis of 32 MEDs with paired normal DNA samples was performed with four microsatellite markers flanking the GSK-3b locus; loss of heterozygosity with at least one marker was identified in 7 tumors. Sequencing of the remaining GSK-3b allele in these cases failed to identify any mutations. The authors concluded that activating mutations in the β-catenin gene may be involved in the development of a subset of MEDs, and that the GSK-3b gene does not appear to be a target for inactivation in MEDs.

7. EPIDERMAL GROWTH FACTOR (EGF) AND ITS RECEPTOR (EGFR)

> "The growth and proliferation of cells are usually tightly regulated processes that are activated by stimuli from their environment. Epidermal growth factor (EGF)-related peptides represent a class of molecules that can trigger cell proliferation, among several cellular processes, such as differentiation, migration, and survival. Binding of EGF-like peptides to the EGF receptor (EGFR) at the cell surface leads to a cascade of intracellular reactions that transduce signals to the nucleus, resulting in particular gene expression patterns. However, in many tumor cells, the regulation of EGFR activity is lost, due to increased or aberrant expression of the receptor or its ligands, and this contributes to many processes important for tumor growth, including cell proliferation, survival, angiogenesis, invasion, and metastasis."
>
> Janmaat and Giaccone, 2003 (296)

Epidermal growth factor (EGF) and its receptor (EGFR) were originally isolated from the *"nerve growth-promoting substance"* by Cohen in 1960 and the search for tissue-specific factors involved in the regulation of cell growth and division began its dynamic advance (297). The human homologue of EGF was originally named urogastrone, and was found to inhibit gastric acid secretion, demonstrating that growth factors are able to regulate some basic physiological processes (298). The members of the type I family of growth factor receptors, the erbB family of receptor tyrosine kinases, consists of the epidermal growth factor receptor (EGFR), also known as c-erbB-1; c-erbB-2/HER-2/neu; c-erbB-3/HER-3; and c-erbB-4/HER-4 (299-304). The sequence identity between these polypeptides is about 40% to 60% in their extracellular domains and 60% to 80% in their intracellular tyrosine kinase domains (305). The EGFR, a 170 kD transmembrane glycoprotein, contains an intracellular domain with intrinsic protein kinase activity, important in the "vertical" pathway of post-receptor signal transduction (306). It is well known that EGF is a potent mitogen for a number of human tissues (307-309). During the past decade, a number of ligands to EGFR have been identified: EGF, transforming growth factor α (TGF-α), amphiregulin (AR), heparin-binding EGF (HB-EGF), epiregulin and βcellulin (310-314). The distribution of EGFR in normal human tissues, as detected by immunohistochemistry, has been demonstrated by Damjanov and co-workers (315).

The human c-erbB-2 oncogene encodes a 185 kD transmembrane protein (p185-HER2) which shares a significant sequence homology with the intracellular portion of the EGFR and the v-erbB oncogene of the avian erythroblastosis virus (316-318). None of the fully characterized ligands bind directly to the c-erbB-2 oncogene protein, but observations of its biology have led to the identification of a family of ligands, the so-called neuregulins (319, 320), which include the neu-differentiation factors (321, 322), heregulins (323), which are ligands with acetylcholine receptor-inducing activity (324), and glial growth factor (325). The EGFR and c-erbB-2 receptors, when expressed at high density, may have cell transforming properties, suggesting that they can act as dominant oncogenes.

The c-erbB-3 oncogene receptor is expressed in normal human tissues, with a high density on mature and differentiated cells of the gastrointestinal tract and in the neurons of the central nervous system (313). The 6.2 Kb c-erbB-3 specific mRNA has been observed predominantly in epithelial tissues. It is now clear that the members of the NDF/heregulin family of EGF-like molecules bind to the c-erbB-3/HER-3 protein (322-324, 326, 327). Significantly elevated levels of its normal transcript were detected in several breast cancer cell lines *in vitro* with neither amplification nor gene rearrangement (328).

It is well established that selected growth factors, proto-oncogenes and some chemicals play pivotal roles in normal human intrauterine ontogenesis and are also quite important in embryonal organogenesis, as well as neoplastic transformation (329). EGFRs were found to be intensely expressed in the villous cytotrophoblasts in human placenta during the first trimester (340). The c-erbB-2/HER-2 oncoprotein has also been detected during the first and third trimesters along the apical membrane of the syncytiotrophoblast, in cells which lacked Ki-67 expression. In advanced pregnancy, EGFR immunoreactivity was localized on the proliferative and differentiated villous trophoblast. In contrast, c-erbB-2 was located only in more differentiated trophoblasts.

Another embryological study in mammals has reported that mutant mice embryos carrying a c-erbB-2 null allele died before day 11 of intrauterine ontogenesis, probably as a result of dysfunctions associated with a complete absence of cardiac trabeculae. The development of the neural crest (NC) derived cranial sensory ganglia and the ontogenesis of motor nerves was also markedly reduced (320). Both c-erbB-2 and neuregulin were detected in NC cells migrating away from the neural tube suggesting that both molecules act *via* an autocrine mechanism and play a key role in the development and differentiation of NC related structures. Both molecules were also found in the developing heart, c-erbB-2 was detected in cardiac myocytes, whereas neuregulins were expressed in the adjacent endocardium. These results suggest that both molecules promote heart development *via* a paracrine mechanism (331). During the past two years, neuregulin-deficient, as well as c-erbB-2 and c-erbB-4-deficient mice have been bred genetically and experimentally observed (332, 333). Neuregulin-deficient mice display a cranial-neural and cardiac pathology remarkably similar to that of c-erbB-2 deficient mice. The c-erbB-4 deficiency was associated with the lack of cardiac trabeculae development, much like the c-erbB-2 deficiency. In the human thymus, EGF has been shown to promote the acquisition of a neural IP by the cells of the reticulo-epithelial (RE) cellular network and also enhances neuropoietic cytokine secretion (334).

7.1 Results

In our systematic immunohistochemical screening the expression profile of c-erbB-2 (HER-2), c-erbB-3 (HER-3) and c-erbB-4 (HER-4) (internal and external domains) was observed in 22 MEDs/PNETs. 11/22 MEDs/PNETs demonstrated high level of immunoreactivity (A, B), detectable by the immunostaining intensity. In 11/22 MEDs/PNETs 10 to 50% (++) of the tumor cells showed c-erbB-2 (HER-2) and c-erbB-4 (HER-4) positivity, while c-erbB-3 (HER-3) expression was detected only in >10 % (+) of the neoplastically transformed cells. Expression of c-erbB-2 (HER-2) and c-erbB-4 (HER-4) was detectable only in high grade glial tumors, in 9 AAs and 6 GBMs in 10 to 50% (++) of their tumor cells with identified strong immunoreactivity (A, B immunostainings, stronger in GBMs).

Survival rate in the follow up demonstrated that 70% of patients revealed to have c-erbB-2 (HER-2) positive MEDs/PNETs (usually children under 4 years of age) succumbed to their neoplastic disorder. In our opinion, we suggest that c-erbB-2 positivity is a poor prognostic marker in MEDs/PNETs. The Kaplan-Meier estimation revealed a significant correlation between c-erbB-2 expression and survival (p=0.002). In further correlative statistical observation the expression of synaptophysin and GFAP in 11/22 MEDs/PNETs revealed a negative correlation between the expression of c-erbB-2 (HER-2) and these two marker proteins. These results suggest that c-erbB-2 (HER-2), which may be predominantly expressed by more embryonal MEDs, may delineate a poorer prognostic subgroup, especially in children diagnosed with MED prior to their fourth year of life.

7.2 Discussion

EGFR overexpression has been observed in human cancers including breast, ovarian, prostate, bladder, lung, brain, and pancreas (335-340). EGFR appears to contribute to the growth and survival of tumor cells in addition to maintaining normal cellular function. The EGFR signal transduction pathways contribute to the development of malignancies through various processes, such as effects on cell cycle progression, inhibition of apoptosis, angiogenesis, tumor cell motility, and metastases (341-343). Neoplastically transformed cells that overexpress EGFR can subvert the G1 to S transition by modulating the levels of cyclin D1 and the cyclin-dependent kinase (Cdk) inhibitor $p27^{Kip1}$ through both MAPK and PI3K/Akt signaling pathways (344, 345). Members of the EGFR family, especially the EGFR and c-erbB-2 receptor, have frequently been implicated as important prognostic indicators in a variety of neoplasms, including GBM (346). In head and neck squamous cell carcinoma (HNSCC), EGFR expression is reported in approximately 90% of specimens, and it is associated with a poor prognosis. In breast carcinomas, the overexpression of the c-erbB-2 oncogene protein, combined with the presence or absence of gene amplification, possesses significant prognostic value for the evaluation of decreased survival in the lymph node positive stage of the disease, but does not correlate well with neoplastic progression (347-364). Amplification of *c-myc*, but not c-erbB-2, is associated with high proliferative capacity in breast cancers (365). Evidence for tumor biological synergism between c-erbB-2 and c-erbB-3 suggests the possibility that demonstration of simultaneously signaling pathways may improve the predictability of malignancy free survival in BC (347, 363, 366, 367). It is now clear that c-erbB-3 is frequently overexpressed in a variety of primary cancers [oral squamous cell carcinoma (368), ovarian (369)], but to a lesser extent in *in vitro* established tumor cell lines. Both estrogen and EGF down-regulate the expression of the c-erbB-2 oncogene product in BC cells, probably by distinct pathways (370). The overexpression of any members of the EGFR family suggests the presence of IP changes in the neoplastic cells as a result of continuous and rapid cell transformation.

Cutaneous melanomas are malignant neoplasms that originate from epidermal melanocytes (371). Congenital nevi and dysplastic nevi have been determined as premalignant lesions in many malignant melanoma cases (372-374). It has been reported that certain kinds of human melanomas are capable of producing TGF-α and EGF (375-376). In fact, TGF-α was first identified as a polypeptide growth factor synthesized by a human melanoma cell line (377). The pattern of expression of members of the EGFR family in human malignant melanomas, as described in the literature, certainly lacks clarity. A group of researchers have reported that primary and metastatic melanomas express more EGFRs than normal melanocytes, while others described no difference in the density of EGFR expression on benign nevi compared to *in situ* and metastatic melanomas (378, 379). The expression of c-erbB-2 (HER-2) and c-erbB-3 (HER-3) in human malignant melanomas has also been described (380). The unregulated expression (overexpression) of the c-erbB-2 gene, involved in cellular growth, may be mediated by either amplification or post-translational stabilization and has been reported in 25-30% of primary, heterogeneous human BC cells (381-384). The expression of the c-erbB-2 oncoprotein has been associated with poor prognosis and high cellular proliferation activity in a variety of human malignancies, including GBMs (385); lung carcinomas (386); gastric carcinomas (387); bladder carcinoma (388); prostate carcinomas (389, 390); ovarian carcinomas (366), endometrial carcinomas (391-395), and BCs (396-414). In numerous cases, however, no statistically significant correlation between *c-erbB-2* expression the prognostication of disease progression could be established (415-422). Press and co-workers (423) have attributed this variability to differences in tissue processing (fixation), epitope and antigen retrieval methods employed, and the sensitivity of the particular antibody used in detecting the c-erbB-2 oncoprotein.

The c-erbB-3 protein has been reported to be overexpressed in breast (424), gastrointestinal (425-427) and pancreatic (428) cancers, but neither clinical nor prognostic importance of this oncoprotein has yet been described. Cooperative signaling of the c-erbB-2 and c-erbB-3 oncoproteins has been demonstrated *in vitro* (362). Transphosphorylation of c-erbB-3 by c-erbB-2 has been observed in cell lines co-expressing these two oncoproteins, while no such occurrence was detected in cell lines transfected with non-functional c-erbB-2. In addition, the formation of heterodimers of c-erbB-2 and c-erbB-3, with much greater catalytic activity than monomers, has also been established as a way of cooperative signaling by these two oncogene products (362). Thus the detection of the co-expression of these two oncoproteins may have more clinical and prognostic significance than the detection of overexpression of just the c-erbB-3 protein.

The overexpression and aberrant function of the EGFR, c-erbB-1 (HER-1) and its ligands and co-receptors in a wide spectrum of epithelial neoplasms have provided a rationale for targeting this signaling network with novel biological treatment approaches (429). Several antireceptor therapeutic strategies have been pursued, but two stand ahead in their clinical development. One approach has been the generation of small molecules that compete with adenosine triphosphate (ATP) for binding to the receptor's kinase pocket, thus blocking receptor activation and the transduction of postreceptor signals. The second approach utilizes humanized monoclonal antibodies generated against the receptor's ligand-binding extracellular domain. These antibodies block binding of receptor-activating ligands and, in some cases, can induce receptor endocytosis and downregulation. Clinical studies already suggest that both of these approaches, either alone or in combination with standard anti-neoplastic therapies, are well tolerated and can induce favorable clinical responses and tumor stabilization in a variety of common solid neoplasms. The crucial role of the epidermal growth factor receptor (EGFR) in neoplastic proliferation and the overexpression of EGFR and is often associated with an aggressive IP in brain tumors, in over 50% of cases of NSCLC, HNSCC and colon cancer. As it is well established, EGFR is commonly overexpressed in a number of epithelial malignancies and several other malignancies provides the rationale for targeting and interrupting this key signaling network (430). Several EGFR-targeting agents have been recently developed (C225, ABX-EGF, E7.6.3, EMD 55900, ICR62, ZD1839, CP358774, PD168393, CGP75166/PKI166, CGP59326A, BIBX1382). The two most advanced EGFR inhibitors in development are C225 and ZD1839. C225 is an antibody directed against the ligand-binding domain of human EGFR, which competes for receptor binding with EGF and other ligands (431). *In vitro*, C225 inhibits EGFR tyrosine kinase activity and proliferation of EGFR-overexpressing squamous cell carcinoma cell lines. Synergy was observed with doxorubicin, cisplatin and radiation in preclinical animal models.

8. P53, THE GUARDIAN OF THE INTEGRITY OF THE GENOME

> "The main difference between normal cells and tumor cells results from discrete changes in specific genes important for cell proliferation control mechanisms and tissue homeostasis. These genes are mainly proto-oncogenes or tumor-suppressor genes, and their mutation could play a role in cell hyperproliferation and carcinogenesis. Tumor-suppressor genes normally function as a physiological barrier against clonal expansion or mutation accumulation in the genome. They also control and arrest growth of the cells that hyperproliferate due to oncogene activity. Alteration or DNA damage in tumor-suppressor genes and oncogenes are considered key events in human carcinogenesis. Tumor-suppressor protein p53 is an important transcription factor, which plays a central role in the cell cycle regulation mechanisms and cell proliferation control, and its inactivation is considered a key event in human carcinogenesis."
>
> Batinac, 2003 (432)

The most important aspect of the regulation of the cell cycle is the assurance of faithful duplication of the genome. The transcription factor and tumor suppressor p53 and its two homologues p63 and p73 form a family of proteins. In recent years, it has become increasingly clear that the p53 gene, originally mistaken for an oncogene and presently considered a tumor suppressor gene, produces a protein which is absolutely critical in guarding cells against replication with altered genomes. The tumor suppressor activity of p53 plays a major role in limiting abnormal proliferation, and inactivation of the p53 response is becoming increasingly accepted as a hallmark of cellular neoplastic transformation (433). The nuclear localization of p53 is important for its tumor suppressor function. In contrast, both p63 and p73, which are close relatives of p53, are rarely mutated in neoplastic cells. p63 and p73 show much greater molecular complexity than p53 because they are expressed both as multiple alternatively spliced C-terminal isoforms, and as N-terminally deleted dominant-negative proteins that show reciprocal functional regulation (434). In addition, several other factors, such as post-translational modifications and specific and common family regulatory proteins result overall in subtle modulation of their biological effects. Although all p53, p63, and p73 family members are regulators of the cell cycle and apoptosis, the developmental abnormalities of p73- and p63-null mice do not show enhanced neoplastic susceptibility of p53 knockouts, suggesting that complex regulatory processes modulate the functional effects of this family of proteins. High levels of p53 are associated with the arrest of the cell cycle in G1, which allows for necessary repairs to be carried out following DNA damage and hypoxia (65,435). If the damage cannot be repaired, cell death by apoptosis is triggered. At a theoretical level, therapeutic approaches that reinstate p53 activity, or augment p63 and p73, provide plausible and potentially efficacious routes towards new developments of anti-neoplastic treatments.

The p53 gene is located on the short arm of chromosome 17, in a region commonly affected by allelic deletions and point mutations (436, 437). 17p13 has been identified as the site of the p53 gene and this site is altered in over 50% of all human cancers. The p53 gene product was originally identified as being complexed to the large T antigen of simian virus 40 (438). In recent years, the ability of p53 to form complexes with other DNA tumor virus oncoproteins such as adenovirus E1B (439) and human papilloma virus-16 (HPV-16) and HPV-18 E6 (440) has been established as a mode of inactivating wild-type p53 function. It is well documented that carcinogenesis is a complex process which requires multiple significant genetical changes. Mutations in the p53 gene are the most common lesions in a variety of human malignancies identified to date and have been strongly associated with the molecular pathogenesis of the disease (441-443). Over

98% of these mutations are clustered in evolutionarily conserved blocks of amino acids between codons 110 and 310 covering exons 5 and 8 (444). Within this stretch of amino acids, domain IV spanning codons 236 through 258 has been determined to be the site of the majority of p53 gene mutations. Mutations in this highly conserved stretch of the p53 gene are associated with nearly 50% of all childhood sarcomas, including OSs (445, 446), and a large proportion of Burkitt's lymphomas and L3-type B cell acute lymphoblastic leukemias (ALLs) (447), ASTRs and other brain tumors (53,448-453), malignant melanomas (454-457), from irreversible premalignant, squamous metaplasia or dysplasia to advanced non-small cell lung cancers (458-464), hepatocellular carcinomas (465-468), ovarian cancers (469), and breast carcinomas (63,469-478) among many other types of cancer (479). Due to the critical cell-cycle regulatory role of p53, most dysfunctions of this phosphoprotein are associated with neoplastic cell subpopulations with high proliferative indices.

The wild-type form of p53 has now been termed and recognized as the "guardian of the genome" by one of its discoverers Lane (480). It is well documented that p53 is also a potent transcriptional regulator and that it carries out this activity by sequence-specific double-stranded DNA binding (481, 482) as a tetramer. It has been shown that normal p53 regulates not only the procession through the G1 phase of the cell cycle and into the S phase (*i.e.* the G1 checkpoint control for DNA damage), but that it also activates the DNA-repair machinery if a mismatch or mutation is found in the DNA (483). p53 protein levels are quite low immediately after mitosis, but increase in G1. During normal procession through the cell cycle, p53 is phosphorylated by cdc2 kinase (484) and casein kinase II (485) during S phase, thereby inactivating the protein.

In human neoplasms, p53 inactivation occurs frequently by mutation, and possibly also by nuclear exclusion of wt p53. First reported by Moll (486), the "cytoplasmic sequestration" of wild type p53 has been thoroughly studied to elucidate the molecular mechanism of this process, employing neuroblastoma cell lines as the model. At Columbia University, the research group of Nikolaev (487) has just isolated the cytoplasm anchor protein PARC, a Parkin-like ubiquity ligase which specifically binds to p53 and anchors it so that the "guardian of the genome" cannot play its tumor suppressor role in the nucleus. PARC, as a member of p53 associated protein complexes, directly interacts forming an approximately 1 Mda complex with p53 in the cytoplasm of unstressed cells. Overexpression of PARC protein promotes the so-called cytoplasm sequestration of ectopic p53. Anti-PARC siRNA-manipulation relocates p53 into the nucleus, restitutes a function to p and chemo-radiosensitivity to malignant neutralists (488). The PARC protein has been suggested as a potential molecular target for anti-neoplastic immunotherapy (489).

p53 also acts to integrate multiple stress signals into a series of diverse antiproliferative responses. One of the most important p53 functions is its ability to activate apoptosis, with disruption of this process promoting tumor progression and chemoresistance (490). p53 apparently promotes apoptosis through transcription-dependent and -independent mechanisms that act in concert to ensure that the PCD program proceeds efficiently. Moreover, the apoptotic activity of p53 is tightly controlled and is influenced by a series of quantitative and qualitative events that influence the outcome of p53 activation. Other p53 family members can also promote PCD, either in parallel or together with p53. Although incomplete, our current understanding of p53 illustrates how PCD can be integrated into a larger tumor suppressor network controlled by different signals, environmental factors, and cell type. Further understanding of this network in more detail will provide insights into neoplastic and other diseases, and will identify new strategies to improve therapy options.

Immunocytochemical p53 overexpression is a morphological demonstration of an alteration in the p53 gene and signals a progression toward a neoplasm associated IP (437, 491, 492). Most of the p53 alterations are mutations clustered in exon 8, in evolutionarily conserved blocks of amino acids (465,493-495), and are usually G to T type transversions (496). The type of mutation reflects the mutagen involved because of their characteristic mutational spectra. Spontaneous point mutations are predominantly G to A transitions (497), whereas the G to T transversions are chemically induced (498). The mutation of the p53 gene results in the synthesis of an abnormal protein, which, being metabolically stable accumulates in the nucleus, reaching the concentration needed for immunocytochemical detection (493). Altered p53 proteins can be detected not only in the nucleus, but also in the cytoplasm, perhaps by virtue of their association with a 70 kD heat shock protein (HSP 70) (499, 500).

The accumulation of p53 protein may be the earliest step of the progression of a pilocytic (relatively "benign") subtype ASTR to diffuse, low-grade ("malignant") ASTR, since it provides an opportunity for further genetic alterations, which can be passed on through cell division. By not having the ability to induce DNA-damage related G1 arrest, repair the genome, or trigger apoptosis, the cell is rendered genetically instable and can further dedifferentiate its IP.

8.1 Results

Immunocytochemical presence and cellular localization of the p53 gene product protein employing MoAB PAb1801 was observed in ten out of the ten ASTRs (100%) in this study. The immunoreactivity was always heterogeneous, and loosely grouped cells with similar staining characteristics were detected within the cellular and hormonal microenvironment of the ASTRs.

We observed p53 expression in five pilocytic ASTR cases. The intensity of the immunoreactivity was high; (B) in two cases, (A,B) in two cases and (A) in one case. However, only a low percentage (usually under 10%) of the total cell number reacted positively (+), although in one case, the number of cells stained was under 1% (±). Most of the staining was localized to the nucleus, although in the three cases with higher intensity staining, some cytoplasmic immunoreactivity was detected.

The two cases of *pure* AAs also accumulated p53 protein in their nucleus and cytoplasm. The immunoreactivity (staining intensity of overexpression) in this ASTR subtype was very high (A) and the number of total cells positively stained ranged from (+++) to (++++). The case of the *mixed* ASTR with primitive neuroectodermal tumor elements also reacted positively with MoAB PAb1801, but with a lower intensity (B) although a similar proportion of tumor cells exhibited immunoreactivity (+++). As would be expected in this case of lower intensity staining with antisera directed against p53, no cytoplasmic staining was observed. High intensity staining (A, B) in well over 90% of the total cell number (++++) was observed in the anaplastic oligo-astrocytoma; with mostly nuclear, but occasional cytoplasmic reactivity.

p53 immunoreactivity was of greatest intensity (A) in the single case of GBM observed by us. In this case, a very high proportion (well over 90%) of the total cell number reacted positively with MoAB PAb1801 (++++). The staining pattern was nuclear, but strong cytoplasmic immunoreactivity (overexpression) was also observed. The immunoreactivity pattern was mostly heterogeneous, with cells groups of similar intensity clustered within the ASTRs. The number of neoplastic cells stained and the intensity of the immunoreactivity correlated directly with the known degree of malignancy of the various subtypes of ASTRs: lowest in the pilocytic ASTR cases and highest in GBM. Low-grade human ASTRs possess an intrinsic tendency for cell dedifferentiation toward the embryonic cell immunophenotype (IP). Loss of p53 function is associated with most, if not all, human malignancies. Mutation of p53 has yet to be demonstrated in pilocytic ASTRs. The accumulation of p53 in some of pilocytic ASTR cells, demonstrated in our study, suggests that the mere dysfunction of the p53 protein may be involved in the early stages of ASTR progression from the grade I pilocytic subtype to the more "malignant" *pure* ASTR, which is characterized by p53 gene mutations. The loss of p53 provides the necessary genetic instability needed for further IP changes and further progression towards more malignant IPs, *e.g.* AA and GBM. Such facts make the employment of p53 in the diagnostic and prognostic assessment of ASTRs indispensable. p53 levels may be used in identifying cell clones within pilocytic ASTR microenvironments, which have a clear tendency for progression toward more malignant IPs and the establishment of the alteration of the p53 gene in more advanced ASTR subtypes (grades II to IV). As expected, the breast carcinomas used as positive tissue controls in this study reacted positively with the anti-p53 MoAB, while the normal, human, postnatal thymus tissue showed no detectable levels of p53 protein (negative tissue control).

8.2 Discussion

p53 mutations have, however, been detected in a high-frequency of low-grade (grade II) and AAs (grade III) and thus the mutation of p53 has been implicated in the early stages of malignant ASTR tumor progression (501). We identified p53 accumulation in all four such ASTRs. This again may be related to the initial dysfunction of the p53 protein observed in pilocytic ASTRs. The lack of p53 function in the grade I ASTRs provides a way for the neoplastically transformed cells to alter their genome and still be able to proliferate. This allows for the alteration of any of a number of genes implicated in cell-cycle suppression including the most important "target", p53. Genetic alterations in grade II and grade III ASTRs include the loss of a variety of loci: 1p, 1q, 9p, 9q, 10p, 10q, 11p, 13q, 17p, 19p, 19q, and 22q which include the sites for critical cell cycle regulatory genes, such as RB1 and p16, and of course p53 (502). The loss of function of these crucial genes has been implicated in the formation or progression of ASTRs (503-509). This large increase in the number of significant genetic alterations is probably the direct effect of the initial dysfunction of the p53 protein observed in pilocytic (grade I) ASTRs and the subsequent and consequent mutation in the p53 gene due to the loss of the quality control mechanism during DNA replication.

Through genetic analyses it has become apparent that there are two ways that the WHO grade IV, GBM can arise. The first pathway is characterized by the loss of the short arm of chromosome 17 and it is through this route that lower-grade ASTRs may progressively dedifferentiate and advance to GBM. The other, *de novo* pathway involves the upregulation of the epidermal growth factor receptor (EGFR), as well as loss of chromosome 10 (501). The detection of high levels of p53 protein or EGFR in glioblastoma cells may be used as part of a broad panel of factors in assessing this ASTR subtype. Thus, in our opinion, a single molecule cannot reliably be used in distinguishing between such closely related human malignancies.

p53 expression levels may also be used to evaluate the malignant tendencies of grade I, II, and III ASTRs. A high level of p53 protein in pilocytic ASTR cells may be used to establish a tendency of those cells to further dedifferentiate to low-grade (grade II) ASTRs, while an accumulation of p53 in grade II ASTR cells may be used in predicting the likelihood of mutations of the genome and further dedifferentiation towards even more malignant AA and GBM IPs.

The loss of normal DNA mismatch repair is thought to promote tumorigenesis by accelerating the accumulation of mutations in oncogenes and tumor suppressor genes. Defective DNA mismatch repair results from genetic or epigenetic alterations that inactivate the DNA mismatch repair genes hMLH1 or hMSH2, and rarely hMSH6. Inactivation of hMLH1, hMSH2, and hMSH6 is observed as a loss of expression of these proteins by immunohistochemistry.

Germline mutations in hMLH1 and hMSH2 are associated with susceptibility to hereditary nonpolyposis colorectal cancer (HNPCC) (510, 511). The methylation of a CpG site in hMLH1 has been shown to inhibit the binding of the CBF transcription factor to the corresponding CCAAT box in the hMLH1 gene, thus silencing its expression in colon cancer cells (512). It has also been shown that missense variations in hMLH1 do not necessarily result in microsatellite instability of the corresponding tumor DNA (513). Additionally, mutations in the promoter region of hMSH2 have been found to have a limited role in development of suspected HNPCC and sporadic early onset colorectal cancer (514). A reduction of mismatch repair protein levels has been described in cutaneous malignant melanomas (515), as well as prostate cancer (516), cervical cancer (517), and endometrial carcinoma (518). However, in certain malignancies, such as bronchioloalveolar carcinoma (519) and hepatocellular carcinoma (520), lack of defects in mismatch repair genes have been shown, suggesting that mutations in these genes are not always involved in generalized carcinogenesis or cancer progression.

9. APOPTOSIS IN BRAIN TUMORS

> "Apoptosis or programmed cell death is a distinct mode of cell destruction and represents a major regulatory mechanism in eliminating abundant and unwanted cells during embryonic development, growth and differentiation.

> So, apoptosis is a mode of cell death that occurs under normal physiological conditions and not only when the cell is damaged or attacked by viruses." (521)

Normal development (pre- and postnatal), cell maturation and differentiation in a multicellular mammalian organism, as well as the maintenance of cellular homeostasis, require a perfect, dynamic co-regulation of cell proliferation and cell death (522, 523). Furthermore, the balance between cell proliferation and cell death determines the growth and differentiation of every complex multicellular tissue (524). At the same time, programmed cell death (PCD) is a highly conserved mechanism that plays an essential role not only in numerous normal developmental and regulatory processes and disease states. It is mediated by a variable interaction among several components of the cell, including cell surface death receptors, the caspase cascade, mitochondrial metabolism and energetics, and the cytoskeleton (525).

Three decades ago, Kerr and co-authors proposed the introduction of the scientific term "apoptosis" from the Greek *apo* (away from) and *ptosis* (falling) for the morphological nomenclature (526). Apoptosis is thus described as a process of falling leaves from the trees or the shedding of the petals from flowers. It is a form of suicidal cell death that requires the active cellular synthesis of molecules involved in processes that provide the energy necessary for programmed cell death to occur. PCD is a distinct mode of cellular suicide and represents one of the most important regulatory mechanisms of homeostasis because it eliminates abundant, cells with altered genetic information and unwanted, damaged cells (521,527).

Specifically, apoptosis is a physiological process wherein the cell initiates a sequence of events culminating in the fragmentation of its DNA, nuclear collapse and finally disintegration of the cell into small, membrane-bound apoptotic bodies. Dramatic changes in the morphological structure of cells such as: 1) condensation of the nuclear (chromatin) and cytoplasmic structures (especially the mitochondria); 2) blebbing of the cell membrane; 3) characteristic swelling of the endoplasmic reticulum; and 4) fragmentation of the cells in membrane bound apoptotic bodies, are the signs of total cell destruction. Complete breakdown of the cell membrane and cytoplasm is associated with the advanced stage of apoptotic changes, during secondary necrosis and this can be identified only in *in vitro* experiments (528, 529).

Mammalian cells are capable of committing "active" suicide in response to specialized pathological actions employing a phylogenetically developed intrinsic program of death, triggered by signal transduction through specific receptors (526).

Genetic studies have provided new insights into the developmental control of PCD. During the development of the nematode *Caenorhabditis elegans*, three genes related to apoptosis were identified: ced3 and ced4 (two potential tumor suppressor genes), as well as an inhibitor, a cell protector gene named ced9 (71,530-532). The ced9 anti-apoptotic gene is 23% identical at the amino acid level to the bcl-2 proto-oncogene, which has been discovered as a gene associated with a frequent translocation breakpoint in some B cell leukemias (532). These essential apoptotic molecules have proven to have mammalian homologues. The ced3 gene encodes a cysteine protease that is homologous to the interleukin-1β-converting enzyme (ICE). During the last two to three years, a number of novel genes involved in apoptosis, belonging to the ICE/ced-3 family and others, have been discovered: mouse Nedd 2, Ich-1, an ICE/ced-3-related gene which encodes both positive and negative regulators, TX/Ich-2/ICErel-II, the gene encoding CPP32, a novel human apoptotic protein with homology to the mammalian IL-1β-converting enzyme, Mch2 and Mch3, ICErel II and ICErel III, ICE-LAP3 and ICE-LAP6 and MACH, a novel MORT1/FADD-interacting protease involved in CD95/*FasR*/APO-1 and TNF receptor-induced PCD. There is evidence that ICE family members are also auto-proteolytic and are able to process one another's substrates, such as poly-ADP ribose polymerase (PARP), lamins and so forth. In most cases, the apoptotic mechanism is diminished or delayed if low quantities of extracellular Ca^{++} and Mg^{++} are present; implying that intracellular influx of Ca^{++} and Mg^{++} (evidence of Ca- and Mg-dependent endonuclease involvement) is a basic requirement for PCD. Tumor necrosis factor-related apoptosis-inducing ligand or Apo 2 ligand (TRAIL/Apo2L) is a member of the tumor necrosis factor (TNF) family of ligands capable of initiating apoptosis through engagement of its death receptors (533). TRAIL selectively induces apoptosis in a variety of neoplastically transformed cells, but not most normal cells, and therefore has garnered intense interest as a promising target or agent for anti-neoplastic therapy. TRAIL is expressed on different cells of the immune system and plays a role in both T lymphocyte and natural killer (NK) cell mediated tumor surveillance and suppression of suppressing tumor metastasis. Some mismatch repair deficient tumors evade TRAIL-induced apoptosis and acquire TRAIL resistance through different mechanisms. Death receptors, members of the TNF receptor family, signal apoptosis independently of the p53 tumor suppressor gene. TRAIL treatment in combination with chemo- or radio-therapy enhances TRAIL sensitivity or reverses TRAIL resistance by regulating the downstream effectors. Efforts to identify agents that activate death receptors or block specific effectors may improve therapeutic design.

9.1 The family of cysteine aspartyl proteases or caspases

Caspases are the executioners in the apoptotic or programmed cell death (PCD) pathway (534). Molecular biological research revealed that the caspase family functions as a well directed orchestra of at least 14 proteases, all containing a common active site of cysteine within the conserved peptide sequence QACXG. Their cleavage target is the peptide bond C-terminal to asparate residues. Caspases are divided into three subgroups upon their tetrapeptide sequence recognition: Group I caspases (caspase-1, caspase-4 and caspase-5) generally prefer the (W/L)EHD sequence; Group II caspases (caspase-3 and caspase-7) prefer DEXD; and Group III caspases (caspase-6, caspase-8, caspase-9 and caspase-10) recognize (L/V)EXD. The VAD tripeptide is targeted by all caspases; therefore, VAD is the structural base for pan caspase reagents (535).

All caspases are present in mammalian cells in the form of enzymatically inert zymogens (named procaspases). The zymogens undergo cleavage, resulting in four large and small subunits. These active or mature caspases represent a functional tetramer containing two distinguished active sites and two heterodimer subunits, one 20 kD (p20) and an other 10 kD (p10) (536). The prodomain is typically short for the so-called effector caspases (caspase-3 and caspase-7) and longer for the rest of them. The longer prodomain can include recognizable motifs, such as the CARD (caspase activation and recruitment domain: in caspase-1, caspase-2, caspase-4 and caspase-9) and the DED (death effector domain: in caspase-8 and caspase-10).

Recently, it has been shown that caspases target more than 100 various proteins belonging to diverse groups, such as signaling, regulatory, structural, *etc*. The activities of caspase family members as part of the complex PCD process are regulated by a number of complex interactions between them, adapter proteins and target substrates and different types of receptor families (537-539).

The intrinsic part of the PCD program arises from the mitochondria, the so-called "killer organelles" (540). The release of cytochrome c from the mitochondrial intermembrane space into the cytosol forms the caspase-activating complex or apoptosome (541). The apoptosome is completed from cytochrome c, ATP, Apaf-1 and caspase-9 zymogen, the last two of which interact *via* their CARDs. Smac (or DIABLO) is a recently identified, novel proapoptotic molecule that is also released from mitochondria into the cytosol during apoptosis. Smac functions by eliminating the caspase-inhibitory properties of the inhibitors of apoptosis proteins (IAP), particularly XIAP (542). After the initial caspase-9 activation and the activation of the following effector caspases, the so-called terminal caspase-cascade is built up, a two step process that is essential to PCD. The high level of effectiveness of PCD is ensured after autocatalytic activation, processing a number of caspases and regulatory proteins, which strongly amplifies the apoptotic process.

9.1.1 Caspase-3

Caspase-3 is ubiquitously expressed in the human cells and, like other caspases, is synthesized as an inactive, 32 kDa proenzyme. Upon activation, caspase-3 is cleaved at Asp28-Ser29 and Asp175-Ser176, thereby generating two subunits of 17 kDa and 12 kDa, respectively. Activation of caspase-3 occurs in response to a wide variety of PCD inducers including *FasL-FasR*. The *in vivo* patterns of CPP32 (caspase-3) gene expression were determined by Krajewska and co-workers employing an immunohistochemical approach and formalin-fixed, paraffin wax-embedded normal human tissues (543). Krajewska and co-workers used a rabbit polyclonal antiserum against recombinant human CPP32 protein, which was proven specific by immunoblot analysis of various human tissues and cell lines. CPP32 immunoreactivity was selectively found in certain cell types and was typically present within the cytosol, although occasional cells also contained nuclear immunostaining. CPP32-positive normal cells included epidermal keratinocyes, cartilage chondrocytes, bone osteocytes, heart myocardiocytes, vascular smooth muscle cells, bronchial epithelium, hepatocytes, thymocytes, plasma cells, renal tubule epithelium, spermatogonia, prostatic secretory epithelial cells, uterine endometrium and myometrium, mammary ductal epithelial cells and the gastrointestinal epithelium of the stomach, intestine and colon. In contrast, weak or absent CPP32 immunoreactivity was observed in endothelial cells, alveolar pneumocytes, kidney glomeruli, mammary myoepithelial cells, Schwann cells and most types of brain and spinal cord neurons. Consistent with a role for CPP32 in apoptotic cell death, clear differences in the relative intensity of CPP32 immunostaining were noted in some shorter-lived types of cells compared to longer-lived, including (a) germinal center (high) *versus* mantle zone (low) B lymphocytes within the secondary follicles of lymph nodes, spleen and tonsils; (b) mature neutrophils (high) *versus* myeloid progenitor cells (low) in bone marrow; (c) corpus luteal cells (high) *versus* follicular granulosa cells (low) in the ovary; and (d) prostate secretory epithelial cells (high) *versus* basal cells (low). CPP32 was also found to be highly expressed in atherosclerotic plaques and to be colocalized with apoptotic cells (544).

9.1.2 Caspase-6

The observation that the nematode cell death effector gene product Ced-3 is homologous to human interleukin-1β-converting enzyme (caspase-1) has led to the discovery of at least nine other human caspases, many of which are implicated as mediators of apoptosis (530, 531, 543). The activity of ICE-like proteases or caspases is essential for apoptosis (544). Multiple caspases participate in apoptosis in mammalian cells but how many caspases are involved and what their relative contribution to cell death is poorly understood. To identify caspases activated in apoptotic cells, an approach to simultaneously detect multiple active caspases was developed. Employing neoplastically transformed cells as a model, it has been established that CPP32 (caspase-3) and Mch2 (caspase-6) are the major active caspases in cells undergoing PCD, being activated in response to distinct apoptosis-inducing stimuli and in all cell lines analyzed. Both CPP32 and Mch2 are present in apoptotic cells as multiple active species. In a given cell line, these species remained the same irrespective of the apoptotic stimulus used. However, the species of CPP32 and Mch2 detected varied between cell lines, indicating differences in caspase processing. The influence of a number of environmental parameters, including pH, ionic strength, detergent and specific ion concentrations, on the activity and stability of four caspases involved in death receptor-mediated, apoptosis have been observed (543). Based on these observations, the following buffer was recommended as optimal for observation of their characteristics *in vitro*: 20 mM piperazine-N,N'-bis(2-ethanesulfonic acid) (PIPES), 100 mM NaCl, 10 mM dithiothreitol, 1 mM EDTA, 0.1 3-[(3-cholamidopropyl)dimethylammonio]-2-hydroxy-1-propanesulfonic acid (CHAPS), 10% sucrose, pH 7.2. Caspase activity is not affected by concentrations of Ca^{2+} below 100 mM, but is abolished by Zn^{2+} in the submicromolar range, a common characteristic of cysteine proteases. Optimal pH values vary from 6.8 for caspase-8 to 7.4 for caspase-3 and activity of all is relatively stable between 0 and 150 mM NaCl. Consequently, changes in the physiologic pH and ionic strength would not significantly alter the activity of the enzymes, in as much as all four caspases are optimally active within the range of these parameters found in the cytosol of living and dying human cells.

Two novel synthetic tetrapeptides, VEID-CHO and DMQD-CHO, could selectively inhibit caspase-6 and caspase-3, respectively (545). These inhibitors were used to dissect the pathway of caspase activation in *Fas*-stimulated Jurkat cells and identify the roles of each active caspase in apoptotic processes. Affinity labeling techniques revealed a branched protease cascade in which caspase-8 activates caspase-3 and -7 and caspase-3, in turn, activates caspase-6. Caspase-6 cleaves nuclear mitotic apparatus protein (NuMA) and mediates the shrinkage and fragmentation of cell nuclei. Caspase-3 cleaves NuMA at sites distinct from caspase-6 and mediates DNA fragmentation and chromatin condensation. It is also involved in extranuclear apoptotic events: cleavage of PAK2, formation of apoptotic bodies and exposure of phosphatidylserine on the cell surface. In contrast, a caspase(s) distinct from caspase-3 or -6 mediates the disruption of mitochondrial membrane potential (permeability transition) and the shrinkage of cytoplasm. These findings demonstrate that caspases are organized in a protease cascade and that each activated caspase plays a distinct role(s) in the execution of *Fas* and TNF receptor 1, the so-called death receptor-mediated PCD.

As already alluded to above, caspases play a major role in the transduction of extracellular apoptotic signals and execution of PCD in mammalian cells (546). Ectopic overexpression of the short prodomain caspase-3 and -6 precursors in mammalian cells does not induce apoptosis, which is due to their inability to undergo autocatalytic processing/activation, suggesting that they depend on the long prodomain caspases for activation. To investigate directly the apoptotic activity of these two caspases *in vivo*, constitutively active recombinant precursors of caspase-3 and -6 were generated, which was achieved by making contiguous precursor caspase-3 and -6 molecules with these molecules have their small subunits preceding their large subunits. Unlike their wild-type counterparts, these recombinant molecules were capable of autocatalytic processing in an *in vitro* translation reaction, suggesting that they are catalytically active. They were also capable of autoprocessing and inducing apoptosis *in vivo* independent of the upstream caspases. Furthermore, their autocatalytic and apoptotic activities were inhibited by the pancaspase inhibitor z-VAD-fluoromethylketone, but not by CrmA or Bcl-2, thus directly demonstrating that the targets of inhibition of apoptosis by CrmA and Bcl-2 are upstream of caspase-3 and -6. Since caspase-3 and -6 are the most downstream executioners of apoptosis, the constitutively active versions of these caspases could be used at very low concentrations in gene therapy model systems to induce apoptosis in target tissues or neoplastically transformed cells.

The apical procaspases, -8 β and -10, were efficiently processed and activated in yeast (547). Although protease activity, *per se*, was insufficient to drive cell death, caspase-10 activity had little effect on cell viability, whereas expression of caspase-8 β was cytotoxic. This lethal phenotype was abrogated by co-expression of the pan-caspase inhibitor, baculovirus p35 and by mutation of the active site cysteine of procaspase-8β. In contrast, autoactivation of the executioner caspase-3 and -6 zymogens was not detected. Procaspase-3 activation required co-expression of procaspase-8 or -10. Surprisingly, activation of procaspase-6 required proteolytic activities other than caspase-8, -10, or -3. Caspase-8 β or -10 activity was insufficient to catalyze the maturation of procaspase-6. Moreover, a constitutively active caspase-3, although cytotoxic in its own right, was unable to induce the processing of wild-type procaspase-6 and *vice versa*. These results distinguish sequential modes of activation for different caspases *in vivo* and establish a yeast model system to examine the regulation of caspase cascades. Moreover, the distinct terminal phenotypes induced by various caspases attest to differences in the cellular targets of these apoptotic proteases (548), with some of these proteases also possibly participating in the granzyme B apoptotic pathways (549).

Exit of cytochrome c from mitochondria into the cytosol has been implicated as an important step in apoptosis (550). In the cytosol, cytochrome c binds to the CED-4 homologue, Apaf-1, thereby triggering Apaf-1-mediated activation of caspase-9. Caspase-9 is thought to propagate the death signal by triggering other caspase activation events, the details of which remain obscure. Slee and co-workers reported that six additional caspases (caspase-2, -3, -6, -7, -8 and -10) are processed in cell-free extracts in response to cytochrome c and that three others (caspase-1, -4 and -5) failed to be activated under the same conditions. Association assays confirmed that caspase-9 selectively bound to Apaf-1 *in vitro*, whereas caspases-1, -2, -3, -6, -7, -8 and -10 did not. Depletion of caspase-9 from cell extracts abrogated cytochrome c-inducible activation of caspase-2, -3, -6, -7, -8 and -10, suggesting that caspase-9 is required for all of these downstream caspase activation events. Immunodepletion of caspases-3, -6 and -7 from cell extracts enabled us to order the sequence of caspase activation events downstream of caspase-9 and reveal the presence of a branched caspase cascade. Caspase-3 is required for the activation of four other caspases (-2, -6, -8 and -10) in this pathway and also participates in a feedback amplification loop involving caspase-9.

9.1.3 Caspase-8 (Mch5)

Emerging evidence suggests that an amplifiable protease cascade consisting of multiple aspartate specific cysteine proteases (ASCPs) is responsible for the apoptotic changes observed in mammalian cells undergoing PCD. Two ASCPs were cloned by Fernandes-Alnemri and co-workers from human Jurkat T-lymphocytes (549). Like other ASCPs, the new proteases, named Mch4 and Mch5 (caspase-8), are derived from single chain proenzymes. However, their putative active sites contain a QACQG pentapeptide instead of the QACRG present in all known ASCPs. Also, their N termini contain *Fas*-associated DD protein (FADD)-like death effector domains, suggesting possible interaction with FADD. In fact, caspase-8 binds to the death effector domain (DED) of FADD. Expression of Mch4 in *Escherichia coli* produced an active protease that, like other ASCPs, was potently inhibited (Kj = 14 nM) by the tetrapeptide aldehyde DEVD-CHO. Interestingly, both Mch4 and the serine protease granzyme B cleave recombinant proCPP32 and proMch3 at a conserved IXXD-S sequence to produce the large and small subunits of the active proteases. Granzyme B also cleaves proMch4 at a homologous IXXD-A processing sequence to produce mature Mch4. These observations suggest that CPP32 and Mch3 are targets of mature Mch4 protease in apoptotic cells. The presence of the FADD-like domains in Mch4 and Mch5 suggests a role for these proteases in the *Fas*-apoptotic pathway. In addition, these proteases could participate in the granzyme B apoptotic pathways.

The *Fas*/APO-1-receptor-associated cysteine protease, caspase-8 or Mch5 (MACH/FLICE) is believed to be the enzyme responsible for activating a protease cascade after *Fas*-receptor ligation, which leads to cell death (551). Binding of caspase-8 to the *Fas* receptor results in oligomerization of the caspase-8 protein, which in turn drives its autoactivation through self-cleavage. Once activated, the caspase-8 activates the downstream caspases, committing the cell to undergo PCD. The *Fas*-apoptotic pathway is potently inhibited by the cowpox serpin CrmA, suggesting that Mch5 could be the target of this serpin. Bacterial expression of proMch5 generated a mature enzyme composed of two subunits, which are derived from the precursor proenzyme by processing at Asp-227, Asp-233, Asp-391 and Asp-401. Recombinant Mch5 is able to process/activate all known ICE/Ced-3-like cysteine proteases and is potently inhibited by CrmA. This contrasts with the observation that Mch4, the second FADD-related cysteine protease that is also able to process/activate all known ICE/Ced-3-like cysteine proteases, is poorly inhibited by CrmA. These data suggest that Mch5 is the most upstream protease that receives the activation signal from the *Fas*-receptor to initiate the apoptotic protease cascade that leads to activation of ICE-like proteases (TX, ICE and ICE-relIII), Ced-3-like proteases (CPP32, Mch2, Mch3, Mch4 and Mch6) and the ICH-1 protease. On the other hand, Mch4 could be a second upstream protease that is responsible for activation of the same protease cascade in CrmA-insensitive apoptotic pathways.

Phytohemagglutinin-activated peripheral $CD95^+$ T cells (day 1 T cells) are resistant to CD95-mediated apoptosis (552). After prolonged interleukin-2 treatment, these T cells become CD95-mediated apoptosis-sensitive (day 6 T cells). To elucidate the molecular mechanism of apoptosis resistance, day 1 and day 6 T cells were tested for the formation of the CD95 death-inducing signaling complex (DISC). DISC-associated active FADD-like interleukin-1 β-converting enzyme-like protease (FLICE), also referred to as

MACH/caspase-8, was only found in apoptosis-sensitive day 6 T lymphocytes. Further analysis of mRNA and protein expression levels of apoptosis-signaling molecules FADD, receptor interacting protein, hematopoietic cell protein tyrosine phosphatase, *Fas*-associated phosphatase-1, FLICE, bcl-2, Bcl-xL and Bax-α showed that only the expression level of Bcl-xL correlated with T cell resistance to CD95-mediated apoptosis (day 1 T cells: Bcl-xhiL, day 6 T cells: Bcl-XloL). In T cells activated *in vitro*, up-regulation of Bcl-xL, has been correlated with general apoptosis resistance. However, the experiments presented suggest that resistance to CD95-mediated apoptosis in T cells can also be regulated at the level of recruitment of FLICE to the DISC.

Cells of the monocyte/macrophage lineage play a central role in both innate and acquired immunity of the host (553). However, the acquisition of functional competence and the ability to respond to a variety of activating or modulating signals require maturation and differentiation of circulating monocytes and entail alterations in both biochemical and phenotypic profiles of the cells. The process of activation also confers survival signals essential for the functional integrity of monocytes, enabling the cells to remain viable in microenvironments of immune or inflammatory lesions that are rich in cytotoxic inflammatory mediators and reactive free-radical species. Investigations on activation-induced resistance to apoptosis in human monocytes at the molecular level yielded the following observations: (a) activation results in selective resistance to apoptosis, particularly to that induced by signaling *via* death receptors and DNA damage, (b) concurrent with activation, the most apical protease in the death receptor pathway, caspase-8/FLICE, is rapidly down-regulated at the mRNA level representing a novel regulatory mechanism and (c) activation of monocytes also leads to dramatic induction of the Bfl-1 gene, an anti-apoptotic member of the Bcl-2 family.

Tumor necrosis factor-alpha (TNF-α) binding to the TNF receptor (TNFR) potentially initiates apoptosis and activates the transcription factor nuclear factor kappa B (NF-kappaB), which suppresses PCD by an unknown mechanism (554). The activation of NF-kappaB was found to block the activation of caspase-8. TRAF1 (TNFR-associated factor 1), TRAF2 and the inhibitor-of-apoptosis (IAP) proteins c-IAP1 and c-IAP2 were identified as gene targets of NF-kappaB transcriptional activity. In cells in which NF-kappaB was inactive, all of these proteins were required to fully suppress TNF-induced apoptosis, whereas c-IAP1 and c-IAP2 were sufficient to suppress etoposide-induced apoptosis. Thus, NF-kappaB activates a group of gene products that function cooperatively at the earliest checkpoint to suppress TNF-α-mediated apoptosis and function more distally to suppress genotoxic agent-mediated apoptosis.

9.1.4 Caspase-9

Recent progress in studies on apoptosis has revealed that cytochrome c is a pro-apoptotic factor (555). It is released from its places on the outer surface of the inner mitochondrial membrane at early steps of apoptosis and, combining with some cytosolic proteins, activates conversion of the latent apoptosis-promoting protease pro-caspase-9 to its active form. Cytochrome c release can be initiated by the pro-apoptotic protein Bax. This process is blocked by the anti-apoptotic proteins Bcl-2 and Bcl-xL. The role of cytochrome c in apoptosis may be understood within the framework of the concept assuming that the evolutionary primary function of PCD was to purify tissues from ROS-overproducing cells. In this context, the pro-apoptosis activity of cytochrome c might represent one of the anti-oxidant functions inherent in this cytochrome. Among other cytochrome c-linked antioxidant mechanisms, the following systems can be indicated: (1) Cytochrome c released from the inner mitochondrial membrane to the intermembrane space can operate as an enzyme oxidizing O_2- back to O_2. The reduced cytochrome c is oxidized by cytochrome oxidase (or in yeasts and bacteria, by cytochrome c peroxidase). (2) The intermembrane cytochrome c can activate the electron transport chain in the outer mitochondrial membrane. This bypasses the initial and middle parts of the main respiratory chain, which produce, as a rule, the major portion of ROS in the cell. (3) The main respiratory chain losing its cytochrome c is inhibited in such a *Fas*hion that antimycin-like agents fail to stimulate ROS production.

Caspase-9 was initially purified and identified as the third protein factor, Apaf-3, that participates in caspase-3 activation *in vitro*. Procaspase-9 and Apaf-1 bind to each other *via* their respective NH_2-terminal, CED-3 homologous domains in the presence of cytochrome c and dATP, an event that leads to caspase-9 activation. Activation of procaspase-9 by Apaf-1 in the cytochrome c / dATP-dependent pathway requires proteolytic cleavage to produce the mature caspase-9 molecule (556). On the other hand, deletion of Apaf-1's WD-40 repeats leaves Apaf-1 constitutively active and capable of processing procaspase-9 independently of cytochrome c and dATP (557). Apaf-1-mediated processing of procaspase-9 occurs at Asp-315 by an intrinsic autocatalytic activity of procaspase-9 itself (558). Apaf-1 can form oligomers and may facilitate procaspase-9 autoactivation by oligomerizing its precursor molecules. Once activated, caspase-9, also called initiator caspase, can start the caspase cascade involving the downstream activation of caspase-3,-6 and -7 PCD executioners. Depletion of caspase-9 from S-100 extracts diminished caspase-3 activation. Mutation of the active site of caspase-9 attenuated the activation of caspase-3 and cellular PCD response *in vivo*, indicating that caspase-9 is the most upstream member of the apoptotic protease cascade that is triggered by cytochrome c and dATP.

9.1.5 Results

The presence and cellular localization of caspase-3, -8 and -9 employing specific, anti-caspases MoABs was observed high-grade ASTRs, in AAs, and GBMs, employing the usual alkaline phosphatase conjugated antigen detection technique. The immunoreactivity of caspase-3 demonstrated a dominant cytoplasmic pattern in more than 50 per cent of the tumor cells with high intensity staining (+++, A,B). In about 10 per cent of the neoplastic cells, caspase-3 (CPP32) was translocated from the cytoplasm into the nuclei in regressing, apoptotic cells. A similar antigen expression pattern was observed in TILs. Human prenatal thymus was employed as both positive and negative tissue controls.

The presence and cellular localization of the caspase-3, a group II executioner of the apoptotic or cell death pathway, was also observed in childhood PNETs/MEDs. The immunoreactivity demonstrated a dominant cytoplasmic pattern in more than 20 per cent of the tumor cells with high intensity staining (+++, A,B). In about 5 per cent of the neoplastic cells, caspase-3 (CPP32) was translocated from the cytoplasm into the nuclei in regressing, apoptotic cells. The great majority of the MED/PNET tissue did not demonstrate any caspase-3 immunoreactivity and was therefore employed as a negative tissue control.

The presence and cellular localization of the caspase-6, another group II executioner of the apoptotic or cell death pathway, employing anti-caspase-6 MoAB was observed in anaplastic, high grade ASTRs and in GBMs. The immunoreactivity demonstrated a dominant cytoplasmic pattern in more than 50 per cent of the tumor cells with high intensity staining (+++, A,B). In about 10 per cent of the neoplastic cells, caspase-6 was translocated from the cytoplasm into the nuclei in regressing, apoptotic cells.

The immunoreactivity of caspase-8 also demonstrated a dominant cytoplasmic pattern in more than 10 per cent of the ASTR tumor cells with high intensity staining (++, A,B). In about 1 to 5 per cent of the neoplastic cells, caspase-8 was translocated from the cytoplasm into the nuclei in regressing, apoptotic cells. Presence of active or mature caspase-8 was also detected in the tumor infiltrating leukocytes (TIL), a possible escape mechanism favorable for the further tumor development and local, *in situ* invasion.

Caspase-9 results revealed similar immunoreactivity to caspase-8 with a dominant cytoplasmic pattern in more than 10 per cent of the neoplastically transformed ASTR cells, showing high intensity staining (++, A,B). In about

1 to 5 per cent of the neoplastic cells, caspase-9 was translocated from the cytoplasm into the nuclei in regressing, apoptotic cells. Presence of active or mature caspase-9 was also detected in the tumor infiltrating TIL cells.

9.1.6 Discussion

Until our systematic studies and according to the literature, brain tumor immunocytochemistry of caspases has only been carried out only in neuroblastomas, but not in ASTRs. Our studies were actually the first ones done in ASTR. Frequently, neuroblastomas demonstrate spontaneous regression and cellular differentiation, which may at least partly be regulated by signaling done through the nerve growth factor and its receptors, TRK-A and p75LNTR (541). Nakagawara and co-workers carried out an immunohistochemical study of 52 neuroblastic tumors to test whether the cell death-related proteases, interleukin-1 β converting enzyme (ICE), CPP32 and Ich-1 were involved in the regression of the tumors. High levels of expression of ICE and CPP32 were significantly correlated with a high level of TRK-A expression, single copy of N-myc, younger age, lower stages and better prognosis. The immunohistochemical studies and Western blot analyses as well as the terminal dUTP-biotin nick end labeling (TUNEL) method revealed that both ICE and CPP32 were translocated from the cytoplasm into the nuclei in regressing, apoptotic tumor cells. These results suggest that ICE and CPP32 cysteine proteases play an important role in regulating the PCD mechanism of the favorable neuroblastic tumors.

Cell death in the core of human brain tumors is triggered by hypoxia and lack of nutrients, but the mode of cell death, whether necrosis or apoptosis, is not clearly defined (560). To identify the role of PCD in brain tumor cells, macromolecular (RNA and protein) synthesis and activity in the central to peripheral region of benign [desmoplastic infantile ganglioglioma (DIG) and transitional meningioma (TMG)] and malignant [ependymoma (END), anaplastic astrocytoma (AA) and GBM] tumors were observed by Ray and co-workers in tumors which had not previously received radiotherapy or chemotherapy. Normal brain tissue (NBT) served as the control. RT-PCR analysis of tumor tissues covering central to peripheral regions detected mRNA overexpression of pro-apoptotic gene Bax in malignant tumors, indicating a commitment to apoptosis. The mRNA expression of calpain (a Ca^{2+}-dependent cysteine protease) and calpastatin (endogenous calpain inhibitor) was altered resulting in an elevated calpain/calpastatin ratio. The calpain content and activity were increased, suggesting a role for calpain in cell death. In the mitochondria-dependent death pathway, caspase-9 and caspase-3 were also overexpressed in tumors. The increased caspase-3 activity cleaved poly(ADP-ribose) polymerase (PARP). Agarose gel electrophoresis detected a mixture of random and internucleosomal DNA fragmentation in malignant brain tumor cells. Overexpression of pro-apoptotic Bax, up-regulation of calpain and caspase-3 and occurrence of internucleosomal DNA fragmentation were presented indicating that one mechanism of cell death in malignant brain tumors is the PCD and that enhancement of this process therapeutically may promote decreased tumor growth.

9.2 Apoptosis mediated by CD95/*FasR*/APO-1

It was recently well established that in mammalian cells, several gene products have been implicated as participating in or regulating PCD: CD95/*FasR*/APO-1, bcl-2, c-myc, p53, and $p21^{WAF1}$ (561-568). *Fas*/APO-1/CD95 mediated apoptosis is one of the major mechanisms of programmed cell death (PCD) (569-571). CD95/*FasR*/APO-1 is a type I transmembrane receptor with a molecular weight of 48 kD, and a cysteine rich extracellular domain of 155 amino acids. Its transmembrane and intracellular domain contain 19 respectively 145 amino acids. CD95/*Fas*/APO-1 belongs to the gene superfamily, which includes such molecules as tumor necrosis factor (TNF) receptors, nerve growth factor (NGF) receptor, CD27, CD30, and CD40 (562,566, 571-575). The CD95/*FasR*/APO-1 molecule has been found on rapidly proliferating cells including thymocytes, immunocompetent T lymphocytes, thymic stromal cells (RE cells, IDCs, Langerhans cells, etc.), and epithelial cells of the skin and gut (576). Anti-*FasR* MoABs or natural *FasL* can induce PCD under a number of conditions (577-582). *FasL* and *FasR* have been interpreted as the death factor and its receptor (574). Recent results have suggested that *FasR* expression in cortical thymocytes is induced at a special developmental stage, characterized by the IP transition from $CD3^-CD4^{low}CD8^{low}$ to $CD3^{low}CD4^+CD8^+$ (575). In mice, loss of function mutation of *FasR* and *FasL* (*lpr*-lymphoproliferation and *gld*-generalized lymphoproliferative disease) causes lymphadenopathy and autoimmune disease (583-585). These animals accumulate a large number of very unusual T lymphocytes with a double negative IP ($CD4^-CD8^-B220^+$) in the secondary lymphatic organs (lymph nodes and spleen), and have profound thymic maturation defects, loss of T lymphocyte tolerance, B cell defects, hypergammaglobulinemia, production of autoreactive antibodies and manifestations of autoimmune diseases such as pneumonitis, synovitis, arthritis, and CNS disease (573-575, 583). Human naive immunocompetent, single positive T lymphocytes cannot be killed by anti-*FasR* antibody, despite the intense presence of *FasR* antigen on their cell surface, because these cells express proteins that are capable of inhibiting the *FasR* mediated PCD mechanism (565, 579, 590). *FasR* induced PCD can be enhanced by metabolical inhibitors such as cycloheximide, demonstrating that the pathway is different from that of glucocorticoids or T lymphocyte activators. Phosphatase (*e.g.* pervanadate) inhibition has been proposed as a key mechanism in the elimination of *FasR* induced apoptosis signaling (591, 592).

In previous observations employing reverse transcriptase-polymerase chain reaction, *Fas* has been shown to be frequently expressed in malignant gliomas (593, 594). *Fas*/APO-1/CD95 expression in ASTRs was revealed using a polyclonal anti-*Fas* antibody. Immunoreactivity to *Fas* was detected in 1/9 (11%) low-grade ASTRs (grade II), 2/11 (18%) anaplastic ASTRs (grade III), and in 13/15 (87%) GBMs (grade IV). In GBMs, *Fas* expression was almost exclusively observed in neoplastically transformed glial cells surrounding foci of necrosis. In these perinecrotic areas, there was also an accumulation of glioma cells undergoing apoptosis, as detected by *in situ* nick-end labeling. These results suggest that *Fas*-mediated apoptosis may be part of the pathogenetic mechanisms of necrosis which constitutes a histological hallmark of GBM. The expression of *FasR* and *FasL* on the surface of neoplastically transformed cells comprising the microenvironment of a given tumor (*e.g.* glioma) has also been described, raising the possibility of autocrine suicide or "fratricide" (595).

9.2.1 Results

During our immunocytochemical screening of 42 childhood ASTRs tissues divided according to WHO classification: 6 WHO grade I or pilocytic ASTRs; 14 WHO grade II or low grade ASTRs; 16 WHO grade III or anaplastic ASTRs and 6 WHO grade IV or glioblastoma multiforme (GBM), we detected strong expression (intensity of staining: "A"-the highest possible; number of stained cells: ++ to ++++, between 20% to 90%) of *FasR* in paraffin sections. *FasR* was present on 70% to 90% of tumor cells in pilocytic ASTRs, in 50% to 60% of the tumor cells in low grade ASTRs, on between 30% and 40% of the tumor cells in anaplastic ASTRs, and on between 20% to 35% of GBM cells. Apoptotic cells were recognized by their nuclear morphology and their percentage was correlated to *bcl-2* expression. Apoptotic ASTR cells were further divided into "early" and "late" stages, according to Cohen (596). The "early" tumor cells undergoing PCD displayed initial signs of nuclear chromatin clumping and the "late" stages were subdivided in 1) ASTR cells with clumped chromatin adhering to their nuclear membrane to form a crescent or ring shape; 2) tumor cells with severe nuclear shrinkage and condensation; and 3) ASTR cells displaying the characteristic apoptotic bodies.

A systematic observation for the presence of apoptosis related markers (especially *FasR*) in MED/PNET cells in PCD was carried out. A strong expression (intensity of staining: "A" - the highest possible; number of stained neoplastic cells: +++ to ++++, between 50% to 90%) of *FasR*, was detected. A decrease in the percentage of PCD affected cells in MEDs and *bcl-2* expression were not in direct correlation with a probably increasing grade of MED malignancy. Thus, we detected an initial increase, followed by decrease in the proportion of apoptotic cells during the early stage of a more malignant (more undifferentiated, more embryonic IP, showing less differentiated cell features IP) form of MEDs/PNETs.

The panel of normal tissues employed as positive and negative tissue controls demonstrated presence of *FasR* in the prenatal thymus, mature tonsils, and colon epithelium. Whereas *FasR* is expressed in every colonocyte of normal colon mucosa, APO-1 or CD95 is down-regulated or lost in the majority of colon carcinoma cells (in accordance to 597, 598).

9.2.2 Discussion

A broad spectrum of neoplastic cells have been identified to express *FasR*: 1) carcinomas of epithelial origin, such as breast (ductal invasive, lobular invasive, mucinous), renal cell, gastric, colorectal, endometrial, prostate, pancreas, hepatocellular and large cell and squamous cell lung carcinomas; 2) non-epithelial neoplasms such as B cell, mediastinal B cell and nodal non-Hodgkin's lymphomas, large granular lymphocytic leukemia of T or NK cell origin, malignant fibrous histiocytoma, malignant mesothelioma, leiomyosarcoma, epitheloid sarcoma and alveolar soft part sarcoma, as well as melanomas. Flow cytometry studies have also detected *FasR* expression on cells of adult T cell, and hairy cell leukemias, as well as in chronic B cell lymphocytic leukemia (BCLL). The co-expression of both *FasR* and *FasL* on several malignant cell types may represent an effective mechanism of tumor escape from the cellular immunological response of the host. It has been well established that brain tumors and melanomas produce their autocrine *FasL*, and even become capable of switching the signal transduction associated with *FasL-FasR* coupling from the PCD pathway to a tumor growth, proliferative pathway.

The CD95 ligand (CD95L) is a cytotoxic cytokine that induces apoptosis in susceptible target cells. CD95L-mediated apoptosis was observed in established human MED cell lines (599, 600). The authors found that DAOY, MED-1 and D-283 cells are susceptible to CD95L-induced apoptosis when RNA and protein synthesis are inhibited. Pre-exposure of D-283, but not DAOY or MED-1 cells, to interferon-γ or tumor necrosis factor-α (TNF-α), enhances CD95R expression and primes these cells for CD95L-mediated apoptosis. Inhibitors of interleukin 1-converting enzyme (ICE)-like protease (caspase) activity block CD95L-induced cytotoxicity, suggesting that caspases mediate the death signal induced by CD95L in human MED cells. Interestingly enough, MED cells belong to an increasing number of neoplastic cell types that coexpress CD95 and CD95L. We strongly support the most important conclusion of this article that *FasR* (CD95, APO-1) may be a promising target of immunochemotherapy for human MED (569,599).

German researchers studied apoptosis in 40 MEDs by *in situ* end labeling (ISEL) of DNA strand breaks. ISEL was performed on paraffin-embedded material from classical and desmoplastic MEDs, and MEDs with glial and combined cell differentiation (601). The number of labeled apoptotic cells varied considerably from tumor to tumor, as well as, between different areas in the same tumor. However, there was no significant difference in the averaged numbers or in the pattern of apoptotic tumor cells between the different differentiations of MED (601). Unlabeled tumor cells displaying features of apoptosis were found in small numbers, indicating that tumor cells may also be able to undergo cell death different from classical apoptosis. A significantly higher number of apoptotic tumor cells in male patients could be demonstrated corresponding to the reported difference in survival rates between male and female patients suffering from MED. A clear cut trend of a negative relation between apoptosis and age by operation was found.

As it was already demonstrated, bcl-2 is an oncoprotein, described to be inactivated by phosphorylation employing the *c-Raf-1* directed signal transduction pathway (67,68,74,75,602-608). Overexpression of bcl-2 protein in cells has been described as an immunocytochemical sign of cell survival (609-611). The ratio of bcl-2 protein to other bcl-2 family members is believed to modulate the apoptotic mechanism. Expression of bcl-2 can render neoplastic cells resistant to several agents that are capable of inducing PCD (physical treatments such as: heat shock, cold shock, UV light, γ-irradiation, or chemical agents often used in chemotherapy, like cisplatin, etoposide, teniposides, DNA alkylating agents, macromolecular synthesis inhibitors and others (612-621). Therefore, drugs that can reopen the apoptotic pathway closed by bcl-2 blocking function can render neoplastic cells extremely susceptible to therapeutic treatments that employ the PCD.

Expression of bcl-2, a PCD-suppressing molecule, and bax, the bcl-2 related PCD-accelerating protein was observed in several histological types of human brain tumors (622). Thirty-six cases of human brain tumors: four ASTRs, three AAs, four GBMs, five MEDs, one ependymoma, two choroid plexus papilloma, one ganglioglioma, one central neurocytoma, four meningotheliomatous meningiomas, three transitional meningiomas, four fibroblastic meningiomas, three acoustic neurinomas, and one craniopharyngioma were analyzed for the localization of bax and bcl-2 proteins. There was no relationship found between the degree of the histological malignancy and the presence of bax or bcl-2 proteins in varied human brain tumors. However, it is suggested that reduced expression of bax protein is necessary for the malignant transformation and progression of the brain tumors, since no histologically malignant bax^{+} brain tumors were present. These findings indicate that the expression pattern of bax and bcl-2 may reflect histogenetic differences of each type of brain tumors.

Similarly to Yew and co-investigators (623) and the group of Schiffer (624, 625), we also registered a decrease in the percentage of apoptotic ASTR cells and bcl-2 expression was evident with the increasing grade of ASTR malignancy. Our data support the idea that a reverse relationship exists between the ASTR malignancy grade and the rate of neoplastic cells involved in PCD. Extensive brain tumor research in Dr. Kleinhaus laboratory in Lyon, France, identified the presence of apoptotic tumor cells in the perinecrotic cell microenvironment of GBM tissues (594,626, 627).

Similar to our immunocytochemical screening for PCD markers, observations of the expression of bcl-2, bcl-X, Mcl-1, and bax were carried out by immunohistochemical methods in 93 CNS tumors, including 49 gliomas (30 ASTRs and 19 GBMs), 16 MEDs, 19 neuroblastomas (NBs: 9 undifferentiated and 10 differentiated), and 9 miscellaneous neuroectodermal neoplasms (628). Among the 49 gliomas, immunopositivity (defined as > or = 10%) was observed for *bcl-2* in 45 (92%), bcl-X in 48 (98%), Mcl-1 in all 49 (100%), and bax in 48 (98%). In 11 (37%) of 30 astrocytomas (WHO grades I to III), the tumor specimens were composed predominantly of malignant cells with strong-intensity bcl-2 immunostaining, whereas none of the 19 GBMs (WHO grade IV) exhibited strong-intensity bcl-2 immunoreactivity ($p = 0.001$) (628). Similarly, Mcl-1 immunoreactivity was strong in 15 (50%) of 30 ASTRs, compared with only 2 (11%) of 19 GBMs ($p = 0.005$). The percentage of Mcl-1 immunopositive tumor cells was also higher in ASTRs than GBMs ($p<0.002$). Thus, contrary to a priori expectations, the expression of the anti-apoptotic proteins *bcl-2* and *Mcl-1* was significantly higher in ASTRs than in GBMs. Of the 16 MEDs, immunopositivity was found for bcl-2 in 4 (25%), bcl-X in 9 (56%), Mcl-1 in 8 (50%), and bax in all 16 (100%) of the cases. The intensity of immunostaining was strong for bcl-2 in only 1 (6%) specimen, for bcl-X in 3 (19%), and for Mcl-1 in 2 (12.5%), in contrast to bax immunostaining, which was strong in 12 (75%) MED tumor cells. Significantly higher percentages of bax-immunopositive tumor cells were also found in MEDs, compared with bcl-2, bcl-X, and Mcl-1 ($p<0.0001$). All 19 NBs were immunopositive for bcl-2, bcl-X, Mcl-1, and bax. Higher percentages of bcl-X and Mcl-1 immunopositive neoplastic cells were observed in cellularly well differentiated tumors (p = 0.04 and 0.004, respectively). The intensity of Mcl-1 immunostaining was also generally higher in differentiated than undifferentiated NBs (strong immunointensity in 7 of 10 versus 0 of 9; p = 0.002). Conversely, strong-intensity bax immunoreactivity was associated with undifferentiated histology (5 of 9 (56%) versus 1 of 10 (10%); p = 0.03). In summary, these results begin to delineate trends in the regulation of the relative levels of the bcl-2 family proteins, bcl-2, bcl-X, Mcl-1, and bax in gliomas, MEDs, NBs, and some of their histological subtypes. We agree with the conclusions that the expression of some of these bcl-2 family genes may be differentially regulated in association with tumor progression and intratumoral cell differentiation, which provides insights into the diverse biology and clinical behavior of these CNS tumors (628).

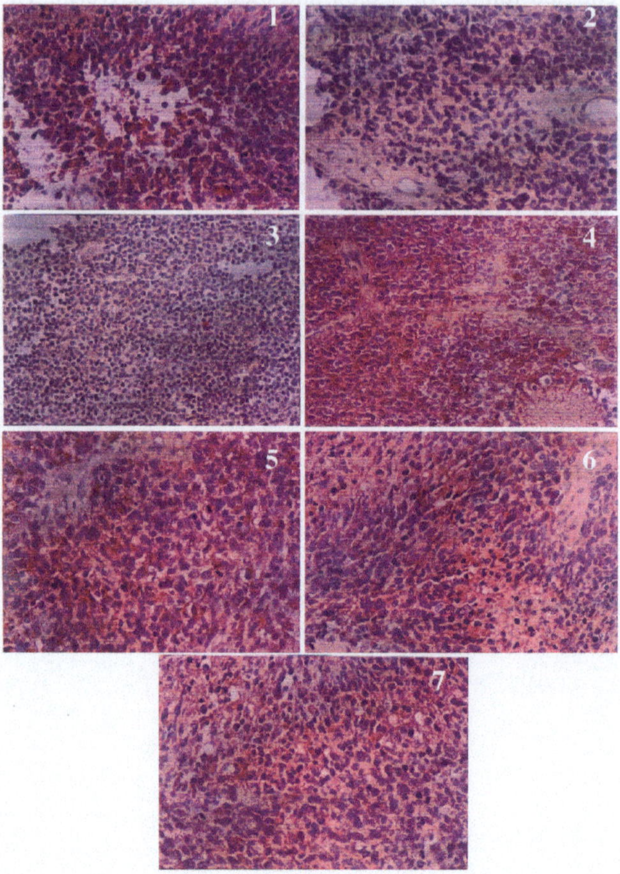

Figure 2-3.

1. Anaplastic ASTR. The expression pattern of caspase-3 is dominantly cytoplasmic in great number of tumor cells (> 50%). In some tumor cells caspase-3 is translocated to the cell nuclei. Alkaline phosphatase conjugated, streptavidin-biotin enhanced immunohistochemical antigen detection technique. Magnification: 200x.
2. GBM. Similar to the anaplastic ASTR expression pattern of caspase-3 is identified. The tumor infiltrating leukocytes (TIL) demonstrated mostly cytoplasmic immunoreactivity. Alkaline phosphatase conjugated, streptavidin-biotin enhanced immunohistochemical antigen detection technique. Magnification: 200x.
3. Human prenatal thymus (16 weeks of intrauterine life). Employed as positive and negative tissue control for caspase-3 immunohistochemistry. Alkaline phosphatase conjugated antigen detection technique. Magnification: 200x.
4. Anaplastic ASTR. The number of tumor cells expressing caspase-8 is significantly lower (about 10%). Alkaline phosphatase conjugated antigen detection technique. Magnification: 200x.

5 GBM. Caspase-8 immunohistochemistry demonstrated similar results to the observation in anaplastic ASTR. The expression pattern is dominantly cytoplasmic. Alkaline phosphatase conjugated antigen detection technique. Magnification: 200x.

6 Anaplastic ASTR. Caspase-9 presence in the dedifferentiated tumor cells. Alkaline phosphatase conjugated antigen detection technique. Magnification: 200x.

7 GBM. Caspase-9 immunohistochemistry identified its strong, cytoplasmic expression in the tumor cells. The TIL cells demonstrated similar immunoreactivity. Alkaline phosphatase conjugated antigen detection technique. Magnification: 200x.

10. SURVIVIN

Survivin, a protein that inhibits apoptosis and regulates cell division (629), was discovered in 1997 as an anti-apoptotic protein on the basis of its baculovirus inhibitor of apoptosis repeat (BIR) domain (630). It was recently identified by hybridization screening of human genomic libraries with the cDNA of a factor Xa receptor, effector cell protease receptor-1 (EPR-1). The survivin gene has been found to span 15 kb and is located on chromosome 17 at band q25. Its expression is highly cell-cycle regulated and is detectable in the nucleus selectively at the G2/M phase (631). Transcription of survivin has even been shown to be directly repressed by p53, another cell cycle checkpoint-regulating protein that induces apoptosis (632). Survivin also plays a role in cell cycle progression as demonstrated when it was disrupted by antisense targeting in Hela cells and resulted in spontaneous apoptosis and aberrant mitosis (633) and an increase in caspase-3 activity (631). Survivin is required for cell division *in vivo* as well.

Survivin belongs to the Inhibitor of Apoptosis Protein (IAP) family and as such, is pathologically over-expressed in most human cancers and functions at the cross-roads of cell death and cell cycle regulation. Survivin has actually been detected in most cancers in humans. Survivin thus appears to have an important role in regulating apoptosis at cell cycle checkpoints. Survivin controls a checkpoint associated with chromosome segregation and cell division. Specifically, survivin inhibits apoptosis *via* its BIR domain by either directly or indirectly interfering with the function of the caspases (634), which are responsible for inducing apoptosis.

The structure of the survivin protein is greatly related to its function as an inhibitor of apoptosis. The amino terminal portion of survivin consists of three α helices and three β-sheets, which closely resemble the BIR domain that is part of the IAP family (630, 635). It is the BIR domain of the IAP family that is involved in the function of these proteins as inhibitors of apoptosis (636). Survivin was shown to exist as dimer with the two BIR domain forming a "bow-tie" shape, but how the dimeric structure is involved in the inhibition of apoptosis and cytokinesis requires further investigation (635).

The structure of survivin has been compared to another member of the IAP family, XIAP, which contains three BIR domains (637). XIAP inhibits caspase-3 and caspase-7 *via* a linker region between the first two BIR domains. It also binds to and inhibits caspase-9 through its third BIR (BIR3) domain. The BIR domain of survivin appears to be closely related to the three-dimensional structure of the BIR3 domain of XIAP, which suggests that survivin may bind to caspase-9 (638). In fact, the survivin-caspase-9 complex has been shown to have an effect on the mitotic apparatus, which allowed for caspase-9 dependent apoptosis to occur (639). Specifically, the interaction between survivin and caspase-9 led to the loss of phosphorylation at threonine 34 on the T34A mutant of survivin resulting in the dissociation of an immunoprecipitable survivin-caspase-9 complex on the mitotic apparatus.

Interestingly, the coding strand of the Survivin gene is entirely complementary (antisense) to a previously characterized gene encoding the effector cell protease receptor-1 (EPR-1) (640). Separate promoters, oriented in opposing directions, control expression of ERP-1 and survivin in an exclusionary *Fas*hion, wherein transcripts produced for one of these appear to inhibit the expression of the other through an antisense RNA-based mechanism (640). It remains to be determined whether any of these gene loci are directly involved in genetic alterations found in tumors. Ambrosini and co-workers also discovered that the down-regulation of survivin by forced expression of EPR-1 increased apoptosis and inhibited growth of transformed cells.

Direct binding between survivin and the caspases has not been confirmed, but survivin may also inhibit caspase activity indirectly. There is evidence of an indirect regulation of caspase activity by survivin in the mitochondrial pathway of apoptosis. When *Fas* is stimulated in a cell culture, survivin has been found to interact with Cdk4, which releases p21 to complex with caspase-3, which is the initial step in the inactivation of capase-3 in the mitochondria (641, 642). While several reports also demonstrate that purified survivin directly binds to caspase-3 and inhibits its activity *in vitro* (640, 643, 644); at best, the mechanism of caspase-3 inhibition by survivin remains, nonetheless, controversial.

10.1 Results

In this immunohistochemical study, we observed the presence and tissue localization of survivin, employing the specific MoAB against it in MEDs/PNETs. A sensitive, four-step, alkaline phosphatase conjugated antigen detection technique was used. The immunoreactivity demonstrated a distribution pattern in 10 to 50 per cent of the tumor cells with medium intensity immunoreactivity (++, B) in the tissue.

10.2 Discussion

Survivin has been shown to be expressed by neoplasms originating from different cell lineages. There are also cumulative evidences that spontaneous immune response against survivin derived epitopes may occur. Katoh and co-workers used RT-PCR, Western-blot analysis and immunohistochemistry to show that survivin is widely expressed by gliomas, meningiomas and schwannomas, both *in vitro* and *in vivo* (645). Their data indicate that survivin may serve as an attractive target for immunotherapies designed for brain tumors.

Survivin expression is considered an important prognostic factor of many tumors. Kajiwara and co-workers investigated 43 astrocytic tumors (8 diffuse astrocytomas; 15 anaplastic astrocytomas; 20 glioblastoma) for survivin mRNA expression by reverse transcriptase-polymerase chain reaction amplification (646). Thirty-four of 43 (79.1%) astrocytic tumors expressed survivin, with the distributions specifically including 3 of 8 (37.5%) diffuse astrocytomas, 13 of 15 (86.7%) anaplastic astrocytomas, and 18 of 20 (90.0%) glioblastomas. Expression of survivin (p = 0.0057) and EGFR (p = 0.0112) was significantly associated with malignant grade of astrocytic tumors. It was found that patients with survivin-positive astrocytic tumors had significantly shorter overall survival times compared with patients who had survivin-negative tumors (p = 0.0271). The authors concluded that survivin expression in astrocytic tumors varies with the grade of histologic malignancy and may play an important role in the oncogenesis and progression of astrocytic tumors. These data thus suggest that survivin also has great potential as a target in anti-neoplastic biological therapy of astrocytic tumors. Our results indicate that survivin is also present in MEDs and PNETs and may thus, prove to be a useful tool in prognosis and another possible immunotherapy option.

While survivin has been detected in most tumors, it is not by any expressed in all tumors. For example, low grade non-Hodgkin's lymphomas, which are known for their activation of another type of anti-apoptotic gene, Bcl-2 (649), rarely express this IAP family protein (630). These lymphomas are also tumors with very low growth fractions, a characteristic that could have bearing on the apparent cell cycle-dependent expression of survivin. Moreover, even within a given type of cancer, heterogenity in survivin expression may be observed. Immunohistochemical assessments of survivin expression in tumors where immunointensity, percentage immunopositivity, or both have been measured for purposes of segregating survivin-negative from survivin-positive (survivin low from high) tumors suggest that expression of survivin (or higher levels of survivin expression) is associated with worse clinical outcome or other unfavorable prognostic features in neuroblastomas, colon and gastric cancers (630, 640, 648, 649). Although preliminary, assessments of survivin expression may be of prognostic significance for patients with some types of cancer.

11. TUMOR-RELATED NEOANGIOGENESIS IN CHILDHOOD BRAIN TUMORS

The growth of solid neoplasms, even at clinically undetectable sizes (a few mm^3), depends upon the continuous formation of new blood capillaries (*i.e.* neo-vascularization [NV]). The generation of an invasive cellular IP in neoplastic cells (and distant metastases) is also an NV dependent process. Endothelial cells undergo rapid proliferation and express upregulated quantities of integrins ($\alpha v\beta 3$ and $\alpha v\beta 5$) during neoplasm related and induced NV. These integrins bind to a short sequence of amino acids (so-called RGD fragment) in extracellular collagen and the enzyme MMP-2 (matrix metalloproteinase 2), allowing the newly organized blood capillaries to advance through the extracellular matrix. The involvement of MMPs and urokinase in NRA and consequently in tumor growth and progression has also been well established (650).

Two significant papers have reported that bFGF, which was the first endogenous angiogenic factor discovered, may play a crucial role in the switch of neoplastic cellular IP to angiogenic *via* a 17 kD protein product designated bFGF binding protein (FGF-BP), which is secreted by the neoplastic cells themselves (651, 652). In 1994, the same authors reported that neoplastic cell growth and tumor related angiogenesis can be upregulated by gene transfection to increase FGF-BP secretion by neoplastic cells (653). Furthermore, a ribozyme targeting approach was employed to suppress FGF-BP expression in malignant colon carcinoma cells. A hypothesis has been developed that states that upregulated FGF-BP in neoplastically transformed cells is secreted into the extracellular microenvironment to bind inactive bFGF molecules. The soluble, displaced from the extracellular matrix, activated bFGF molecules are then able to mediate several functions, among them the induction of NRA. Czubayko and co-investigators (651) suggest that secretion of FGF-BP by neoplastic cells is in fact "the angiogenic switch" in human neoplastic diseases. A relatively small (about 20%) reduction in FGF-BP expression by the ribozyme targeting method resulted in disregulation and size reduction of neoplastic growth *in vivo*. This phenomenon has been described earlier as a three-fold reduction of VEGF/PF expression that correlated with almost complete obliteration of xenotransplanted human GBM growth in nude mice. bFGF has been shown to modulate malignant glioma growth and is required for the clonogenic expansion of human glioma cells during the progression of malignant disease (654 ,655). Continued observations of NRA are necessary to find adequate answers for the numerous questions that still remain unanswered; including what is responsible for the upregulation of FGF-BP expression in neoplastic cells in the first place. It is quite probable that genetic alterations, such as activation of the oncogene *ras* and inactivation of the tumor suppressor gene p53 are associated with FGF-BP expression or other even earlier events leading up to NRA. Ras and other dominant oncogenes activate phosphatidylinositol-3-kinase, representing one possible way of angiogenic switch regulation. The anti-neoplastic adjuvant therapeutic implications and possibilities of the genetics of NRA are obviously very significant. Brain tumors, especially GBM, are among the most vascularized human neoplasms, and thus are candidates for anti-angiogenic therapy. Vascular endothelial growth factor/vascular permeability factor receptors (VEGF/PF-R1 or flt-1 and VEGF/PF-R2 or flk-1) are formed *de novo* in a glioma progression-dependent manner.

High grade brain tumors are associated with prominent neovascularization, composed mostly of vascular smooth muscle cells (VSMC). Stiles and co-workers observed the expression of the potent smooth muscle mitogen endothelin-1 (ET-1) and one of its secretagogues, transforming growth factor-β1 (TGF-β1) in a series of astrocytic tumors (656). TGF-β1 is also of interest due to its known activity as an angiogenic factor. Using immunohistochemical methods, 30 surgical cases were examined by the authors including, 10 GBM, 10 anaplastic ASTRs, and 10 low-grade ASTRs. Employing a MoAB to TGF-β1 and a polyclonal to ET-1, both growth factors were detected in all cases of GBM examined. In cases of anaplastic ASTR, 4 tumors were positive for both factors; 2 contained only ET-1; 2 contained only TGF-β1; and 2 exhibited no tumor cell immunoreactivity for either factor. In low-grade ASTR, 4 of 10 tumors demonstrated weak ET-1 immunoreactivity; 2 of those contained TGF-β1 immunoreactive neoplastically transformed astrocytes; 6 tumors were negative for both factors. In all tumors that expressed both factors, serial sections showed that regions of ET-1 immunopositivity also tended to be positive for TGF-β1. Endothelial cells within all tumors were positive for ET-1. The authors concluded that ET-1 and TGF-β1 are present in human ASTRs and their expression correlates with tumor vascularity and malignancy grade, and suggested roles for both ET-1 and TGF-β1 in the growth and progressive angiogenesis of human glial tumors.

11.1 Endoglin (EDG/CD105)

Human endoglin (EDG/CD105) is a homodimeric transmembrane component of the TGF-β type I receptor complex. Endoglin was initially characterized as an endothelial cell membrane glycoprotein involved in intercellular recognition in the early 1990s (657, 658). EDG, a TGF-β1 and TGF-β3 binding protein, received the CD105 designation at the 5th International Conference on Human Leukocyte Differential Antigens in 1993 (659). EDG has also been defined as a proliferation-associated antigen (PAA), expressed at high, upregulated density on actively proliferating endothelial cells during neoplasm related and induced angiogenesis (660, 661). EDG expression also has been reported on proerythroblasts, activated monocytes, and lymphoblasts in childhood leukemia (662-670). The human EDG gene (END) has been mapped to chromosome 9 by fluorescent *in situ* hybridization coupled with Distamicin A (DA)/4',6-diamidino-2-phenylindole (DAPI) banding on human chromosomes.

11.2 Vascular Endothelial Growth Factor (VEGF)

Vascular endothelial growth factor (VEGF) is a disulfide-linked homodimeric glycoprotein of about 40 kD that promotes fluid and protein leakage from blood vessels being a potent microvascular permeability enhancing cytokine and a selective mitogen for endothelial cells (671, 672). It has been shown to be a potent mediator of brain tumor angiogenesis, vascular permeability, and glioma growth (673). Of the four known isoforms, the smaller two $VEGF_{121}$ and $VEGF_{165}$ are secreted proteins acting as diffusible agents, whereas the larger two $VEGF_{189}$ and $VEGF_{206}$ remain cell-associated (674).

Implantation and growth of the placenta requires extensive angiogenesis to establish the vascular structures involved in exchange (675). Failure to establish adequate blood supply to the fetus may have serious clinical consequences such as intrauterine growth retardation. VEGF is an identified growth factor with significant angiogenic properties. There are four species of mRNA encoding VEGF in both first trimester and term placenta. *In situ* hybridization was used to localize the sites of expression of VEGF mRNA in these tissues. VEGF expression was seen in villous trophoblast in the first trimester and in extravillous trophoblast at term, and in both fetal macrophages within the villi and maternal macrophages in the decidua. Glandular epithelium in maternal decidua also expressed VEGF mRNA. The strongest site of expression was in maternal macrophages adjacent to Nitabuch's stria, a zone of necrosis at the site of implantation. This complex pattern of expression suggests that VEGF is involved in angiogenesis on both maternal and fetal sides of the placenta and that macrophages are the primary source of VEGF. However, VEGF may also play a role in term placenta, when extensive angiogenesis has diminished, possibly regulating vascular permeability.

Vascular endothelial growth factor B (VEGF-B) is structurally closely related to VEGF and binds one of its receptors, VEGFR-1 (676). *In situ* hybridization and immunohistochemistry were employed to localize VEGF-B mRNA and protein in embryonic mouse tissues. In 8.5-17.5 day embryos, VEGF-B was most prominently expressed in the developing myocardium, but not in the cardiac cushion tissue. The strong expression in the heart persisted at later developmental stages, while weaker signals were obtained from several other tissues, including developing muscle, bone, pancreas, adrenal gland, and from the smooth muscle cell layer of several larger vessels, but not from endothelial cells. VEGF-B is likely to act in a paracrine *Fas*hion, as its receptor is almost exclusively present in endothelial cells. VEGF-B may have a role in vascularization of the heart, skeletal muscles and developing bones, and in paracrine interactions between endothelial and surrounding muscle cells.

VEGF is one of the most important mediators of tumor-associated neoangiogenesis. Vascular growth factor mRNA levels (TGF-α, TGF-β, bFGF, VEGF) were screened employing Northern blot analysis in human gliomas and meningiomas to examine their correlation with neoplastic transformation related angiogenesis (677). The number of blood capillaries was scored by counting their numbers after employing von Willebrand factor immunocytochemistry. The authors normalized the growth factor mRNA levels versus the glyceraldehyde phosphate dehydrogenase mRNA level. In the 17 gliomas and 16 meningiomas, only the mRNA level of VEGF correlated significantly with vascularity ($r=0.499$; $p < 0.05$ in gliomas; $r=0.779$; $p < 0.001$ in meningiomas). These results clearly suggest an important regulatory role for VEGF in tumor related angiogenesis in human gliomas and meningiomas. A number of *in situ* hybridization and Northern blot observations have identified strong expression of VEGF m-RNA in glioblastomas, meningiomas, and hemangioblastomas. It has been suggested that VEGF may be considered as the most important stimulatory factor of endothelial cell proliferation in CNS malignancies.

VEGF demonstrated a temporal and spatial expression pattern during neural ontogenesis that is compatible with an inducer function for blood vessel growth from the perineural vascular plexus (678-681). There is considerable experimental evidence that VEGF is strongly involved in neoplasm provoked neoangiogenesis and, consequently the growth and progression of primary neoplasms (*i.e.* gliomas), including the formation of an invasive and metastatic IP (673, 678-681). The *de novo* expression of

VEGF on endothelial cells depends on activated oncogenes and inactivated tumor-suppressor genes, as well as additional complex regulatory mechanisms involving several other factors (*e.g.* growth factors, hypoxia, and neoplastic promoters).

11.3 Results

Immunocytochemically, childhood brain tumor related angiogenesis was observed in 62 cases. As an important marker, strong expression (A; +++ to ++++) of EDG on endothelial cells was demonstrated in all 62 childhood brain tumor cases. MEDs/PNETs (n=34) were characterized by the expression of CD105 on the endothelial cells of blood capillaries located within the neuroectodermally derived, neoplastically transformed tissue. The tumor related blood capillaries were much thicker in their morphological appearance. The endothelial cells of the blood capillaries within the neuroectodermal in origin brain tumors demonstrated an oversized appearance, suggesting an increase in cytoplasmic complexity and high proliferative activity. Special attention was given to the immunocytochemical analysis of ASTRs (n=28) due to the well-characterized progression established for these tumors. The newly formed tumor blood capillaries' most striking feature was the presence of markedly enlarged and disorganized perivascular space, measuring at least 2-5 mm. This alteration in the perivascular space was detected in all observed childhood brain tumors and was independent of the tumor's size, as well as its morphological and cellular differentiation features. Perivascularly located pericytes express MHC class II and adhesion molecules, and are strong candidates for intratumoral antigen presentation to both $CD4^+$ and $CD8^+$, tumor infiltrating T lymphocytes. The normal brain (part of the panel of normal control tissues) displayed a relationship between the blood capillaries and the adjacent parenchyma in which the capillaries seemed to be in much closer apposition with glial cells, which resulted in the absence of perivascular spaces in the normal CNS. In necrotic areas, cell debris was detected within the perivascular space. In larger spaces, monocyte/macrophages and collagen fibrils were also observed. Endothelial cell proliferations represented the great majority of capillaries within astrocytic glial tumors, especially in the anaplastic ASTR and all six glioblastoma multiforme cases. Morphologically, the pattern of endothelial cell proliferations can be divided into three subtypes: 1) solid-glomeruloid; 2) channeled-branching; and 3) channeled-telangiectatic. The first group is characterized by its cohesive cells; the second, by the single layer of endothelial cells that lines the capillaries; and the third (along with the second), by the regular presence of macrophages in the surrounding tissue.

The first two subtypes represented the majority of endothelial cell proliferations. Blood vessels in several normal human tissues (cortex, cerebellum, thymus, tonsil, spleen, lymph node, skin) employed as control tissues contained significantly lower levels of CD105 or EDG (B and mostly C; ± to +), in accordance with the extremely slow turnover rate of normal endothelial cells. We identified the presence and tissue localization of $VEGF_{121}$, employing the specific MoAB against it in anaplastic, high-grade anaplastic ASTRs (AAs) and in GBMs. The immunoreactivity demonstrated cytoplasmic, cell surface and extracellular matrix distribution pattern in more than 90 per cent of the tumor cells with high intensity immunostaining (++++, A,B) in every high-grade glioma tissue. The most characteristic $VEGF_{121}$ expression was identified in the cytoplasms of tumor cells, a specificity of an embryonic and more malignant cellular IP. Neoplastically transformed astrocytes with certain IP are capable of provoking neoangiogenesis and proliferation of the endothelial cells. Both of these tumor developmental characteristics that are typical for high-grade gliomas were detected. Human prenatal and postnatal thymus served as both positive and negative tissue controls in which the VEGF immunoreactivity was located in the thymic medulla whereas the undifferentiated masses of cortical thymocytes demonstrated negative primary antigen-antibody reaction.

11.4 Discussion

A very recent publication focused on the molecular aspects underlying the differences in terms of the sets of genes that control pathogenesis of the different subtypes of astrocytic glioma (682). By performing cDNA-array analysis of 53 patient biopsies, comprising low-grade ASTR, secondary GBMs (respective recurrent high-grade tumors), and newly diagnosed primary GBMs, it was clear that human gliomas can be differentiated according to their gene expression. Low-grade ASTR have the most specific and similar expression profiles, whereas primary GBM exhibit much larger variation between tumors. Secondary GBM display features of both other groups. The authors identified several sets of genes with relatively highly correlated expression within groups that: (a) can be associated with specific biological functions; and (b) effectively differentiate tumor class. One prominent gene cluster discriminating primary versus non-primary GBM comprises mostly genes involved in angiogenesis, including VEGF fms-related tyrosine kinase 1 but also IGFBP2, which has not yet been directly linked to angiogenesis. The *in situ* hybridization studies revealed a co-expression of IGFBP2 and VEGF in pseudopalisading cells surrounding tumor necrosis, which provided further evidence for a possible involvement of IGFBP2 in angiogenesis. The separating groups of genes were found by the unsupervised coupled two-way clustering method, and their classification power was validated by a supervised construction of a nearly perfect glioma classifier.

Basic fibroblast growth factor (bFGF) and VEGF are two growth factors which have been established as potent angiogenic factors and endothelial cell mitogens in human ASTRs, during NRA (683-685). Growth factors levels and staining patterns of tumor cells were evaluated immunocytochemically in seven cases of WHO grade II brain ASTR (686). Four (group A) were diagnosed to express anaplastic progression at their second operation, and three (group B) did not. The proliferation index was measured by immunostaining with an anti-Ki-67 MoAB (MIB1). Immunostaining for bFGF was localized to both the nucleus and cytoplasm of ASTR cells, whereas VEGF reactivity was mainly confined to the cytoplasm. The mean cell count parameters reported in this study were: 1) for bFGF: 20.08 ± 6.38 in group A and 0.87 ± 0.90 ($p < 0.01$) for group B; 2) for VEGF: 43.75 ± 17.09 in group A and 0.8 ± 1.06 ($p < 0.05$) for group B; and 3) proliferation index: 3.20 ± 0.81 in group A and 0.77 ± 1.03 in group B ($p < 0.05$). These data strongly support a significant regulatory role for bFGF and VEGF in the development of a more aggressive astrocytoma IP and behavior. In SCID (severe combined immunodeficiency) mice, experimental inhibition of VEGF alone has been found to be sufficient for preventing the growth of primary neoplasms and their dissemination *in vivo*. The inhibitory effect on metastasis formation appeared to be distinct from that on primary tumor progression (687).

An experimental observation studied whether expression of VEGF is correlated with *in vivo* measurements of the capillary permeability and vascular volume of primary human brain tumors (688). During the observations, 13 brain tumor samples (seven GBMs, one ANA-ASTR, two low-grade ASTRs, one pilocytic ASTR, and three primary cerebral lymphomas) were stereotactically obtained from 14 patients. A semiquantitative polymerase chain reaction was used to quantify the relative expression of VEGF messenger RNA in the tumors. VEGF protein was also demonstrated in tissue sections by immunohistochemical techniques. A two-compartment dynamic computed tomographic method was employed to quantitatively measure the aforementioned parameters in the regions from which the biopsies were obtained. In glial tumors, there was significant correlation of VEGF messenger RNA levels with capillary permeability ($p < 0.05$) and vascular volume ($p < 0.01$). Although all primary cerebral lymphomas showed considerable increases in capillary permeability and vascular volume, VEGF expression was only slightly up-regulated in these neoplasms.

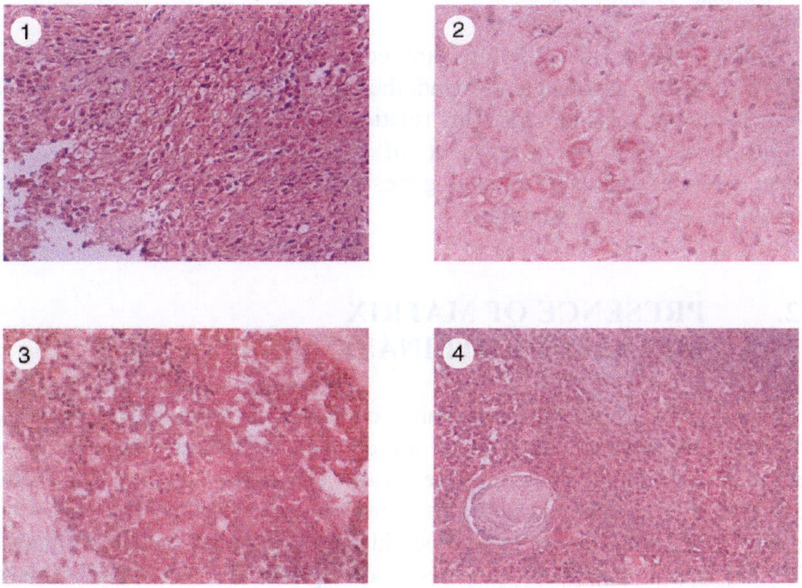

Figure 2-4.

1. Anaplastic ASTR. Cytoplasmic VEGF121 expression in the neoplastically transformed astrocytes. Alkaline phosphatase conjugated, streptavidin-biotin enhanced immunohistochemical antigen detection technique. Magnification: 200x.
2. GBM. VEGF121 presence in the cytoplasms of tumor cells and strong endothelial involvement during the tumor related capillary proliferation and neoangiogenesis. Alkaline phosphatase conjugated, streptavidin-biotin enhanced immunohistochemical antigen detection technique. Magnification: 200x.

3. MED/PNET. Strong, cytoplasmic VEGF121 expression in the tumor cells. Alkaline phosphatase conjugated, streptavidin-biotin enhanced immunohistochemical antigen detection technique. Magnification: 200x.
4. Human thymus (5 years of age). No presence of VEGF121 in medullary, already immunocompetent T lymphocytes and in one giant Hassall's body. Alkaline phosphatase conjugated, streptavidin-biotin enhanced immunohistochemical antigen detection technique. Magnification: 200x.

These experimental findings are consistent with the hypothesis that VEGF may be responsible for endothelial cell proliferation and vascular permeability in glial tumors. This relationship has implications for clinical applications, such as assessment of delivery of water-soluble drugs, treatment of edema and anti-angiogenesis directed biological therapy based on inhibition of VEGF function.

12. PRESENCE OF MATRIX METALLOPROTEINASES (MMPS)

Growth, development, maintenance of homeostasis, as well as processes of tissue regeneration and repair all depend on the precise spatial and temporal remodeling of connective tissue. The proper regulation of tissue remodeling is accomplished by the complex control of the expression and activity of a large number of endopeptidases. The matrix metalloproteinases (MMPs) are a family of enzymes that degrade the extracellular matrix (ECM) and are considered to be important in neoplastic cell invasion and metastasis. Structural changes in the extracellular matrix are necessary for cell migration during tissue remodeling and neoplastic invasion. The MMPs), which are members of the matrixin subfamily of the metalloproteinase family of enzymes, are critical modulators of extracellular matrix (ECM) composition, and may also be involved in protein processing and activation. Assessment of the expression of the various MMPs provides important data on two levels: 1) expression of these proteins signals ongoing, *in situ* modulation of ECM composition and tissue remodeling; and 2) in the case of neoplastically transformed tissues, the expression of MMPs may be indicative of changes in the behavior of certain colonies of cells (if expression is focal) or involvement of the particular MMPs in generalized tissue remodeling during the establishment and expansion of the neoplastic mass.

Gelatinase A (MMP-2) was originally identified as a type IV collagenase secreted by a murine tumor (689). MMP-2 is constitutively expressed by many cells, is present in a wide variety of tissues and is activated while on the cell surface. Its activation has been implicated in processes of cell proliferation, adhesion and migration, and this has lent MMP-2 added importance in investigations of tumor cell-ECM interactions. Substrates for MMP-2 include gelatin and collagens type IV, V, and VII (690-692), cartilage type X collagen (693, 694), and elastin (695). Gelatinase B (MMP-9) has been shown to cleave pepsin-solubilized type IV and V collagen, as well as type XI collagen (696-699). Unlike MMP-2, MMP-9 does not cleave type I collagen. MMP-9 also possesses endopeptidase activity against the cartilage proteoglycan, aggrecan (700), link protein which stabilizes the interaction between aggrecans and hyaluronate (701), as well as elastin (695) of the ECM.

Stromelysin 1 (MMP-3) was originally isolated from a medium of human rheumatoid synoviocytes stimulated with macrophage-conditioned medium (702) Stromelysin 2 (MMP-10) was initially identified in a cDNA library derived from human mammary carcinoma cells and has been established to be 78% identical in sequence to MMP-3 (703). The substrate specificity of the two stromelysins is very similar, but MMP-10 possesses a significantly lower catalytic efficiency. The diversity of the substrates is striking, ranging from aggregan (704) and cartilage (705), to a variety of collagens (types II, IV, IX, XI) (706, 707), fibronectin (708), α_2-macroglobulin (709), ovostatin (709), proteinase and chymotrypsin inhibitors (710), antithrombin III (710), substance P (711), insulin B chain (707), IGF-BP-3 (712), nidogen (713), SPARC (osteonectin) (714), fibulin-2 (715), fibrin γ-chain (716), as well as proenzymes of the matrixin family including proMMP-1, -3, -7, -8, -9, -13 (717-722).

Human collagenase 3 (MMP-13), initially isolated from a breast carcinoma cell cDNA library (723), possesses nearly 90% homology to the rodent mesenchymal collagenase. Through analyses of its substrate specificity, MMP-13 has been established to cleave type II collagen, the predominant interstitial collagen of cartilage, with much greater efficiency than types I and III, and thus has been designated a selective type II collagenase (724-726).

The expression patterns of 15 MMPs and three tissue inhibitors of metalloproteinase have been observed in gliomas, MEDs, and normal brain tissue (727). Increased levels of mRNAs encoding gelatinase A, gelatinase B, two membrane-type MMPs (mt1- and mt2-MMP), and tissue inhibitors of metalloproteinase-1 in GBMs and MEDs. A significant increase in mt1-MMP, gelatinase A, gelatinase B, and tissue inhibitors of metalloproteinase-1 was observed in glioblastomas as compared with low-grade astrocytomas, anaplastic astrocytomas, and normal brain. In MEDs, the expression of mt1-MMP, mt2-MMP, and gelatinase A were also increased, but to a lesser extent than that observed in glioblastomas. Substrate gel electrophoresis showed that the activated forms of gelatinases A and B were present in glioblastomas and MEDs. These results suggest that increased expression of mt1-MMP/gelatinase A is closely related to malignant progression observed in gliomas (727).

12.1 Results

Positive and negative control tissues for each of the antibodies were: MMP-2 (pos.: thyroid and ovarian carcinoma, Ewing's sarcoma, leiomyoma and carcinoid; neg.: hepatocellular, gastric and renal cell carcinoma, mesothelioma, undifferentiated carcinoma, lymphoma, rhabdomyosarcoma, malignant fibrous histiocytoma, and epithelioid sarcoma); MMP-3 (all tumors observed were positive: hepatocellular, gastric, thyroid, renal cell and ovarian carcinoma, Ewing's sarcoma, leiomyoma, carcinoid, carcinoma, mesothelioma, undifferentiated carcinoma, lymphoma, rhabdomyosarcoma, malignant fibrous histiocytoma, and epithelioid sarcoma); MMP-9 (pos.: focal positivity observed in ovarian carcinoma, mesothelioma, Hodgkin's and other lymphomas, and Ewing's sarcoma; neg.: hepatocellular, gastric, thyroid, and renal cell carcinoma, leiomyoma, carcinoid, undifferentiated carcinoma, rhabdomyosarcoma, malignant fibrous histiocytoma, and epithelioid sarcoma); MMP-10 (all tumors observed were positive: hepatocellular, gastric, thyroid, renal cell and ovarian carcinoma, Ewing's sarcoma, leiomyoma, carcinoid, carcinoma, mesothelioma, undifferentiated carcinoma, lymphoma, rhabdomyosarcoma, malignant fibrous histiocytoma, and epithelioid sarcoma); and MMP-13 (pos.: focal and perivascular positivity observed in renal cell and ovarian carcinoma, and lymphoma, neg.: hepatocellular, gastric, and thyroid carcinoma, mesothelioma, undifferentiated carcinoma, rhabdomyosarcoma, malignant fibrous histiocytoma, epithelioid sarcoma, leiomyoma, and carcinoid).

In our immunohistochemical screening, we sought to characterize the MMP expression profile of spontaneous childhood MEDs/PNETs employing a chosen library of MoABs directed against epitopes on the MMP-2, -3, -9, -10, and -13 enzymes and a sensitive antigen detection technique.

Strong overall expression of MMP-3 and -10 was found in MEDs/PNETs, especially in the ECM adjacent to blood vessels (728). Positive immunoreactivity could be seen for these two MMPs in the ECM surrounding over 90% of the neoplastically transformed cells, and the staining intensity was also the strongest possible (A,B). Focal (surrounding less than 10% of the neoplastically transformed cells) but strong (A,B) immunoreactivity was determined for collagenase-3 (MMP-13), an endopeptidase characterized by a potent degrading activity against a wide spectrum of substrates. Weak (surrounding anywhere between 10% and 90% of the neoplastically transformed cells, of B and B,C intensity) expression of

MMP-2 (gelatinase A) and MMP-9 (gelatinase B), two cytokine-induced MMPs, was also observed.

During our systematic screening, strong overall expression of MMP-3 and -10 was found in all 24 ASTRs, especially in the ECM adjacent to blood vessels (729). Positive immunoreactivity could be seen for these two MMPs in the ECM surrounding over 90% of the neoplastically transformed cells (++++), and the staining intensity was also the strongest possible (A,B). No immunoreactivity could be determined for neither the gelatinases (MMP-2 and -9) nor for collagenase-3 (MMP-13) in the ASTR cases observed by us. These results are somewhat different than that previously determined in ASTRs *in vitro* and glioblastomas *in vivo*. Thus, it is very likely that the different stages of glioma progression, from low-grade pilocytic ASTRs, through AAs, and finally to GBM, are each associated with particular, unique patterns of MMP and tissue inhibitors of metalloproteinases expression and utilization.

12.2 Discussion

Proteases such as matrix metalloproteinases (MMPs), cysteine- and serine-proteinases are capable of degrading extracellular matrix and basement membranes and have been implicated in human brain tumors. MMPs are a homologous family of zinc-dependent proteases. Within this group, attention has been focused on the gelatinases (MMP-2 and -9), which are thought to play an important role in brain tumor progression (730).

The expression of MMP-2 and 9 and membrane type metalloproteinase (MMP-14) has been observed in intrinsic human primary brain tumors of various histological type and grade. Zymography results showed that MMP-2 was the most prominent proteolytic enzyme in all the cell lines studied (with one exception) while MMP-9 was only faintly expressed. However, the corresponding paraffin sections showed no expression of either MMP-2, -9 or -14 within the neoplastically transformed cells, positivity being confined to haematogenous cells and the vascular endothelium. Fluorescence immunocytochemical studies, using monoclonal antibodies to MMP-2, 9 and 14, showed granular cytoplasmic reactivity *in vitro*. In addition, there was strong focal positivity at the cell membrane with MMP-14 in some high grade gliomas suggesting that MMPs are produced at the leading edge of the cell by individual subpopulations of invading glia, in small quantities and on demand *in vivo*. It can be concluded that local microenvironmental conditions *in vitro* appear to stimulate such MMP activity (731).

We established overall expression of MMP-3 and -10, focal immunoreactivity for MMP-13, and weak expression of MMP-2 and -9. Such data in PNETs/MEDs are very limited. Most of the work on ECM modulation, MMP and tissue inhibitors of metalloproteinases expression has been done on gliomas, due to the well characterized stages of progression of astrocytic neoplasms, from pilocytic astrocytoma to glioblastoma multiforme, expression of MMP-1 (732), -2 (733-735), and -9 (736). In a recent study by Vince and co-workers (737), the differential intra- and intertumoral heterogeneity and patterns of MMP expression was assessed in human glioblastomas *in vivo*. Twelve glioblastoma samples were analyzed for MMP expression by semi-quantitative RT-PCR. A total of 56 samples (8 adjoining regions of 6 GBMs) were immunohistochemically examined for the expression and regional distribution of MMP-2, -3, -7, and -9. MMP-2 mRNA was detected in all samples while MMP-9 was found in numerous samples. Correspondingly, strong expression levels of both of these cytokine-induced gelatinases were confirmed by immunohistochemistry. MMP-2 was expressed by both neoplastic cells and endothelium while MMP-9 was found to be restricted to endothelial cells. MMP-3 was not detected in any of the samples while MMP-7 was found around tumor cells of three samples from one patient only. The strong immunoreactivity seen for MMP-2 around tumor cells and blood vessels suggests a role in both tissue degradation and tumor neoangiogenesis.

The role of MMPs in the growth and invasion of human glioma cells (T98G and A172 cell lines) cultured on various ECM components including type I, IV and V collagens, fibronectin, laminin, and reconstituted basement membrane (Matrigel) has been observed. T98G glioma cells grew well on these ECM components and invaded the reconstituted basement membrane. In contrast, A172 glioma cells showed growth inhibition on collagen types IV and V and Matrigel without invasion of the Matrigel. Gelatin zymography and enzyme immunoassays demonstrated that T98G glioma cells, but not A172 cells, secrete a large amount of MMP-2, and this was confirmed by immunoblotting and immunohistochemistry (738). Of the two different tissue inhibitors of metalloproteinases (TIMP-1 and TIMP-2), T98G cells produced only TIMP-1 during culture on Matrigel, whereas A172 cells secreted both. Although both human recombinant TIMP-1 and TIMP-2 stimulated T98G cell growth slightly on Matrigel, the *in vitro* invasiveness was significantly reduced by only recombinant TIMP-2. These results suggest that MMP-2 plays an important role in the ECM invasion of T98G human glioma cells *in vitro* (738).

The neural cell adhesion molecules (NCAMs) of neuronal and glial cells provide a calcium-independent mechanism for cell-cell and cell-ECM adhesion. NCAMs are downregulated to promote cell disaggregation during cell migration in the developing nervous system, whereas MMPs facilitate migration. Recent studies have shown downregulation of MMP secretion in rat glioma cells transfected with NCAM cDNA. An inverse correlation has been described between the expression of NCAM-A and that of MMP-2 and -9 *in vitro*, although the patterns of expression showed no obvious correlation with histological type or grade of the parent tumors. These results suggest that downregulation of NCAM-A may contribute to neoplastic cell invasiveness by promoting both cell disaggregation and protease secretion (119,739).

In tissue specimens of 27 primary and 17 secondary GBMs and the precursor lesions, the immunocytochemical expression patterns of the membrane protein CD44s, the basal lamina proteins laminin, collagen IV, and fibronectin, the lectin galectin-3 recognizing tenascin and NCAM as well as of the matrix-degrading enzymes matrix metalloproteinase MMP-2 and MMP-9, and cathepsin D were observed (740). Besides expression of basal lamina proteins in vessels, all glioblastomas and the precursor lesions showed strong immunoreactivity for CD44s, tenascin, galectin-3, and NCAM which were restricted to solid tumor masses. Present in solid tumor areas, MMP-2, MMP-9 and cathepsin D were also strongly expressed by single neoplastic cells invading adjacent brain tissue at the infiltrative margin. Neither the expression pattern in primary and secondary glioblastomas nor in the precursor tumors revealed significant differences. There was also no intraindividual constant expression pattern during glioma progression or correlation with malignancy. Restricted expression of CD44s, galectin-3, tenascin and NCAM in solid tumor masses seems to contribute to homotypic neoplastic cell adhesion while single tumor cells abolish this expression profile and acquire invasive activities by expression of cathepsin D, MMP-2 and MMP-9 (740).

The differential intra- and intertumoral heterogeneity and patterns of MMP expression have been described in human primary brain tumors, including AAs and GBMs (741-744). MMP-2 mRNA was detected in all samples while MMP-9 was found in numerous samples. Correspondingly, strong expression levels of both gelatinase proteins were detected by immunohistochemistry. MMP-2 was expressed by both tumor cells and endothelium (730) while MMP-9 was found to be restricted to proliferating endothelial cells and selected neoplastically transformed astrocytes (731). MMP-3 protein was not detected in any of the samples, while MMP-7 was found around tumor cells in one case. The strong immunoreactivity seen for MMP-2 around tumor cells and blood vessels suggests a role in both tissue degradation and brain tumor related neoangiogenesis. The marked localization of MMP-9 to the endothelium and its presence in non-infiltrative benign lesions, however, makes a direct proteolytic role of this MMP on ECM components during glioma invasion unlikely. Its close association with vascular structures, however, may indicate a possible link to neoangiogenesis. The significance of MMP-7 which was only identified in tumor cells in three samples remains unclear. MMP-3, although strongly expressed in cell lines, does not appear to play a role in GBM pathogenesis *in vivo*.

Three different membrane-type matrix metalloproteinases (MT1-, MT2-, and MT3-MMPs) are known to activate the zymogen of MMP-2 (pro-MMP-2, progelatinase A) *in vitro*, which, as noted above, is one of the key MMPs in invasion and metastasis of various cancers, including glioblastomas. The production and activation of pro-MMP-2, expression of MT1-, MT2-, and MT3-MMPs and their correlation with pro-MMP-2 activation, and localization of MMP-2, MT1-MMP, and MT2-MMP have also been observed in human astrocytic tumors. Pro-MMP-2 production was found to be upregulated in AAs and GBMs as compared with low-grade ASTRs, metastatic brain tumors, or normal brains. MT1-MMP and MT2-MMP are expressed predominantly in glioblastoma tissues, and their expression levels increase significantly as neoplastic grade increases. MT3-MMP is detectable in both astrocytic tumor and normal brain tissues, but the mean expression level is approximately 50-fold lower compared with that of MT1-MMP and MT2-MMP in the GBMs. The activation ratio of pro-MMP-2 correlates directly with the expression levels of MT1-MMP and MT2-MMP but not MT3-MMP. Neoplastically transformed astrocytes of GBM express MT1-MMP and MT2-MMP transcripts, and immunocytochemical observations have established co-expression of MT1-MMP and MT2-MMP in MMP-2 immunopositive neoplastically transformed astrocytes in glioblastoma, suggesting a critical role for cell surface MMPs in GBM pathogenesis (732-734).

In conclusion, it seems likely that the expression of MMPs is related to stage of disease and agressiveness of the neoplastically transformed cells. Expression of the stromelysins, for instance, which was so pronounced in ASTRs as described in our study (729), is nonexistent in high-grade glioblastoma multiforme, while the gelatinases are not expressed in ASTRs, but seem to be critical in GBM growth and development, suggesting stage-specific activation of the various MMPs during the malignant progression of the neoplastically transformed astrocytic IP.

In conclusion, thorough observations on the expression, production and role of ECM remodeling proteases should be conducted in PNETs/MEDs in order to clarify the role of such enzymes in the formation, growth and progression of these intracranial malignancies. The possibility of the application of specific protease inhibitors in the treatment of minimal residual disease following stereotactic resection should lend a further motivation to such efforts.

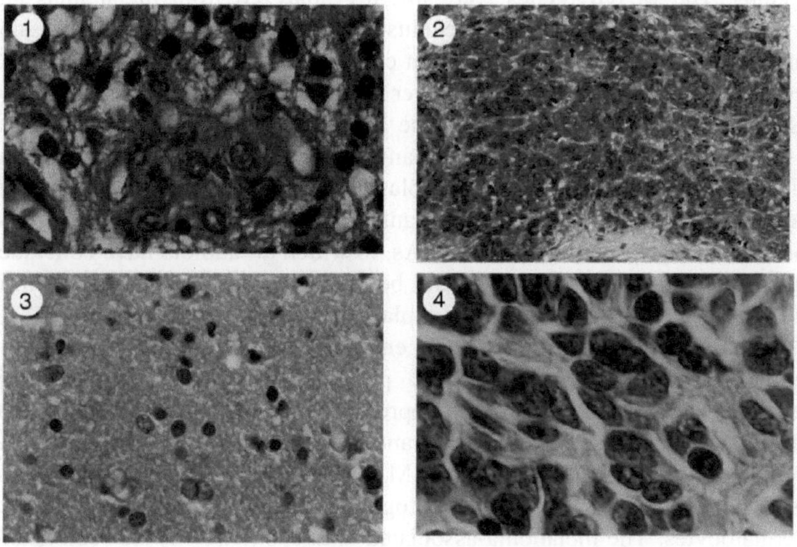

Figure 2-5.

1. Childhood anaplastic ASTR. Expression of *FasR* on neoplastically transformed cells. Alkaline phosphatase conjugated, streptavidin-biotin enhanced immunohistochemical antigen detection technique. Magnification: 1000x.
2. Childhood MED/PNET. Strong presence of *FasR* on tumor cells. Alkaline phosphatase conjugated, streptavidin-biotin enhanced immunohistochemical antigen detection technique. Magnification: 500x.
3. ASTR. Strong immunohistochemical evidence of stromelysin/MMP-10 expression in the extracellular matrix (ECM), surrounding the neoplastically transformed astrocytes. Alkaline phosphatase conjugated, streptavidin-biotin enhanced immunohistochemical antigen detection technique. Magnification: 500x.
4. Childhood MED/PNET. Expression of MMP-10 around the tumor cells. Alkaline phosphatase conjugated, streptavidin-biotin enhanced immunohistochemical antigen detection technique. Magnification: 1000x.

13. THE MAGE GENE FAMILY

During the last decade, the aberrant expression of normal testicular proteins in neoplastically transformed cells became common knowledge. Cancer/testis-antigens (CTAs) represent a novel family of immunogenic proteins. The MAGE genes were initially analyzed from melanomas and turned out to have an almost exclusively neoplasm specific expression pattern. This expression pattern might contribute to the genetic instability of neoplastically transformed cells. Over the past few years, more than 200 research articles have appeared in the literature concerning CTAs (745). It has become obvious that CTAs can and will be used for the diagnosis and microstaging of various human neoplasms (746). Thus far, we have never seen authors writing about specific antigen families all arrive at the same conclusion, except in the case of CTAs, in which all authors have concluded that these antigens could and should be employed for more individualized, antigen targeted and specific anti-neoplastic immunotherapy.

MAGE peptides are already being employed for CT antigen directed anti-neoplastic immunotherapy. MAGE proteins affect only neoplastically transformed cells (activation or derepression of normally silent CT genes) but not normal cells, because testis and placenta lack HLA expression. In normal adult tissues, most 23 human MAGE genes are expressed only in the testis, but only in the mitotic spermatogonia (germ cells) and in the primary spermatocytes. The melanoma-associated antigens (MAGE) are regarded as inducing strong tumor-specific immune response (747-749). During the last decade, 23 human and 12 mouse MAGE genes have been isolated and characterized in various mammalian neoplasms (750). The first member, MAGE-A1, of the MAGE gene family was discovered and characterized as genes encoding melanoma antigens (751). Numerous gene homologues to MAGE-A1 have been identified on the Xq28, Xp21,3 Xq26, and Xp11.23 and are classified as MAGE-A, -B, -C and -D (752-756). The MAGE-C1 gene appears to be located on band Xq26, whereas the MAGE-A and MAGE-B genes are located on Xq28 and Xp21, respectively. Like other MAGE genes, MAGE-C1 is expressed in a significant proportion of neoplastic cells of various histological types, whereas it is silent in normal tissues except testis (755). Previous observations have characterized the MAGE-A subfamily as: (a) they were not present in normal cells except for testis (primitive germ cells, spermatogonia) and placenta (757); and (b) some antigens encoded by the gene family are presented by the human leukocyte antigen (HLA) (758). These results suggest that immunoreaction against MAGE proteins could be expected to affect only neoplastically transformed cells (activation or derepression of normally silent CT genes) but not normal cells because testis and placenta lack HLA expression (759, 760). Hence, experimental vaccination with MAGE-A peptide provoked neoplasm regression in melanoma patients without significant side effects (761). Recently, it was reported that the MAGE-D genes are well conserved between man and mouse, suggesting that these genes have important functions. The COOH-terminal domain of MAGE-D3 is identified to be trophinin, a previously detected protein believed to be involved in embryo implantation (749). The genomic structure of the MAGE-D genes indicate that one of them corresponds to the founder member of the family and that all of the other MAGE genes are retrogenes derived from that common ancestral gene. It is important to emphasize that the MAGE-D genes are universally present in normal tissues and the scientific literature has not classified them as CTAs.

13.1 Results

Immunocytochemical presence and cellular localization of the MAGE-1 CT antigen, employing anti-MAGE-1 MoAB was observed only in anaplastic, high grade ASTRs (100%) and in GBMs in this study. The immunoreactivity was always heterogeneous, showing a cytoplasmic pattern and loosely grouped cells with similar staining characteristics being detected within the cellular and hormonal microenvironment of the ASTRs.

We never observed MAGE-1 expression in the lowest grade of malignancy: pilocytic ASTR cases.

The two cases of *pure* anaplastic ASTRs demonstrated expression of MAGE-1, located in the cytoplasm of neoplastically transformed cells. The staining intensity in this ASTR subtype was high (A and B) and the number of total cells positively stained ranged from (+) to (++). In the case of the *mixed* ASTR with primitive neuroectodermal tumor (PNET) elements, MAGE-1 was also present, but with a lower intensity (B) although a similar proportion of ASTR tumor cells exhibited immunoreactivity (++). No staining for PNET elements was identified. High intensity staining (A,B) in well over 50% of the total cell number (+++) was observed in the anaplastic oligo-astrocytoma and other anaplastic ASTRs. MAGE-1 immunoreactivity was of greatest intensity (A) in the four cases of GBM observed by us. A high proportion (over 50%) of the total cell number reacted positively (+++).

As expected, the normal testis employed as positive tissue control in this study reacted positively with the anti-MAGE-1 MoAB, while the other normal human tissues, including the postnatal thymus tissue demonstrated no detectable presence of MAGE-1 CT antigen (negative tissue controls).

13.2 Discussion

In the literature only few observations reported presence of CTAs in human brain tumors (762). Meningiomas were found to express only HOM-TES-14/SCP-1. SSX-4 was found to be the only CT gene expressed in oligodendrogliomas and it was also expressed in oligoastrocytomas and astrocytomas. ASTR cells proved to be the most positive for HOM-TES-14/SCP-1 and SSX-4, with expression of HOM-TES-85, SSX-2 and MAGE-3 also defined in these tumors. MAGE-3 was detected only in grade IV GBMs, while the expression of the other CT genes showed no clear correlation with histological grade. Sixty percent of ASTR analyzed were found to express at least one CT gene, 21% expressed two CT genes, and 8% coexpressed three CT genes.

In the present study of MAGE-1 CT antigen, expression in childhood ASTR subtypes using an antigen specific MoAB, we demonstrated the presence of this CTA only in each of the high grade ASTR cases. The greatest intensity of immunoreactivity and proportion of antigen positive cells was observed in glioblastomas, the subtype of ASTR associated with the greatest malignant potential. The anaplastic ASTRs also revealed presence of MAGE-1 antigen in their cytoplasm. No significant difference between the immunoreactivity of pure and mixed anaplastic ASTRs was observed. It is significant that no presence of MAGE-1 antigen was detectable in any case of pilocytic and low grade ASTRs.

MAGE-1 CT antigen expression levels may be used to evaluate the malignant and dedifferentiation tendencies of grade I, II, and III ASTRs. No expression in pilocytic ASTR cells may be used to establish a starting grade tendency of those cells to possible further IP dedifferentiation to low-grade (grade II) ASTRs, while an accumulation of this antigen in grade II and III ASTR cells may be used in predicting the likelihood of mutations of the genome and further dedifferentiation towards even more malignant anaplastic ASTR and glioblastoma multiforme IPs.

REFERENCES

1. Miescher S, Whiteside TL, de Tribolet N, von Fliedner V: *In situ* characterization, clonogenic potential, and anti-tumor cytolytic activity of T lymphocytes infiltrating human brain cancers. J Neurosurg *68:* 438-448, 1988.
2. Ruiter DJ, Bhan AK, Harrist TJ, Sober AJ, Mihm JC Jr: Major histocompatibility antigens and mononuclear inflammatory infiltrates in benign nevomelanocytic proliferations and malignant melanoma. J Immunol *129:* 2808-2815, 1982.
3. Shimokawara I, Imamura M, Yamanaka N, Ishii Y, Kikuchi K: Identification of lymphocyte subpopulations in human breast cancer tissue and its significance: An

immunoperoxidase study with anti-human T- and B-cell sera. Cancer *49:* 1456-1464, 1982.

4. Kornstein MJ, Brooks JS, Elder DE: Immunoperoxidase localization of lymphocyte subsets in the host response to melanoma and nevi. Cancer Res *43:* 2749-2753, 1983.
5. von Hanwehr RI, Hofman FM, Taylor CR, Apuzzo ML: Mononuclear lymphoid populations infiltrating the microenvironment of primary CNS tumors: Characterization of cell subsets with monoclonal antibodies. J Neurosurg *60:* 1138-1147, 1984.
6. Hiratsuka H, Imamura M, Kasai K, Kamiya H, Ishii Y, Kohama G, Kikuchi K: Lymphocyte subpopulations and T-cell subsets in human oral cancer tissues: Immunohistologic analysis by monoclonal antibodies. Am J Clin Pathol *81:* 464-470, 1984.
7. Rowe CJ, Beverley PC: Characterization of breast cancer infiltrate using monoclonal antibodies to human leukocyte antigens. Br J Cancer *49:* 149-159, 1984.
8. Paine JT, Handa H, Yamasaki T, Yamashita J, Miyatake S: Immunohistochemical analysis of infiltrating lymphocytes in central nervous system tumors. Neurosurg *18:*
9. Stevens A, Kloter I, Roggendorf W: Inflammatory infiltrates and natural killer cell presence in human brain tumors. Cancer *61:* 738-743, 1988.
10. Chin Y, Janseens J, Vandepitte J, Vandenbrande J, Opdebeek L, Raus J: Phenotypic analysis of tumor-infiltrating lymphocytes from human breast cancer. Anticancer Res *12:*1463-1466, 1992.
11. Dietl J, Horny HP, Ruck P, Kaiserling E: Dysgerminoma of the ovary. An immunohistochemical study of tumor-infiltrating lymphoreticularcells and tumor cells. Cancer *71:* 2562-2568, 1993.
12. Finke JH, Rayman P, Hart L, Alexander JP, Edinger MG, Tubbs RR, Klein E, Tuason L, Bokowski RM: Characterization of tumor-infiltrating lymphocyte subsets from human renal cell carcinoma: specific reactivity defined by cytotoxicity, interferon-gamma secretion, and proliferation. J Immunother Emph Tumor Immunol *15:* 91-104, 1994.
13. Maccalli C, Mortarini R, Parmiani G, Anichini A: Multiple sub-sets of $CD4^+$ and $CD8^+$ cytotoxic T-cell clones directed to autologous human melanoma identified by cytokine profiles. Int J Cancer *57:* 56-62, 1994.
14. Zhang BX: Obsevation on the phenotypic changes of tumor-infiltrating lymphocytes (TIL) during cultivation *in vitro*. Chinese J Pathol *21:* 281-283, 1992.
15. Shilyansky J, Nishimura MI, Yannelli JR, Kawakami Y, Jacknin LS, Charmley P, Rosenberg SA: T-cell receptor usage by melanoma-specific clonal and highly oligoclonal tumor-infiltrating lymphocyte lines. Proc Natl Acad Sci USA *91:* 2829-2833, 1994.
16. Kuppner MC, Hamou MF, de Tribolet N: Immunohistological and functional analyses of lymphoid infiltrates in human glioblastomas. Cancer Res *48:* 6926-6932, 1988.
17. Rosenberg SA, Spiess P, Lafreniere R: A new approach to the adoptive immunotherapy of cancer with tumor-infiltrating lymphocytes. Science *233:* 1318-1321, 1986.
18. Sensi M, Salvi S, Castelli C, Maccali C, Mazzocchi A, Mortarini R, Nicolini G, Herlyn M, Parmiani G, Anichini A: T cell receptor (TCR) structure of autologous melanoma-reactive cytotoxic T lymphocyte (CTL) clones: tumor-infiltrating lymphocytes overexpress *in vivo* the TCR β chain sequence used by an HLA-A2-restricted and melanocyte-lineage-specific CTL clone. J Exp Med *178:* 1231-1246, 1993.
19. Peoples GE, Yoshino I, Douville CC, Andrews JV, Goedegebuure PS, Eberlein TJ: TCR $V\beta 3^+$ and $V\beta 6^+$ CTL recognize tumor-associated antigens related to HER2/neu expression in HLA-A2+ ovarian cancers. J Immunol *152:* 4993-4999, 1994.

20. Rodolfo M, Castelli C, Bassi C, Accornero P, Sensi M, Parmiani G: Cytotoxic T lymphocytes recognize tumor antigens of a murine colonic carcinoma by using different T-cell receptors. Int J Cancer *57:* 440-447, 1994.
21. Zocchi MR, Ferrarini M, Migone N, Casorati G: T-cell receptor V delta gene usage by tumour reactive gamma delta T lymphocytes infiltrating human lung cancer. Immunol *81:* 234-239, 1994.
22. Schrier PI, Bernards R, Vaessen RTMJ, Houweling A, van der Eb AJ: Expression of class I major histocompatibility antigens switched off by highly oncogenic adenovirus 12 in transformed rat cells. Nature *305:* 771-775, 1983.
23. Becker JC, Termeer C, Schmidt RE, Brocker EB: Soluble intercellular adhesion molecule-1 inhibits MHC-restricted specific T cell/tumor interaction. J Immunol *151:* 7224-7232, 1993.
24. Mule JJ, Schwarz SL, Roberts AB, Sporn MB, Rosenberg SA: Transforming growth factor-beta inhibits the *in vitro* generation of lymphokine-activated killer cells and cytotoxic T cells. Cancer Immunol Immunother *26:* 95-100, 1988.
25. Sporn MB, Roberts AB, Wakefield LM, Assoian RK: Transforming growth factor-beta: biological function and chemical structure. Science *233:* 532-534, 1986.
26. Rivoltini L, Arienti F, Orazi A, Cefalo G, Gasparini M, Gambacorti-Passerini C, Fossati-Bellani F, Parmiani G: Phenotypic and functional analysis of lymphocytes infiltrating paediatric tumours, with a characterization of the tumour phenotype. Cancer Immunol Immunother *34:* 241-251, 1992.
27. Cepek KL, Parker CM, Madara JL, Brenner MB: Integrin aEb7 mediates adhesion of T lymphocytes to epithelial cells. J Immunol *150:* 3459-3470, 1993.
28. Gelb AB, Smoller BR, Warnke RA, Picker LJ: Lymphocytes infiltrating primary cutaneous neoplasms selectively express the cutaneous lymphocyte-associated antigen (CLA). Amer J Pathol *142:* 1556-1564, 1993.
29. Roberts AI, O'Connell SM, Ebert EC: Intestinal intra-epithelial lymphocytes bind to colon cancer cells by HML-1 and CD11a. Cancer Res *53:* 1608-1611, 1993.
30. Ward PA, Varani J: Mechanisms of neutrophil-mediated injury. 5th International ANCA Workshop, 1993.
31. Yamamoto H, Hirayama M, Genyea C, Kaplan J: TGF-beta mediates natural suppressor activity of IL-2-activated lymphocytes. J Immunol *152:* 3842-3847, 1994.
32. Chatani Y, Tanimura S, Miyoshi N, Hattori A, Sato M, Kohno M: Cell type-specific modulation of cell growth by transforming growth factor beta 1 does not correlate with mitogen-activated protein kinase activation. J Biol Chem *270:* 30686-30692, 1995.
33. Tamada K, Harada M, Ito O, Takenoyama M, Mori T, Matsuzaki G, Nomoto K: The emergence of non-cytolytic NK1.1+ T cells in the long-term culture of murine tumour-infiltrating lymphocytes: a possible role of transforming growth factor-beta. Immunology *89:* 627-635, 1996.
34. Shemesh J, Ehrlich R: Aberrant biosynthesis and transport of class I major histocompatibility complex molecules in cells transformed with highly oncogenic human adenoviruses. J Biol Chem *268:* 15704-15711, 1993.
35. Becker JC, Brocker EB: Lymphocyte-melanoma interaction: role of surface molecules. Recent Results Cancer Res *139:* 205-214, 1995.
36. Kleinman GM, Zagzag D, Miller DC: Diagnostic use of immunohistochemistry in neuropathology. Neurosurg Clin N Am *5:* 97-126, 1994.
37. Szymas J: Diagnostic immunohistochemistry of tumors of the central nervous system. Folia Neuropathol *32:* 209-214 1994.

38. Taylor CR: Quality assurance and standardization in immunohistochemistry. A proposal for the annual meeting of the Biological Stain Commission, June, 1991. Biotech Histochem *67:* 110-117, 1992.
39. Stemmer-Rachamimov AO, Louis DN: Histopathologic and immunohistochemical prognostic factors in malignant gliomas. Curr Opin Oncol *9:* 230-234, 1997.
40. Jay V, Edwards V, Halliday W, Rutka J, Lau R: "Polyphenotypic" tumors in the central nervous system: problems in nosology and classification. Pediatr Pathol Lab Med *17:* 369-389, 1997.
41. Beckmann MJ, Prayson RA: A clinicopathologic study of 30 cases of oligoastrocytoma including p53 immunohistochemistry. Pathology *29:* 159-164, 1997.
42. Krouwer HG, van Duinen SG, Kamphorst W, van der Valk P, Algra A: Oligoastrocytomas: a clinicopathological study of 52 cases. J Neurooncol *33:* 223-238, 1997.
43. Hirose T, Schneithauer BW, Lopes MB, Gerber HA, Altermatt HJ, VandenBerg SR: Ganglioglioma: an ultrastructural and immunohistochemical study. Cancer *79:* 989-1003, 1997.
44. Baker DL, Molenaar WM, Trojanowski JQ, Evans AE, Ross AH, Rorke LB, Packer RJ, Lee VM, Pleasure D: Nerve growth factor receptor expression in peripheral and central neuroectodermal tumors, other pediatric brain tumors, and during development of the adrenal gland. Am J Pathol *139:* 115-122, 1991.
45. Kokunai T, Sawa H, Tamaki N: Functional analysis of trk proto-oncogene product in medulloblastoma cells. Neurol Med Chir *36:* 796-804, 1996.
46. Pruim J, Willemsen AT, Molenaar WM, van Waarde A, Paans AM, Heesters MA, Go KG,Visser GM, Franssen EJ, Vaalburg W: Brain tumors: L-[1-C-11]tyrosine PET for visualization and quantification of protein synthesis rate. Radiology *197:* 221-226, 1995.
47. de Wolde H, Pruim J, Mastik MF, Koudstaal J, Molenaar WM:Proliferative activity in human brain tumors: comparison of histopathology and L-[1-(11)C]tyrosine PET. J Nucl Med *38:* 1369-1374, 1997.
48. Sallinen P, Miettinen H, Sallinen SL, Haapasalo H, Helin H, Kononen J: Increased expression of telomerase RNA component is associated with increased cell proliferation in human astrocytomas. Am J Pathol *150:* 1159-1164, 1997.
49. Kyritsis AP, Bondy ML, Hess KR, Cunningham JE, Zhu D, Amos CJ, Yung WK, Levin VA, Bruner JM: Prognostic significance of p53 immunoreactivity in patients with glioma. Clin Cancer Res *1:* 1617-1622, 1995.
50. Korshunov AG, Sycheva RV: An immunohistochemical study of the expression of the oncoprotein p53 in astrocytic gliomas of the cerebral hemispheres. Arkh Patol *58:* 37-42, 1996.
51. Bhattacharjee MB, Bruner JM: p53 protein in pediatric malignant astrocytomas: a study of 21 patients. J Neurooncol *32:* 225-233, 1997.
52. Ng HK, Lo SY, Huang DP, Poon WS: Paraffin section p53 protein immunohistochemistry in neuroectodermal tumors. Pathology *26:* 1-5, 1994.
53. Bodey B, Gröger AM, Bodey B Jr, Siegel SE, Kaiser HE: Immunocytochemical detection of p53 protein expression in various childhood astrocytoma subtypes: Significance in tumor progression. Anticancer Res *17:* 1187-1194, 1997.
54. Biernat W, Kleihues P, Yonekawa Y, Ohgaki H: Amplification and overexpression of MDM2 in primary (de novo) glioblastomas. J Neuropathol Exp Neurol *56:* 180-185, 1997.

55. Watanabe K, Tachibana O, Sata K, Yonekawa Y, Kleihues P, Ohgaki H: Overexpression of the EGF receptor and p53 mutations are mutually exclusive in the evolution of primary and secondary glioblastomas.Brain Pathol *6:* 217-223, 1996.
56. Korkolopoulou P, Christodoulou P, Kouzelis K, Hadjiyannakis M, Priftis A, Stamoulis G, Seretis A, Thomas-Tsagli E: MDM2 and p53 expression in gliomas: a multivariate survival analysis including proliferation markers and epidermal growth factor receptor. Br J Cancer *75:* 1269-1278, 1997.
57. Hagel C, Laking G, Laas R, Scheil S, Jung R, Milde-Langosch K, Stavrou DK: Demonstration of p53 protein and TP53 gene mutations in oligodendrogliomas. Eur J Cancer *32A:* 2242-2248, 1996.
58. Packham G, Cleveland J: c-Myc and apoptosis. Biochim Biophys Acta *1242:* 11-28, 1995.
59. Clarke AR, Purdie CA, Harrison DJ, Morris RG, Bird CC, Hooper ML, Wyllie AH: Thymocyte apoptosis induced by p53-dependent and independent pathways. Nature *362:* 849-852, 1993.
60. Strasser A, Harris AW, Jacks T, Cory S: DNA damage can induce apoptosis in proliferating lymphoid cells via p53-independent mechanisms inhibitable by Bcl-2. Cell *79:* 329-339, 1994.
61. Glickman JN, Yang A, Shahsafaei A, McKeon F, Odze RD: Expression of p53 related protein p63 in the gastrointestinal tract and in esophageal metaplastic and neoplastic disorders. Hum Pathol *32:* 1157-1165, 2001.
62. Matsumoto T, Fujii T, Yabe M, Oka K, Hoshi T, Sato K: MIB-1 and p53 immunocytochemistry for differentiating pilocytic astrocytomas and astrocytomas from anaplastic astrocytomas and glioblastomas in children and young adults. Histopathology *33:* 446-452, 1998.
63. Bodey B, Bodey B JR, Gröger AM, Luck JV, Siegel SE, Taylor CR, Kaiser HE: Immunocytochemical detection of the p170 multidrug resistance (MDR) and the p53 tumor suppressor gene proteins in human breast cancer cells: Clinical and therapeutical significance. Anticancer Res *17:*1311-1318, 1997.
64. Fuchs EJ, McKenna KA, Bedi A: p53-dependent DNA damage-induced apoptosis requires *Fas*-APO-1-independent activation of CPP321. Cancer Res *57:* 2550-2554, 1997.
65. Graeber TG, Osmanian C, Jacks T, Housman DE, Koch CJ, Lowe SW, Giaccia AJ: Hypoxia-mediated selection of cells with diminished apoptotic potential in solid tumors. Nature *379:* 88-91, 1996.
66. Naumowski L, Clearly ML: Bcl-2 inhibits apoptosis associated with terminal differentiation of HL-60 myeloid leukemia cells. Blood *83:* 2261-2266, 1994.
67. Reed JC: Regulation of apoptosis by bcl-2 family proteins and its role in cancer and chemoresistance. Curr Opin Oncol *7:* 541-546, 1995.
68. Haldar S, Basu A, Croce CM: Bcl-2 is the guardian of microtubule integrity. Cancer Res *57:* 229-233, 1997.
69. Oltvai Z, Milliman C, Korsmeyer SJ: Bcl-2 heterodimerizes *in vivo* with a conserved homolog Bax that accelerates programmed cell death. Cell *74:* 609-619, 1993.
70. Pillai MR, Kesari AL, Chellam VG, Madhavan J, Nair P, Nair MK: Spontaneous programmed cell death in infiltrating duct carcinoma: association with p53, BCL-2, hormone receptors and tumor proliferation. Pathol Res Pract *194:* 549-557, 1998.
71. Yuan J, Horvitz HR: The *Caenorhabditis elegans* genes ced-3 and ced-4 act cell autonomously to cause programmed cell death. Dev Biol *138:* 33-41, 1990.
72. Hengartner MO, Ellis RE, Horvitz HR: *Caenorhabditis elegans* gene ced-9 protects cells from programmed cell death. Nature *356:* 494-499, 1992.

73. Steller H: Mechanisms and genes of cellular suicide. Science *267:* 1445-1448, 1995.
74. Reed JC: Bcl-2 and the regulation of programmed cell death. J Cell Biol *124:* 1-6, 1994.
75. Pourzand C, Rossier G, Reelfs O, Borner C, Tyrrell RM: The overexpression of bcl-2 inhibits UVA-mediated immediate apoptosis in rat 6 fibroblasts: evidence for the involvement of bcl-2 as an antioxidant. Cancer Res *57:* 1405-1411, 1997.
76. Alnermi ES, Livingston DJ, Nicholson DW, Salvesen G, Thornberry NA, Wong WW, Yuan J: Human ICE/CED-3 protease nomenclature. Cell *87:* 171, 1996.
77. Kumar S, Kinoshita M, Noda M, Copeland NG, Jenkins NA: Induction of apoptosis by the mouse Nedd 2 gene, which encodes a protein similar to the product of the *Caenorhabditis elegans* cell death gene ced-3 and the mammalian IL-1β-converting enzyme. Genes Dev *8:* 1613-1626, 1994.
78. Sinkovics JG: Malignant lymphoma arising from natural killer cells: report of the first case in 1970 and newer developments in the *FasL→FasR* system. Acta Microbiol Immunol Hung *44:* 295-303, 1997.
79. Wang L, Miura M, Bergeron B, Zhu H, Yuan J: Ich-1, an ICE/Ced-3 related gene, encodes both positive and negative regulators of programmed cell death. Cell *78:* 739-750, 1994.
80. Kamens J, Paskind M, Hugunin M, Talanian RV, Allen H, Banach D, Bump N, Hackett M, Johnston CG, Li P, Mankovich JA, Terranova M, Ghayur T: Identification and characterization of ICH-2, a novel member of the interleukin-1 beta-converting enzyme family of cysteine proteases. J Biol Chem *270:* 15250-15256, 1995.
81. Fernandes-Alnemri T, Litwack G, Alnemri ES: CPP32, a novel human apoptotic protein with homology to *Caenorhabditis elegans* cell death protein Ced-3 and mammalian interleukin-1 beta-converting enzyme. J Biol Chem *269:* 30761-30764, 1995.
82. Fernandes-Alnemri T, Litwack G, Alnemri ES: Mch2, a new member of the apoptotic Ced-3/ICE cysteine protease gene family. Cancer Res *55:* 2737-2742, 1995.
83. Munday NA, Vaillancourt JP, Ali A, Casano FJ, Miller DK, Molineaux SM, Yamin TT, Yu VL, Nicholson DW: Molecular cloning and pro-apoptotic activity of ICErelII and ICErelIII members of the ICE/Ced-3 family of cysteine proteases. J Biol Chem *270:* 15870-15876, 1995.
84. Duan H, Chinnaiyan AM, Hudson P, Wing JP, He WW, Dixit VM: ICE-LAP3, a novel mammalian homologue of of the *Caenorhabditis elegans* cell death protein Ced-3 is activated during *Fas*- and tumor necrosis factor-induced apoptosis. J Biol Chem *271:* 1621-1625, 1996.
85. Duan H, Orth K, Chinnaiyan AM, Poirier GG, Froelich CJ, He WW, Dixit VM: ICE-LAP6, a novel member of the ICE-Ced-3 gene family, is activated by the cytotoxic T cell protease granzyme B. J Biol Chem *271:* 16720-16724, 1996.
86. Boldin MP, Goncharov TM, Goltsev YV, Wallach D: Involvement of MACH, a novel MORT1/FADD-interacting protease, in *Fas*/APO-1 and TNF receptor-induced cell death. Cell *85:* 803-815, 1996.
87. Lazebnik YA, Kaufmann SH, Desnoyers S, Poirier GG, Earnshaw WC: Cleavage of poly(ADP-ribose) polymerase by a proteinase with properties like ICE. Nature *371:* 346-347, 1994.
88. Nicholson DW, Ali A, Thornberry NA, Vaillancourt JP, Ding CK, Gallant M, Gareau Y, Griffin PR, Labelle M, Lazebnik YA: Identification and inhibition of the ICE/CED-3 protease necessary for mammalian apoptosis. Nature *376:* 37-43, 1995.
89. Sorensen CM: Apoptosis or planning a death. Biomedical Products (September, 1996 issue), pp 38-39, 1996.

90. Asano K, Kubo O, Tajika Y, Huang MC, Takakura K, Ebina K, Suzuki S: Expression and role of cadherins in astrocytic tumors. Brain Tumor Pathol *14:* 27-33, 1997.
91. Huber O: Structure and function of desmosomal proteins and their role in development and disease. Cell Mol Life Sci *60:* 1872-1890, 2003.
92. Strelkov SV, Herrmann H, Aebi U: Molecular architecture of intermediate filaments. Bioessays *25:* 243-251, 2003.
93. Hasegawa K, Yoshida T, Matsumoto K, Katsuta K, Waga S, Sakakura T: Differential expression of tenascin-C and tenascin-X in human astrocytomas. Acta Neuropathol (Berl) *93:* 431-437, 1997.
94. Gilbertson RJ, Perry RH, Kelly PJ, Pearson AD, Lunec J: Prognostic significance of HER2 and HER4 coexpression in childhood medulloblastoma. Cancer Res *57:* 3272-3280, 1997.
95. Aaronson SA: Growth factors and cancer. Science *254:* 1146-1153; 1991.
96. Engebraaten O, Bjerkvig R, Humphrey PA, Bigner SH, Bigner DD, Laerum OD: Effect of EGF, bFGF, NGF and PDGF(bb) on cell proliferative, migratory and invasive capacities of human brain-tumour biopsies *in vitro*. Int J Cancer *53:* 209-214, 1993.
97. Chicoine MR, Madsen CL, Silbergeld DL: Modification of human glioma locomotion *in vitro* by cytokines EGF, bFGF, PDGFbb, NGF, and TNFα. Neurosurg *36:* 1165-1171, 1995.
98. U HS, Espiritu OD, Kelley PY, Klauber MR, Hatton JD: The role of the epidermal growth factor receptor in human gliomas: I. The control of cell growth. J Neurosurg *82:* 841-846, 1995.
99. Hitotsumatsu T, Iwaki T, Kitamoto T, Mizoguchi M, Suzuki SO, Hamada Y, Fukui M, Tateishi J: Expression of neurofibromatosis 2 protein in human brain tumors: an immunohistochemical study. Acta Neuropathol *93:* 225-232, 1997.
100. Chang F, Li R, Noon K, Gage D, Ladisch S: Human medulloblastoma gangliosides. Glycobiology *7:* 523-530, 1997.
101. Leung SY, Wong MP, Chung LP, Chan AS, Yuen ST: Monocyte chemoattractant protein-1 expression and macrophage infiltration in gliomas. Acta Neuropathol *93:* 518-527, 1997.
102. Ingber DE: Mechanical signaling and the cellular response to extracellular matrix in angiogenesis and cardiovascular physiology. Circ Res *91:* 877-887, 2002.
103. Moffett JR, Els T, Espey MG, Walter SA, Streit WJ, Namboodiri MA: Quinolinate immunoreactivity in experimental rat brain tumors is present in macrophages but not in astrocytes. Exp Neurol *144:* 287-301, 1997.
104. Yoon Y, Pitts K, McNiven M: Studying cytoskeletal dynamics in living cells using green fluorescent protein. Mol Biotechnol *21:* 241-250, 2002.
105. Osborn M, Weber K: Tumor diagnosis by intermediate filament typing: a novel tool for surgical pathology. Lab Invest *48:* 372-394, 1983.
106. Yung WK, Luna M, Borit A: Vimentin and glial fibrillary acidic protein in human brain tumors. J Neurooncol *3:* 35-38, 1985.
107. Gown AM, Vogel AM: Anti-intermediate filament monoclonal antibodies: tissue-specific tools in tumor diagnosis. Surv Synth Pathol Res *3:* 369-385, 1984.
108. Cooper EH: Neuron specific enolase: a marker of (small cell) cancers of neuronal and neuroendocrine origin. Biomed Pharmacother *39:* 165-166, 1985.
109. Gould VE: Histogenesis and differentiation: a re-evaluation of these concepts as criteria for the classification of tumors. Hum Pathol *17:* 212-215, 1986.
110. Messing A, Brenner M: GFAP: functional implications gleaned from studies of genetically engineered mice. Glia *43:* 87-90, 2003.

111. Eng LF, Ghirnikar RS, Lee YL: Glial fibrillary acidic protein: GFAP-thirty-one years (1969-2000). Neurochem Res *25:* 1439-1451, 2000.
112. Rungger-Brandle E, Achtstatter T, Franke WW: An epithelium-type cytoskeleton in a glial cell: astrocytes of amphibian optic nerves contain cytokeratin filaments and are connected by desmosomes. J Cell Biol *109:* 705-716, 1989.
113. Cosgrove M, Fitzgibbons PL, Sherrod A, Chandrasoma PT, Martin SE: Intermediate filament expression in astrocytic neoplasms. Am J Surg Pathol *13:* 141-145, 1989.
114. Leader M, Collins M, Patel J, Henry K: Vimentin: an evaluation of its role as a tumour marker. Histopathology *11:* 63-72, 1987.
115. Gown AM, Vogel AM: Monoclonal antibodies to human intermediate filament proteins. III. Analysis of tumors. Am J Clin Pathol *84:* 413-424, 1984.
116. Roessmann U, Velasco ME, Gambetti P, Autilio-Gambetti L: Vimentin intermediate filaments are increased in human neoplastic astrocytes (abstract). J Neuropathol Exp Neurol *42:* 309, 1983.
117. Wang E, Cairncross JG, Liem RK: Identification of glial filament protein and vimentin in the same intermediate filament system in human glioma cells. Proc Natl Acad Sci USA *81:* 2102-2106, 1984.
118. Galloway PG, Roessmann U: Anaplastic astrocytoma mimicking metastatic carcinoma. Am J Surg Pathol *10:* 728-732, 1986.
119. Bodey B, Zeltzer PM, Saldivar V, Kemshead J: Immunophenotyping of childhood astrocytomas with a library of monoclonal antibodies. Int J Cancer *45:* 1079-1087, 1990.
120. Perentes E, Rubinstein LJ: Recent applications of immunoperoxidase histochemistry in human neuro-oncology. An update. Arch Pathol Lab Med *111:* 796-812, 1987.
121. Mork SJ, Rubinstein LJ, Kepes JJ, Perentes E, Uphoff DF: Patterns of epithelial metaplasia in malignant gliomas. II. Squamous differentiation of epithelial-like formations in gliosarcomas and glioblastomas. J Neuropathol Exp Neurol *47:* 101-118, 1988.
122. Franke WW, Weber K, Osborn M, Schmid E, Freudenstein C: Antibody to prekeratin. Decoration of tonofilament like arrays in various cells of epithelial character. Exp Cell Res *116:* 429-445, 1978.
123. Franke WW, Appelhans B, Schmid E, Freudenstein C, Osborn M, Weber K: Identification and characterization of epithelial cells in mammalian tissues by immunofluorescence microscopy using antibodies to prekeratin. Differentiation *15:* 7-25, 1979.
124. Steinert PM, Steven AC, Roop DR: The molecular biology of intermediate filaments. Cell *42:* 411-420, 1985.
125. Steinert PM, Roop DR: Molecular and cellular biology of intermediate filaments. Annu Rev Biochem *57:* 593-625, 1988.
126. Kepes JJ, Fulling KH, Garcia JH: The clinical significance of "adenoid" formations of neoplastic astrocytes, imitating metastatic carcinoma, in gliosarcomas. A review of five cases. Clin Neuropathol *1:* 139-150, 1982.
127. Sun TT, Green H: Keratin filaments of cultured human epidermal cells. Formation of intermolecular disulfide bonds during terminal differentiation. J Biol Chem *253:* 2053-2060, 1978.
128. Green H, Fuchs E, Watt F: Differentiated structural components of the keratinocyte. Cold Spring Harb Symp Quant Biol *46:* 293-301, 1982.
129. Moll R, Franke WW, Schiller DL, Geiger B, Krepler R: The catalog of human cytokeratins: patterns of expression in normal epithelia, tumors and cultured cells. Cell *31:* 11-24, 1982.

130. Sun TT, Eichner R, Nelson WG, Tseng SC, Weiss RA, Jarvinen M, Woodcock-Mitchell J: Keratin classes: molecular markers for different types of epithelial differentiation. J Invest Dermatol *8:* 109s-115s, 1983.
131. Steinert PM, Steven AC: Splitting hairs and other intermediate filaments. Nature *316:* 767, 1985.
132. Fuchs E, Green H: Changes in keratin gene expression during terminal differentiation of the keratinocyte. Cell *19:* 1033-1042, 1980.
133. Tseng SC, Jarvinen MJ, Nelson WG, Huang JW, Woodcock-Mitchell J, Sun TT: Correlation of specific keratins with different types of epithelial differentiation: monoclonal antibody studies. Cell *30:* 361-372, 1982.
134. Moll R, von Bassewitz DB, Schulz U, Franke WW: An unusual type of cytokeratin filament in cells of a human cloacogenic carcinoma derived from the anorectal transition zone. Differentiation *22:* 25-40, 1982.
135. Moll R, Krepler R, Franke WW: Complex cytokeratin polypeptide patterns observed in certain human carcinomas. Differentiation *23:* 256-269, 1983.
136. Weiss SW, Langloss JM, Enzinger FM: Value of S-100 protein in the diagnosis of soft tissue tumors with particular reference to benign and malignant Schwann cell tumors. Lab Invest *49:* 299-308, 1983.
137. Debus E, Weber K, Osborn M: Monoclonal antibodies specific for glial fibrillary acidic (GFA) protein and for each of the neurofilament triplet polypeptides. Differentiation *25:* 193-203, 1983.
138. Woodcock-Mitchell J, Eichner R, Nelson WG, Sun TT: Immunolocalization of keratin polypeptides in human epidermis using monoclonal antibodies. J Cell Biol *95:* 580-588, 1982.
139. Damjanov I: Antibodies to intermediate filaments and histogenesis. Lab Invest *47:* 215-217, 1982.
140. Trojanowski JQ, Lee VM, Schlaepfer WW: An immunohistochemical study of human central and peripheral nervous system tumors, using monoclonal antibodies against neurofilaments and glial filaments. Hum Pathol *15:* 248-257, 1984.
141. Sun TT, Shih C, Green H: Keratin cytoskeletons in epithelial cells of internal organs. Proc Natl Acad Sci USA *76:* 2813-2817, 1979.
142. Zipser B, Schley C: Description of two differently distributed central nervous system antigens with single monoclonal antibody and different methods of fixation. Ann NY Acad Sci *420:* 100-106, 1983.
143. Gabbiani G, Kapanci Y, Barazzone P, Franke WW: Immunochemical identification of intermediate-sized filaments in human neoplastic cells. A diagnostic aid for the surgical pathologist. Am J Pathol *104:* 206-216, 1981.
144. Vinores SA, Rubinstein LJ: Simultaneous expression of glial fibrillary acidic (GFA) protein and neuron-specific enolase (NSE) by the same reactive or neoplastic astrocytes. Neuropathol Appl Neurobiol *11:* 349-359, 1985.
145. Lazarides E: Intermediate filaments: a chemically heterogeneous, developmentally regulated class of proteins. Annu Rev Biochem *51:* 219-250, 1982.
146. Kemshead JT, Coakham HB: The use of monoclonal antibodies for the diagnosis of intracranial malignancies and the small round cell tumours of childhood. J Pathol *141:* 249-257, 1983.
147. Kemshead JT: Pediatric Tumors: Immunological and Molecular Markers. Boca Roton, FL, CRC Press, 1989.
148. Molenaar WM, Jansson DS, Gould VE, Rorke LB, Franke WW, Lee VM, Packer RJ, Trojanowski JQ: Molecular markers of primitive neuroectodermal tumors and other pediatric central nervous system tumors. Monoclonal antibodies to neuronal and glial

antigens distinguish subsets of primitive neuroectodermal tumors. Lab Invest *61:* 635-643, 1989.

149. Gould VE, Jansson DS, Molenaar WM, Rorke LB, Trojanowski JQ, Lee VM, Packer RJ, Franke WW: Primitive neuroectodermal tumors of the central nervous system. Patterns of expression of neuroendocrine markers, and all classes of intermediate filament proteins. Lab Invest *62:* 498-509, 1990.
150. Dahl D, Bignami A: Astroglial and axonal proteins in isolated brain filaments. I. Isolation of the glial fibrillary acidic protein and of an immunologically active cyanogen bromide peptide from brain filament preparations of bovine white matter. Biochim Biophys Acta *578:* 305-316, 1979.
151. Sawa H, Takeshita I, Kuramitsu M, Fukui M, Inomata H: Immunohistochemistry of retinoblastomas. J Neurooncol *5:* 351-355, 1987.
152. Dahl D: Isolation of neurofilament proteins and of immunologically active neurofilament degradation products from extracts of brain, spinal cord and sciatic nerve. Biochim Biophys Acta *668:* 299-306, 1981.
153. Wang N, Stamenovic D: Mechanics of vimentin intermediate filaments. J Muscle Res Cell Motil *23:* 535-540, 2002.
154. Helfand BT, Chang L, Goldman RD: The dynamic and motile properties of intermediate filaments. Annu Rev Cell Dev Biol *19:* 445-467, 2003.
155. Bürglin TR: Homeodomain Proteins. *In:* Encyclopedia of Molecular Biology and Molecular Medicine (Meyers RA, ed). Vol 3. Weinheim, VCH Verlagsgesellschaft mbH, p1996, pp 55-76.
156. Stornaiuolo A, Acampora D, Pannese M, D'Esposito M, Morelli F, Migliaccio E, Rambaldi M, Faiella A, Nigro V, Simeone A, Boncinelli E: Human HOX genes are differentially activated by retinoic acid in embryonal carcinoma cells according to their position within the four loci. Cell Differ Dev *31:* 119-127, 1990.
157. Giampaolo A, Acampora D, Zappavigna V, Pannese M, D'Esposito M, Care A, Faiella A, Stornaiuolo A, Russo G, Simeone A, Boncinelli E, Peschle C: Differential expression of human HOX-2 genes along the anterior-posterior axis in embryonic central nervous system. Differentiation *40:* 191-197, 1989.
158. Mavilio F, Simeone A, Boncinelli E, Andrews PW: Activation of four homeobox gene clusters in human embryonal carcinoma cells induced to differentiate by retinoic acid. Differentiation *37:* 73-79, 1988.
159. Simeone A, Acampora D, D'Esposito M, Faiella A, Pannese M, Scotto L, Montanucci M, D'Alessandro G, Mavilio F, Boncinelli E: Posttranscriptional control of human homeobox gene expression in induced NTERA-2 embryonal carcinoma cells.Mol Reprod Dev *1:* 107-115, 1989.
160. Peverali FA, D'Esposito M, Acampora D, Bunone G, Negri M, Faiella A, Stornaiuolo A, Pannese M, Migliaccio E, Simeone A, Della Valle G, Boncinelli E: Expression of HOX homeogenes in human neuroblastoma cell culture lines. Differentiation *45:* 61-69, 1990.
161. Maconochie MK, Nonchev S, Studer M, Chan SK, Popperl H, Sham MH, Mann RS, Krumlauf R: Cross-regulation in the mouse HoxB complex: the expression of Hoxb2 in rhombomere 4 is regulated by Hoxb1. Genes Dev *11:* 1885-1895, 1997.
162. Gould A, Morrison A, Sproat G, White RA, Krumlauf R: Positive cross-regulation and enhancer sharing: two mechanisms for specifying overlapping Hox expression patterns.Genes Dev *11:* 900-913, 1997.
163. Sauvageau G, Thorsteinsdottir U, Eaves CJ, Lawrence HJ, Largman C, Lansdorp PM, Humphries RK: Overexpression of HOXB4 in hematopoietic cells causes the selective

expansion of more primitive populations *in vitro* and *in vivo*. Genes Dev *9:* 1753-1765, 1995.

164. Rancourt DE, Tsuzuki T, Capecchi MR: Genetic interaction between hoxb-5 and hoxb-6 is revealed by nonallelic noncomplementation. Genes Dev *9:* 108-122, 1995.
165. Sun J, Rose JB, Bird P: Gene structure, chromosomal localization, and expression of the murine homologue of human proteinase inhibitor 6 (PI-6) suggests divergence of PI-6 from the ovalbumin serpins. J Biol Chem *270:* 16089-16096, 1995.
166. Hogan BL, Holland PW, Lumsden A: Expression of the homeobox gene, Hox 2.1, during mouse embryogenesis. Cell Diff Dev *25:* 39-44, 1988.
167. Safaei R: A target of the HoxB5 gene from the mouse nervous system. Brain Res Dev Brain Res *100:* 5-12, 1997.
168. Lawrence HJ, Sauvageau G, Humphries RK, Largman C: The role of HOX homeobox genes in normal and leukemic hematopoiesis. Stem Cells *14:* 281-291, 1996.
169. Thorsteinsdottir U, Sauvageau G, Humphries RK: Hox homeobox genes as regulators of normal and leukemic hematopoiesis. Hematol Oncol Clin North Am *11:* 1221-1237, 1997.
170. Shimamoto T, Ohyashiki K, Toyama K, Takeshita K: Homeobox genes in hematopoiesis and leukemogenesis. Int J Hematol *67:* 339-350, 1998.
171. Chiba S: Homeobox genes in normal hematopoiesis and leukemogenesis. Int J Hematol *68:* 343-353, 1998.
172. Moskow JJ, Bullrich F, Huebner K, Daar IO, Buchberg AM: Meis1, a PBX1-related homeobox gene involved in myeloid leukemia in BXH-2 mice. Mol Cell Biol *15:* 5434-5443, 1995.
173. Nakamura T, Largaespada DA, Shaughnessy JD Jr, Jenkins NA, Copeland NG: Cooperative activation of Hoxa and Pbx1-related genes in murine myeloid leukaemias. Nat Genet *12:*149-153, 1996.
174. Lawrence HJ, Sauvageau G, Ahmadi N, Lopez AR, LeBeau MM, Link M, Humphries K, Largman C: Stage- and lineage-specific expression of the HOXA10 homeobox gene in normal and leukemic hematopoietic cells. Exp Hematol *23:* 1160-1166, 1995.
175. Kawagoe H, Humphries RK, Blair A, Sutherland HJ, Hogge DE: Expression of HOX genes, HOX cofactors, and MLL in phenotypically and functionally defined subpopulations of leukemic and normal human hematopoietic cells. Leukemia *13:* 687-698, 1999.
176. Kroon E, Krosl J, Thorsteinsdottir U, Baban S, Buchberg AM, Sauvageau G: Hoxa9 transforms primary bone marrow cells through specific collaboration with Meis1a but not Pbx1b. EMBO J *17:* 3714-3725, 1998.
177. Thorsteinsdottir U, Sauvageau G, Hough MR, Dragowska W, Lansdorp PM, Lawrence HJ, Largman C, Humphries RK: Overexpression of HOXA10 in murine hematopoietic cells perturbs both myeloid and lymphoid differentiation and leads to acute myeloid leukemia. Mol Cell Biol *17:* 495-505, 1997.
178. Ohnishi K, Tobita T, Sinjo K, Takeshita A, Ohno R: Modulation of homeobox B6 and B9 genes expression in human leukemia cell lines during myelomonocytic differentiation. Leuk Lymphoma *31:* 599-608, 1998.
179. Cillo C, Cantile M, Mortarini R, Barba P, Parmiani G, Anichini A: Differential patterns of HOX gene expression are associated with specific integrin and ICAM profiles in clonal populations isolated from a single human melanoma metastasis. Int J Cancer *66:* 692-697, 1996.
180. Lawrence HJ, Stage KM, Mathews CH, Detmer K, Scibienski R, MacKenzie M, Migliaccio E, Boncinelli E, Largman C: Expression of HOX C homeobox genes in lymphoid cells. Cell Growth Differ *4:* 665-669, 1993.

181. Bijl JJ, van Oostveen JW, Walboomers JM, Brink AT, Vos W, Ossenkoppele GJ, Meijer CJ: Differentiation and cell-type-restricted expression of HOXC4, HOXC5 and HOXC6 in myeloid leukemias and normal myeloid cells. Leukemia *12:* 1724-1732, 1998.
182. Meazza R, Faiella A, Corsetti MT, Airoldi I, Ferrini S, Boncinelli E, Corte G: Expression of HOXC4 homeoprotein in the nucleus of activated human lymphocytes. Blood *85:* 2084-90, 1995.
183. Bijl J, van Oostveen JW, Kreike M, Rieger E, van der Raaij-Helmer LM, Walboomers JM, Corte G, Boncinelli E, van den Brule AJ, Meijer CJ: Expression of HOXC4, HOXC5, and HOXC6 in human lymphoid cell lines, leukemias, and benign and malignant lymphoid tissue. Blood *87:* 1737-1745, 1996.
184. Bijl JJ, van Oostveen JW, Walboomers JM, Horstman A, van den Brule AJ, Willemze R, Meijer CJ: HOXC4, HOXC5, and HOXC6 expression in non-Hodgkin's lymphoma: preferential expression of the HOXC5 gene in primary cutaneous anaplastic T-cell and oro-gastrointestinal tract mucosa-associated B-cell lymphomas. Blood *90:* 4116-4125, 1997.
185. Bijl JJ, Rieger E, van Oostveen JW, Walboomers JM, Kreike M, Willemze R, Meijer CJ: HOXC4, HOXC5, and HOXC6 expression in primary cutaneous lymphoid lesions. High expression of HOXC5 in anaplastic large-cell lymphomas. Am J Pathol *151:* 1067-1074, 1997.
186. Alami Y, Castronovo V, Belotti D, Flagiello D, Clausse N: HOXC5 and HOXC8 expression are selectively turned on in human cervical cancer cells compared to normal keratinocytes. Biochem Biophys Res Commun *257:* 738-745, 1999.
187. Shim C, Zhang W, Rhee CH, Lee JH: Profiling of differentially expressed genes in human primary cervical cancer by complementary DNA expression array. Clin Cancer Res *4:* 3045-3050, 1998.
188. Osborne J, Hu C, Hawley C, Underwood LJ, O'Brien TJ, Baker VV: Expression of HOXD10 gene in normal endometrium and endometrial adenocarcinoma. J Soc Gynecol Investig *5:* 277-280, 1998.
189. Boylan JF, Lohnes D, Taneja R, Chambon P, Gudas LJ: Loss of retinoic acid receptor gamma function in F9 cells by gene disruption results in aberrant Hoxa-1 expression and differentiation upon retinoic acid treatment. Proc Natl Acad Sci USA *90:* 9601-9605, 1993.
190. Pratt MA, Langston AW, Gudas LJ, McBurney MW: Retinoic acid fails to induce expression of Hox genes in differentiation-defective murine embryonal carcinoma cells carrying a mutant gene for α retinoic acid receptor. Differentiation *53:* 105-113, 1993.
191. Langston AW, Gudas LJ: Retinoic acid and homeobox gene regulation. Curr Opin Genet Dev *4:* 550-555, 1994.
192. Knoepfler PS, Kamps MP: The Pbx family of proteins is strongly upregulated by a post-transcriptional mechanism during retinoic acid-induced differentiation of P19 embryonal carcinoma cells. Mech Dev *63:* 5-14, 1997.
193. Simeone A, Acampora D, Arcioni L, Andrews PW, Boncinelli E, Mavilio F: Sequential activation of HOX2 homeobox genes by retinoic acid in human embryonal carcinoma cells. Nature *346:* 763-766, 1990.
194. Chang CP, de Vivo I, Cleary ML: The Hox cooperativity motif of the chimeric oncoprotein E2a-Pbx1 is necessary and sufficient for oncogenesis. Mol Cell Biol *17:* 81-88, 1997.
195. Krosl J, Baban S, Krosl G, Rozenfeld S, Largman C, Sauvageau G: Cellular proliferation and transformation induced by HOXB4 and HOXB3 proteins involves cooperation with PBX1. Oncogene *16:* 3403-3412, 1998.

196. Tiberio C, Barba P, Magli MC, Arvelo F, Le Chevalier T, Poupon MF, Cillo C: HOX gene expression in human small-cell lung cancers xenografted into nude mice. Int J Cancer *58:* 608-615, 1994.
197. Flagiello D, Gibaud A, Dutrillaux B, Poupon MF, Malfoy B: Distinct patterns of all-trans retinoic acid dependent expression of HOXB and HOXC homeogenes in human embryonal and small-cell lung carcinoma cell lines. FEBS Lett *415:* 263-267, 1997.
198. Flagiello D, Poupon MF, Cillo C, Dutrillaux B, Malfoy B: Relationship between DNA methylation and gene expression of the HOXB gene cluster in small cell lung cancers. FEBS Lett *380:* 103-107, 1996.
199. Chariot A, Moreau L, Senterre G, Sobel ME, Castronovo V: Retinoic acid induces three newly cloned HOXA1 transcripts in MCF7 breast cancer cells. Biochem Biophys Res Commun *215:* 713-720, 1995.
200. Chariot A, Castronovo V: Detection of HOXA1 expression in human breast cancer. Biochem Biophys Res Commun *222:* 292-297, 1996.
201. Suzuki M, Tanaka M, Iwase T, Naito Y, Sugimura H, Kino I: Over-expression of HOX-8, the human homologue of the mouse Hox-8 homeobox gene, in human tumors. Biochem Biophys Res Commun *194:* 187-193, 1993.
202. De Vita G, Barba P, Odartchenko N, Givel JC, Freschi G, Bucciarelli G, Magli MC, Boncinelli E, Cillo C: Expression of homeobox-containing genes in primary and metastatic colorectal cancer. Eur J Cancer *29A:* 887-893, 1993.
203. Manohar CF, Furtado MR, Salwen HR, Cohn SL: Hox gene expression in differentiating human neuroblastoma cells. Biochem Mol Biol Int *30:* 733-741, 1993.
204. Manohar CF, Salwen HR, Furtado MR, Cohn SL: Up-regulation of HOXC6, HOXD1, and HOXD8 homeobox gene expression in human neuroblastoma cells following chemical induction of differentiation. Tumour Biol *17:* 34-47, 1996.
205. Cillo C, Barba P, Freschi G, Bucciarelli G, Magli MC, Boncinelli E: HOX gene expression in normal and neoplastic human kidney. Int J Cancer *51:* 892-897, 1992.
206. Deschamps J, Meijlink F: Mammalian homeobox genes in normal development and neoplasia. Crit Rev Oncog *3:* 117-173, 1992.
207. Friedmann Y, Daniel CA, Strickland P, Daniel CW: Hox genes in normal and neoplastic mouse mammary gland. Cancer Res *54:* 5981-5985, 1994.
208. Redline RW, Hudock P, MacFee M, Patterson P: Expression of AbdB-type homeobox genes in human tumors. Lab Invest *71:* 663-670, 1994.
209. Cillo C: HOX genes in human cancers. Invasion Metastasis *14:* 38-49, 1994-95.
210. Stuart ET, Yokota Y, Gruss P: PAX and HOX in neoplasia. Adv Genet *33:* 255-274, 1995.
211. Mark M, Rijli FM, Chambon P: Homeobox genes in embryogenesis and pathogenesis. Pediatr Res *42:* 421-429, 1997.
212. Boudreau N, Andrews C, Srebrow A, Ravanpay A, Cheresh DA: Induction of the angiogenic phenotype by Hox D3. J Cell Biol *139:* 257-264, 1997.
213. Care A, Silvani A, Meccia E, Mattia G, Stoppacciaro A, Parmiani G, Peschle C, Colombo MP: HOXB7 constitutively activates basic fibroblast growth factor in melanomas. Mol Cell Biol *16:* 4842-4851, 1996.
214. Kloen P, Visker MH, Olijve W, van Zoelen EJ, Boersma CJ: Cell-type-specific modulation of Hox gene expression by members of the TGF-beta superfamily: a comparison between human osteosarcoma and neuroblastoma cell lines. Biochem Biophys Res Commun *233:* 365-369, 1997.
215. Silverberg E, Boring CC, Squires TS: Cancer statistics, CA-A Cancer J Clinicians *40:* 9-26, 1990.

216. Cooper GM (ed), Elements of human cancer. Jones & Bartlett Publishers, Boston-London, 1992, pp.7-14.
217. Katsetos CD, Krishna L, Frankfurter A, Karkavelas G, Wolfe DE, Valsamis MP, Schiffer D, Vlachos IN, Urich H: A cytomorphological scheme of differentiating neuronal phenotypes in cerebellar medulloblastomas based on immunolocalization of class III β-tubulin isotype (β III) and proliferating cell nuclear antigen (PCNA)/cyclin. Clinical Neuropathol *14:* 72-81, 1995.
218. O'Brien MC, Gupta RK, Lee SY, Bolton WE: Use of a multiparametric panel to target subpopulations in a heterogeneous solid tumor model for improved analytical accuracy. Cytometry *21:* 76-83, 1995.
219. Oda Y, Tsuneyoshi M: A comparative study of nuclear photometry and proliferating activity in neuroectodermal tumors of bone and Ewing's sarcoma of bone. General & Diagnostic Pathol *141:* 121-129, 1995.
220. Ellis PA, Makris A, Burton SA, Titley J, Ormerod MG, Salter J, Powles TJ, Smith IE, Dowsett M: Comparison of MIB-1 proliferation index with S-phase fraction in human breast carcinomas. Brit J Cancer *73:* 640-643, 1996.
221. Oyama T, Take H, Hikino T, Iino Y, Nakajima T: Immunohistochemical expression of metallothionein in invasive breast cancer in relation to proliferative activity, histology and prognosis. Oncol *53:* 112-117, 1996.
222. Enzinger FM, Lattes R, Torloni H: Histological typing of soft tissue tumours. World Health Organization, Geneva, 1971, p. 28.
223. Kury G, Carter HW: Autoradiographic study of human nervous system tumors. Arch Pathol *80:* 38-42, 1965.
224. Tym R: Distribution of cell doubling times in *in vivo* human cerebral tumors. Surg Forum *20:* 445-447, 1969.
225. Hoshino T, Barker M, Wilson CB, Boldrey EB, Fewer D: Cell kinetics of human gliomas. J Neurosurg *37:* 15-26, 1975.
226. Hoshino T, Wilson CB, Rosenblum ML, Barker M: Chemotherapeutic implications of growth fraction and cell cycle time in glioblastomas. J Neurosurg *43.* 127-135, 1975.
227. Böker DK, Stark HJ, Gullotta F, Nadstawek J, Schultheiss R: Immunohistochemical demonstration of the Ki-67-antigen in paraffin-embedded tumor biopsies. Clin Neuropathol *9:* 51-54, 1990.
228. Gerdes J: Ki-67 and other proliferation markers useful for immunohistological diagnostic and prognostic evaluations in human malignancies. Semin Cancer Biol *1(3):* 199-206, 1990.
229. Sledge GW, Eble JN, Roth BJ, Wuhrman BP, Fineberg N, Einhorn LH: Relation of proliferate activity to survival in patients with advanced germ cell cancer. Cancer Res *48:* 3864-3868, 1988.
230. Hall PA, Levison DA, Woods AL, Yu CC, Kellock DB, Watkins JA, Barnes DM, Gillet CE, Camplejohn R, Dover R: Proliferating cell nuclear antigen (PCNA) immunolocalization in paraffin sections: an insex of cell proliferation with evidence of deregulated expression in some neoplasms. J Pathol *162:* 285-294, 1990.
231. Munck-Wikland E, Fernberg JO, Kuylenstierna R, Lindholm J, Aver G: Proliferating cell nuclear antigen (PCNA) expression and nuclear DNA content in predicting recurrence after radiotherapy of glottic cancer. Oral Oncol Eur J Cancer *2913:* 75-79, 1993.
232. Broich G, Lavezzi A-M, Biondo B, Pignataro LD: PCNA - a cell proliferation marker in vocal cord cancer. Part II: recurrence in malignant laryngeal lesions. In Vivo *10:* 175-178, 1996.

233. Visakorpi T: Proliferative activity determined by DNA flow cytometry and proliferate cell nuclear antigen (PCNA) immunohistochemistry as a prognostic factor in prostatic carcinoma. J Pathol *168:* 7-13, 1992.
234. Pignataro LD, Broich G, Lavezzi AM, Biondo B, Ottaviani F: PCNA - a cell proliferation marker in vocal cord cancer. Part I: Premalignant laryngeal lesions. Anticancer Res *15:* 1517-1520, 1995.
235. Fairman MP: DNA polymerase/PCNA: Actions and interactions. J Cell Science *95:* 1-4, 1990.
236. Bravo R, Franke Blundell PA, MacDonald M, Bravo M: Cyclin/PCNA is the auxiliary protein of the DNA polymerase delta. Nature *326:* 517-518, 1987.
237. Cruz-Sanchez FF, Ferreres JC, Figols J, Palacin A, Cardesa A, Rossi ML, Val-Bernal JF: Prognostic analysis of astrocytic gliomas correlating histological parameters with the proliferating cell nuclear antigen labeling index (PCNA-LI). Histol Histopathol *12:* 43-49, 1997.
238. Bodey B: The significance of immunocytochemistry in the diagnosis and therapy of neoplasms. Expert Opinion Biological Therapy *2:* 371-393, 2002.
239. Nurse P: Universal control mechanism regulating onset of M-phase. Nature *344:* 503-508, 1990.
240. Pardee A: G1 events and regulation of cell proliferation. Science *246:* 603-608, 1989.
241. Vulliet PR, Hall FL, Mitchell JP, Hardie DG: Identification of a novel proline-directed serine/threonine protein kinase in rat pheochromocytoma. J Biol Chem *264:* 16292-16298, 1989.
242. Hall FL, Vulliet PR: Proline-directed protein phosphorylation and cell cycle regulation. Current Opinion Cell Biol *3:* 176-184, 1991.
243. Howard A, Pelc SR: Nuclear incorporation of ^{32}P as demonstrated by autoradiographs. Expl Cell Res *2:* 178-187, 1951.
244. Darzynkiewicz Z: Molecular interactions and cellular changes during the cell cycle. Pharmacol Ther *21:* 143-188, 1983.
245. Quastler H, Sherman FG: Cell population kinetics in the intestinal epithelium of the mouse. Exp Cell Res *17:* 420-438, 1959.
246. Miyachi K, Fritzler MJ, Tan EM: Autoantibody to a nuclear antigen in proliferating cells. J Immunol *121:* 2228-2234, 1978.
247. Tan EM: Autoantibodies to nuclear antigens (ANA): their immunobiology and medicine. Adv Immunol *33:* 167-240, 1982.
248. Gerdes J, Lemke H, Baisch H, Wacker H, Schwab U, Stein H: Cell cycle analysis of a cell proliferation associated human nuclear antigen defined by the monoclonal antibody Ki-67. J Immunol *133:* 1710-1715, 1984.
249. Hoshino T, Wilson CB: Cell kinetic analyses of human malignant brain tumours (gliomas). Cancer *44:* 956-962, 1979.
250. Kirkpatrick JP, Marks LB: Modeling killing and repopulation kinetics of subclinical cancer: direct calculations from clinical data. Int J Radiat Oncol Biol Phys *58(2):* 641-654, 2004.
251. Knobler RL, Lublin FD, Streletz LJ, Zimmer M, Joseph J, D'Imperio C, Northrup B, Barolat G, Marcus SG: Intracerebral beta-interferon in brain tumor therapy. Monitoring cerebral function with compressed spectral analysis. Ann NY Acad Sci *540:* 573-575, 1988.
252. Hoshino T, Townsend JJ, Muraoka I, Wilson CB: An autoradiographic study of human gliomas: growth kinetics of anaplastic astrocytoma and glioblastoma multiforme. Brain *103:* 967-984, 1980.

253. Crafts DC, Hoshino T, Wilson CB: Current status of population kinetics in gliomas. Bull Cancer. *64:* 115-124, 1977.
254. Gerdes J: An immunohistological method for estimating cell growth fractions in rapid histopathological diagnosis during surgery. Int J Cancer *35:* 169-171, 1985.
255. Gerdes J, Lemke H, Baisch H, Wacker H, Schwab U, Stein H: Cell cycle analysis of a cell proliferation associated human nuclear antigen defined by the monoclonal antibody Ki-67. J Immunol *133:* 1710-1715, 1984.
256. Gerdes J, Schwab U, Lemke H, Stein H: Production of a mouse monoclonal antibody reactive with a human nuclear antigen associated with cell proliferation. Int J Cancer *31:* 13-20, 1983.
257. Burger PC, Shibata T, Kleihues P: The use of the monoclonal antibody Ki-67 in the identification of proliferating cells: application to surgical pathology. Am J Surg Pathol *10:* 611-617, 1986.
258. Ostertag CB, Volk B, Shibata T, Burger P, Kleihues P: The monoclonal antibody Ki-67 as a marker for proliferating cells in sterotactic biopsies of brain tumours. Acta Neurochirurg (Wien) *89:* 117-121, 1987.
259. Giangaspero F, Doglioni C, Rivano MT, Pileri S, Gerdes J, Stein H: Growth fraction in human brain tumors defined by the monoclonal antibody Ki-67. Acta Neuropathol (Berlin) *74:* 179-182, 1987.
260. Zuber P, Hamou MF, De Tribolet N: Identification of proliferating cells in human gliomas using the monoclonal antibody Ki-67. Neurosurgery *22:* 364-368, 1988.
261. Roggendorf W, Schuster T, Peiffer J: Proliferative potential of meningiomas determined with the monoclonal antibody Ki-67. Acta Neuropathol (Berlin) *73:* 361-364, 1987.
262. Pileri S, Gerdes J, Rivano M, Tazzari PL, Magnani M, Gobbi M, Stein H: Immunohistochemical determination of growth fractions in human permanent cell lines and lymphoid tumors: a critical comparison of the monoclonal antibodies OKT9 and Ki-67. Brit J Haematol *65:* 271-276, 1987.
263. Lloyd RV, Wilson BS, Varani J, Gaur PK, Moline S, Makari JG: Immunocytochemical characterization of a monoclonal antibody that recognizes mitosing cells. Amer J Pathol *121:* 275-283, 1985.
264. Klein G, Steiner M, Wiener F, Klein E: Human leukemia-associated anti-nuclear reactivity. Proc Natl Acad Sci USA *71:* 685-689, 1974.
265. Boker DK, Stark HJ: The proliferation rate of intracranial tumors as defined by the monoclonal antibody KI 67. Application of the method to paraffin embedded specimens. Neurosurg Rev *11:* 267-272, 1988.
266. Murry AW, Kirschner M: Cyclin synthesis drives the early embryonic cell cycle. Nature *339:* 275-280, 1989.
267. Hall FL, Braun RK, Mihara K, Fung YK, Berndt N, Carbonaro-Hall DA, Vulliet PR: Characterization of the cytoplasmic proline-directed protein kinase in proliferative cells and tissues as a heterodimer comprised of p34cdc2 and p58cyclin A. J Biol Chem *266:* 17430-17440, 1991.
268. Doi T, Morita T, Wakabayashi N, Sumi T, Iwai SA, Amekawa S, Sakuda M, Nishimune Y: Induction of instability of p34(cdc2) expression by treatment with cisplatin (CDDP) in mouse teratocarcinoma F9 cells. Cancer Lett *176:* 75-80, 2002.
269. Poggioli GJ, Dermody TS, Tyler KL: Reovirus-induced sigma1s-dependent G2/M phase cell cycle arrest is associated with inhibition of p34(cdc2). J Virol *75:* 7429-7434, 2001.

270. Kanatsu-Shinohara M, Schultz RM, Kopf GS: Acquisition of meiotic competence in mouse oocytes: absolute amounts of p34(cdc2), cyclin B1, cdc25C, and wee1 in meiotically incompetent and competent oocytes. Biol Reprod *63:* 1610-1616, 2000.
271. John S, Workman JL: Bookmarking genes for activation in condensed mitotic chromosomes. Bioessays *20:* 275-279, 1998.
272. Gatter KC, Alcock C, Heryet A, Mason DY: Clinical importance of analysing malignant tumours of uncertain origin with immunohistological techniques. Lancet *1:* 1302-1305, 1985.
273. Morimura T, Kitz K, Budka H: *In situ* analysis of cell kinetics in human brain tumours. Acta Neuropathol *77:* 276-282, 1989.
274. Robbins BA, de la Vega D, Ogata K, Tan EM, Nakamura RM: Immunohistochemical detection proliferating cell nuclear antigen in solid human malignancies. Arch Pathol Lab Med *111:* 841-845, 1987.
275. Hoshino T: A commentary on the biology and growth kinetics of low-grade and high-grade gliomas. J Neurosurg *61:* 895-900, 1984.
276. Schlote W, Lang C, Mobius HJ: Growth fraction and growth pattern of neuroectodermal tumors as determined with the monoclonal antibody Ki-67. Abstract, Clin Neuropathol *7:* 207, 1988.
277. Deckert M, Reifenberger G, Wechsler W: Determination of the proliferative potential of human brain tumors using the monoclonal antibody Ki-67. J Cancer Res Clin Oncol *115:* 179-188, 1989.
278. Detta A, Hitchcock E: Rapid estimation of the proliferating index of brain tumours. J Neuro-Oncol *8:* 245-253, 1990.
279. Ohno S, Nishi T, Kojima Y, Haraoka J, Ito H, Mizuguchi J: Combined stimulation with interferon α and retinoic acid synergistically inhibits proliferation of the glioblastoma cell line GB12. Neurol Res *24:* 697-704, 2002.
280. Wilson CB, Hoshino T, Barker M, Downey R: Kinetics of gliomas in rat and man. Prog Exp Tumor Res *17:* 363-372, 1972.
281. Zatterstrom UK, Kallen A, Wennerberg J: Cell cycle time, growth fraction and cell loss in xenografted head and neck cancer. In Vivo *5:* 137-142, 1991.
282. Stahli C, Staehelin T, Miggiano V, Schmidt J, Haring P: High frequencies of antigen-specific hybridomas: dependence on immunization parameters and prediction by spleen cell analysis. J Immunol Methods *32:* 297-304, 1980.
283. Nishizaki T, Orita T, Saiki M, Furutani Y, Aoki H: Cell kinetics studies of human brain tumours by *in vitro* labelling using anti-BUdR monoclonal antibody. J Neurosurg *69:* 371-374, 1988.
284. Hoshino T, Nagashima T, Cho KG, Murovic JA, Hodes JE, Wilson CB, Edwards MS, Pitis LH: S-phase fraction of human brain tumors *in situ* measured by uptake of bromodeoxyuridine. Int J Cancer *38:* 369-374, 1986.
285. Murovic JA, Nagashima T, Hoshino T, Edwards MS, Davis RL: Pediatric central nervous system tumors: a cell kinetic study with bromodeoxyuridine. Neurosurgery *19:* 900-904, 1986.
286. Silverman CL, Simpson JR: Cerebellar medulloblastoma: the importance of posterior fossa dose to survival and patterns of failure. Int J Radiat Oncol Biol Phys *8:* 1869-1876, 1982.
287. Goz B: The effects of incorporation of 5-halogenated deoxyuridines into the DNA of eukaryotic cells. Pharmacol Rev *29:* 249-272, 1977.
288. Gratzner HG: Monoclonal antibody to 5-bromo- and 5-iododeoxyuridine: A new reagent for detection of DNA replication. Science *218:* 474-475, 1982.

289. Sacchi S, Donelli A, Cocconcelli P, Emilia G, Messerotti A, Piccinini L, Selleri L, Torelli G, Rinaldi G, Torelli U: Monoclonal antibody to 5-bromodeoxyuridine: a sensitive and rapid method for estimating the amount of S-phase cells. In: Biotechnology in Diagnostics (eds. by Koprowsky H, Ferrone S, Albertini A,), pp. 65-70, Elsevier Science Publishers, Amsterdam, 1985.
290. Assietti R, Butti G, Magrassi L, Danova M, Riccardi A, Gaetani P: Cell-kinetic characteristics of human brain tumors. Oncology *47:* 344-351, 1990.
291. Raza A, Preisler HD, Mayers GL, Bankert R: Rapid enumeration of S-phase cells by means of monoclonal antibodies. New Engl J Med *310:* 991, 1984.
292. Pollack IF, Campbell JW, Hamilton RL, Martinez AJ, Bozik ME: Proliferation index as a predictor of prognosis in malignant gliomas of childhood. Cancer *79:* 849-856, 1997.
293. Hamada K, Kuratsu J, Saitoh Y, Takeshima H, Nishi T, Ushio Y: Expression of tissue factor in glioma. Noshuyo Byori *13:* 115-118, 1996.
294. Reyes-Mugica M, Rieger-Christ K, Ohgaki H, Ekstrand BC, Helie M, Kleinman G, Yahanda A, Fearon ER, Kleihues P, Reale MA: Loss of DCC expression and glioma progression. Cancer Res *57:* 382-386, 1997.
295. Zurawel RH, Chiappa SA, Allen C, Raffel C: Sporadic medulloblastomas contain oncogenic β-catenin mutations. Cancer Res *58:* 896-899, 1998.
296. Janmaat ML, Giaccone G: Small-molecule epidermal growth factor receptor tyrosine kinase inhibitors. Oncologist *8:* 576-586, 2003.
297. Carpenter G, Cohen S: Epidermal growth factor. Ann Rev Biochem *48:* 193-216, 1979.
298. Gregory H: Isolation and structure of urogastrone and its relationship to epidermal growth factor. Nature *257:* 325-327, 1975.
299. Ullrich A, Schlessinger J: Signal transduction by receptors with tyrosine kinase activity. Cell *61:* 203-212, 1990.
300. Prigent SA, Lemoine NR: The type I (EGFR-related) family of growth factor receptors and their ligands. Prog Growth Factor Res *4:* 1-24, 1992.
301. Plowman GD, Culouscou J-M, Whitney GS, Green JM, Carlton GW, Foy L, Neubauer MG, Shoyab M: Ligand-specific activation of HER-4/p180erbB4, a fourth member of the epidermal growth factor receptor family. Proc Natl Acad Sci USA *90:* 1746-1750, 1993.
302. Plowman GD, Green JM, Culouscou J-M, Carlton GW, Rothwell VM, Sharon B: Heregulin induces tyrosine phosphorylation of HER-4/p180erbB-4. Nature *366:* 473-475, 1993.
303. Zelada-Hedman M, Werer G, Collins P, Backdahl M, Perez I, Franco S, Jimenez J, Cruz J, Torroella M, Nordenskjold M, Skoog L, Lindblom A: High expression of the EGFR in fibroadenomas compared to breast carcinomas. Anticancer Res *14:* 1679-1688, 1994.
304. Sahin AA: Biologic and clinical significance of HER-2/neu (cerbB-2) in breast cancer. Adv Anat Pathol *7:* 158-166, 2000.
305. Downward J, Yarden Y, Mayes E, Scrace G, Totty N, Stockwell P, Ullrich A, Schlessinger J, Waterfield MD: Close similarity of epidermal growth factor receptor and v-erbB oncogene protein sequences. Nature *307:* 521-527, 1984.
306. Pinkas-Kramarski R, Soussan L, Waterman H, Levkowitz G, Alroy I, Klapper L, Lavi S, Seger R, Ratzkin BJ, Sela M, Yarden Y: Diversification of Neu differentiation factor and epidermal growth factor signaling by combinatorial receptor interactions. EMBO J *15:* 2452-2467, 1996.
307. Stoker MGP, Pigott D, Taylor-Papadimitrious J: Response to epidermal growth factor of cultured human mammary epithelial cells from benign tumours. Nature *264:* 764-767, 1976.

308. Coleman S, Silberstein GB, Daniel CW: Ductal morphogenesis in the mouse mammary gland: evidence supporting a role for epidermal growth factor. Dev Biol *127:* 304-315, 1988.
309. Combes RC, Barret-Lee P, Luqmani Y: Growth factor expression in breast tissue. J Steroid Biochem Mol Biol *37:* 833-836, 1990.
310. Derynck R: Transforming growth factor-beta. Cell *54:* 593-595, 1988.
311. Gottlieb AB, Chang CK, Posnett DN, Fanelli B, Tam JP: Detection of transforming growth factor-beta in normal, malignant, and hyperproliferative human keratinocytes. J Exp Med *167:* 670-675, 1988.
312. Elder JT, Fisher GJ, Lindquist PB, Bennett GL, Pittelkow MR, Coffey RJ Jr, Ellingsworth L, Derynck R, Voorhees JJ: Overexpression of transforming growth factor β in psoriatic epidermis. Science *243:* 811-814, 1989.
313. Rajkumar T, Gullick WJ: A monoclonal antibody to the human c-erbB3 protein stimulates the anchorage-independent growth of breast cancer cell lines. Br J Cancer *70:* 459-465, 1994.
314. Toyoda H, Komurasaki T, Uchida D, Takayama Y, Isobe T, Okuyama T, Hanada K: Epiregulin. A novel epidermal growth factor with mitogenic activity for rat primary hepatocytes. J Biol Chem *270:* 7495-7500, 1995.
315. Damjanov I, Mildner B, Knowles BB: Immunohistochemical localization of the epidermal growth factor receptor in normal human tissues. Lab Invest *55:* 588-592, 1986.
316. Maguire HC, Green MI: The neu (c-erbB-2) oncogene. Semin Oncol *16:* 148-155, 1989.
317. Lupu R, Colomer R, Zugmaier G, Sarup J, Shepard M, Slamon D, Lippman ME: Direct interaction of a ligand for erbB2 oncogene product with the EGF receptor and p185erbB2. Science *249:* 1552-1555, 1990.
318. Volas GH, Leitzel K, Teramoto Y, Grossberg H, Demers L, Lipton A: Serial serum c-erbB-2 levels in patients with breast carcinoma. Cancer *78:* 267-272, 1996.
319. Dougall WC, Qian X, Peterson NC, Miller MJ, Samanta A, Greene MI: The neu-oncogene: signal transduction pathways, transformation mechanisms and evolving therapies. Oncogene *9:* 2109-2123, 1994.
320. Lee K-F, Simon H, Chen H, Bates B, Hung M-C, Hauser C: Requirement for neuregulin receptor erbB2 in neural and cardiac development. Nature *378:* 394-398, 1995.
321. Peles E, Bacus SS, Koski RA, Lu HS, Wen D, Ogden SG, Levy RB, Yarden Y: Isolation of the neu/HER-2 stimulatory ligand: a 44 kd glycoprotein that induces differentiation of mammary tumor cells. Cell *69:* 205-216, 1992.
322. Wen D, Peles E, Cupples R, Suggs SV, Bacus SS, Luo Y, Trail G, Hu S, Silbiger SM, Levy RB: Neu differentiation factor: a transmembrane glycoprotein containing an EGF domain and an immunoglobulin homology unit. Cell *69:* 559-572, 1992.
323. Holmes WE, Sliwkowski MX, Akita RW, Henzel WJ, Lee J, Park JW, Yansura D, Abadi N, Raab H, Lewis GD: Identification of heregulin, a specific activator of p185erbB2. Science *256:* 1205-1210, 1992.
324. Falls DL, Rosen KM, Cor*Fas* G, Lane WS, Fischbach GD: ARIA, a protein that stimulates acetylcholine receptor synthesis, is a member of the neu ligand family. Cell *72:* 801-815, 1993.
325. Marchionni MA, Goodearl AD, Chen MS, Bermingham-McDonogh O, Kirk C, Hendricks M, Danehy F, Misumi D, Sudhalter J, Kobayashi K: Glial growth factors are alternatively spliced erbB2 ligands expressed in the nervous system. Nature *362:* 312-318, 1993.

326. Carraway KL III, Sliwkowski MX, Akita R, Platko JV, Gy PM, Naijens A, Diamonti AJ, Vandlen RL, Cantley LC, Cerione RA: The erbB3 gene product is a receptor for heregulin. J Biol Chem *269:* 14303-14306, 1994.
327. Groenen LC, Nice EC, Burgess AW: Structure-function relationships for the EGF/TGF-β family of mitogens. Growth Factors *11:* 235-257, 1994.
328. Kraus MH, Issing W, Miki T, Popescu NC, Aaronson SA: Isolation and characterization of ERBB3, a third member of the ERBB/epidermal growth factor receptor family: evidence for overexpression in a subset of human mammary tumors. Proc Natl Acad Sci USA *86:* 9193-9197, 1989.
329. Fisher DA: Epidermal growth factor in the developing mammal. Mead Johnson Symp Perinat Dev Med *32:* 33-40, 1988.
330. Muhlhauser J, Crescimanno C, Kaufmann P, Hofler H, Zaccheo D, Castellucci M: Differentiation and proliferation patterns in human trophoblast revealed by c-erbB-2 oncogene product and EGF-R. J Histochem Cytochem *41:* 165-173, 1993.
331. Cor*Fas* G, Rosen KM, Aratake H, Krauss R, Fischbach GD: Differential expression of ARIA isoforms in the rat brain. Neuron *14:* 103-115, 1995.
332. Meyer D, Birchmeier C: Multiple essential functions of neuregulin in development. Nature *378:* 386-390, 1995.
333. Gassmann M, Casagranda F, Orioli D, Simon H, Lai C, Klein R, Lemke G: Aberrant neural and cardiac development in mice lacking the ErbB4 neuregulin receptor. Nature *378:* 390-394, 1995.
334. Screpanti I, Scarpa S, Meco D, Bellaria D, Stuppia L, Frati L, Modesti A, Gulino A: Epidermal growth factor promotes a neural phenotype in thymic epithelial cells and enhances neuropoietic cytokine expression. J Cell Biol *130:* 183-192, 1995.
335. Gullick WJ: Prevalence of aberrant expression of the epidermal growth factor receptor in human cancers. Br Med Bull *47:* 87-98, 1991.
336. Lofts FJ, Gullick WJ: C-erbB2 amplification and overexpression in human tumors. *In:* Oncogenes and Hormones: Advances in Cellular and Molecular Biology of Breast Cancer., ed. RB Dickson and ME Lippman. Boston: GenesBoston, Kluwer Academic Publishers. pp. 161-179, 1991.
337. Salomon DS, Brandt R, Ciardiello F: Epidermal growth factor-related peptides and their receptors in human malignancies. Crit Rev Oncol Hematol *19:* 183-232, 1995.
338. Chow NH, Liu HS, Lee EI, Chang CJ, Chan SH, Cheng HL, Tzai TS, Lin JS: Significance of urinary epidermal growth factor and its receptor expression in human bladder cancer. Anticancer Res *17:* 1293-1296, 1997.
339. Fischer-Colbrie J, Witt A, Heinzl H, Speiser P, Czerwenka K, Sevelda P, Zeillinger R: EGFR and steroid receptors in ovarian carcinoma: comparison with prognostic parameters and outcome of patients. Anticancer Res *17:* 613-619, 1997.
340. Ke LD, Adler-Storthz K, Clayman GL, Yung AW, Chen Z: Differential expression of epidermal growth factor receptor in human head and neck cancers. Head Neck *20:* 320-327, 1998.
341. Grandis JR, Melhem MF, Barnes EL, Tweardy DJ: Quantitative immunohistochemical analysis of transforming growth factor-α and epidermal growth factor receptor in patients with squamous cell carcinoma of the head and neck. Cancer *78:* 1284-1292, 1996.
342. Radinsky R, Risin, Fan, Dong, Bielenberg, Bucana, Fidler: Level and function of epidermal growth factor receptor predict the metastatic potential of human colon carcinoma cells. Clin Cancer Res *1:* 19-31, 1995.

343. de Jong JS, van Diest PJ, van der Valk P, Baak JP: Expression of growth factors, growth-inhibiting factors, and their receptors in invasive breast cancer. II: Correlations with proliferation and angiogenesis. J Pathol *184:* 53-57, 1998.
344. Hackel PO, Zwick E, Prenzel N, Ullrich A: Epidermal growth factor receptors: critical mediators of multiple receptor pathways. Curr Opin Cell Biol *11:* 184-189, 1999.
345. Baselga J, Averbuch SD: ZD1839 ('Iressa') as an anticancer agent. Drugs *60 Suppl 1:* 33-40; discussion 41-42, 2000.
346. Nagane M, Coufal F, Lin H, Bogler O, Cavenee WK, Huang HJ: A common mutant epidermal growth factor receptor confers enhanced tumorigenicity on human glioblastoma cells by increasing proliferation and reducing apoptosis. Cancer Res *56:* 5079-5086, 1996.
347. Slamon DJ, Clark GM, Wong SG, Levin WJ, Ullrich A, McGuire WL: Human breast cancer: correlation of relapse and survival with amplification of the HER-2/neu oncogene. Science *235:* 177-182, 1987.
348. Varley JM, Swallow JE, Brammar WJ, Wittaker JL, Walker RA: Alterations to either c-erbB2 (neu) or c-myc proto-oncogenes in breast carcinomas correlate with poor short-term prognosis. Oncogene *1:* 423-430, 1987.
349. Rios MA, Marcias A, Perez R, Lage A, Skoog L: Receptors for epidermal growth factor and estrogen as predictors of relapse in patients with mammary carcinoma. Anticancer Res *8:* 173-176, 1988.
350. Tauchi K, Hori S, Osamura RY, Tokuda Y, Tajima T: Immunohistochemical studies on oncogene products (c-erbB-2, EGFR, c-myc) and estrogen receptor in benign and malignant breast lesions. With special reference to their prognostic significance in carcinoma. Virchows Arch A Pathol Anat Histopathol *416:* 65-73, 1989.
351. Moller P, Mechtersheimer G, Kaufmann M, Moldenhauer G, Momburg F, Mattfeldt T, Otto HF: Expression of epidermal growth factor receptor in benign and malignant primary tumors of the breast. Virchows Archiv A Pathol Anat Histopathol *414:* 157-164, 1989.
352. Parkes HC, Lillicrop K, Howell A, Craig RK: c-erbB2 mRNA expression in human breast tumours: comparison with c-erbB2 DNA amplification and correlation with prognosis. Br J Cancer *61:* 39-45, 1990.
353. Goldman R, Ben-Levy R, Peles E, Yarden Y: Heterodimerization of the erbB-1 and erbB-2 receptors in human breast carcinoma cells: a mechanism for receptor transregulation. Biochemistry *29:* 11024-11028, 1990.
354. Hainsworth PJ, Henderson MA, Stillwell RG, Bennett RC: Comparison of EGF-R, c-erbB-2 product and ras p21 immunohistochemistry as prognostic markers in primary breast cancer. Eur J Surg Oncol *17:* 9-15, 1991.
355. Allred DC, Clark GM, Tandon AK, Molina R, Torney DC, Osborne CK, Gilchrist KW, Mansour EG, Abeloff M, Eudey L, McGuire WL: HER2/neu in node-negative breast cancer: prognostic significance of overexpression influenced by the presence of *in situ* carcinoma. J Clin Oncol *10:* 599-605, 1992.
356. Gusterson BA, Gelber RD, Goldhirsch A, Price KN, Save-Soderborgh J, Anbazhagan R, Styles J, Rudenstam CM, Golouh R, Reed R, Martinez-Tello F, Tiltman A, Torhorst J, Grigolato P, Bettelheim R, Neville AM, Burki K, Castigione M, Collins J, Lindtner J, Senn HJ: Prognostic importance of c-erbB2 expression in breast cancer. J Clin Oncol *10:* 1049-1056, 1992.
357. Koenders PG, Beex LV, Kienhuis CB, Kloppenborg PW, Benraad TJ: Epidermal growth factor receptor and prognosis in human breast cancer: a prospective study. Breast Cancer Res Treat *25:* 21-27, 1993.

358. Klijn JGM, Berns PMJJ, Schmitz PIM, Foekens JA: The clinical significance of epidermal growth factor receptor (EGFR) in human breast cancer: a review on 5232 patients. Endocrine Rev *13:* 3-17, 1992.
359. Dittadi R, Donisi PM, Brazzale A, Cappellozza L, Bruscagnin G, Gion M: Epidermal growth factor receptor in breast cancer. Comparison with non-malignant breast tissue. Br J Cancer *67:* 7-9, 1993.
360. Allan SM, Fernando IN, Sandle J, Trott PA: Expression of the c-erbB-2 gene product as detected in cytologic aspirates in breast cancer. Acta Cytol *37:* 981-982, 1993.
361. Jardines L, Weiss M, Fowble B, Greene M: neu(c-erbB-2/HER2) and the epidermal growth factor receptor (EGFR) in breast cancer. Pathobiology *61:* 268-282, 1993.
362. Gramlich TL, Cohen C, Fritsch C, De Rose PB, Gansler T: Evaluation of c-erbB-2 amplification in breast carcinoma by differential polymerase chain reaction. Am J Clin Pathol *101:* 493-499, 1994.
363. Alimandi M, Romano A, Curia MC, Muraro R, Fedi P, Aaronson SA, Di Fiore PP, Kraus MH: Cooperative signaling of ErbB-3 and ErbB-2 in neoplastic transformation and human mammary carcinoma cells. Oncogene *10:* 1813-1821, 1995.
364. Ravdin PM, Chamness GC: The c-erbB-2 proto-oncogene as a prognostic and predictive marker in breast cancer: a paradigm for the development of other macromolecular markers. Gene *159:* 19-27, 1995.
365. Kreipe H, Feist H, Fischer L, Felgner J, Heidorn K, Mettler L, Parwaresch R: Amplification of c-myc but not of c-erbB-2 is associated with high proliferative capacity in breast cancer. Cancer Res *53:* 1956-1961, 1993.
366. Slamon DJ, Godolphin W, Jones LA, Holt JA, Wong SG, Keith DE, Levin WJ, Stuart SG, Udove J, Ullrich A, Press MF: Studies of the HER-2/neu proto-oncogene in human breast and ovarian cancer. Science *244:* 707-712, 1989.
367. Quinn CM, Ostrowski JL, Lane SA, Loney DP, Teasdale J, Benson FA: c-erbB-3 protein expression in human breast cancer: comparison with other tumour variables and survival. Histopathology *25:* 247-252, 1994.
368. Shintani S, Funayama T, Yoshihama Y, Alcalde RE, Matsumura T: Prognostic significance of ERBB3 overexpression in oral squamous cell carcinoma. Cancer Lett *95:* 79-83, 1995.
369. Simpson BJ, Weatherill J, Miller EP, Lessells AM, Langdon SP, Miller WR: c-erbB-3 protein expression in ovarian tumours. Br J Cancer *71:* 758-762, 1995.
370. Antoniotti S, Taverna D, Maggiora P, Sapei ML, Hynes NE, De Bortoli M: Oestrogen and epidermal growth factor down-regulate erbB-2 oncogene protein expression in breast cancer cells by different mechanisms. Br J Cancer *70:* 1095-1101, 1994.
371. Herlyn M: Molecular and cellular basis of melanoma. Austin, KG Landers Co, 1993.
372. Clark WH Jr, Elder DE, Guerry D, Epstein ME, Greene MH, van Horn M: A study of tumor progression: the precursor lesions of superficial spreading and nodular melanoma. Hum Pathol *15:* 1147-1165, 1984.
373. Greene MH, Clark WH Jr, Tucker MA, Elder DE, Kraemer KH, Guerry D 4th, Witmer WK, Thompson J, Matozzo I, Fraser MC: Acquired precursors of cutaneous malignant melanoma. The familial dysplastic nevus syndrome. N Engl J Med *312:* 91-97, 1985.
374. Bodey B, Kaiser HE, Goldfarb RH: Immunophenotypically varied cell subpopulations in primary and metastatic human melanomas. Monoclonal antibodies for diagnosis, detection of neoplastic progression and receptor directed immunotherapy. Anticancer Res *16:* 517-531, 1996.
375. Marquardt H, Todaro GJ: Human transforming growth factor. Production by melanoma cell line, purification, and initial characterization. J Biol Chem *257:* 5220-5225, 1982.

376. Delarco JE, Pigott DA, Lazarus JA: Ectopic peptides released by a human melanoma cell line that modulate the transformed phenotype. Proc Natl Acad Sci USA *82:* 5015-5019, 1985.
377. Anisowicz A, Bardwell L, Sager R: Constitutive overexpression of a growth-regulated gene in transformed Chinese hamster and human cells. Proc Natl Acad Sci USA *84:* 7188-7192, 1987.
378. Herlyn M, Clark WH, Rodeck U, Mancianti ML, Jambrosic J, Koprowski H: Biology of tumor progression in human melanocytes. Lab Invest *56:* 461-474, 1987.
379. Ellis DL, Nanney LB, King LE Jr: Increased epidermal growth factor receptors in seborrheic keratoses and acrochordons of patients with dysplastic nevus syndrome. J Am Acad Dermatol *23:* 1070-1077, 1990.
380. Bodey B, Bodey B Jr, Groger AM, Luck JV, Siegel SE, Taylor CR, Kaiser HE: Clinical and prognostic significance of the expression of the c-erbB-2 and c-erbB-3 oncoproteins in primary and metastatic malignant melanomas and breast carcinomas. Anticancer Res *17:* 1319-1330, 1997.
381. Seshadri R, Matthews C, Dobrovic A, Horsfall DJ: The significance of oncogene amplification in primary breast cancer. Int J Cancer *43:* 270-273, 1989.
382. Allred DC, O'Connell P, Fuqua AW: Biomarkers in early breast neoplasia. J Cell Biochem *17G:* 125-131, 1993.
383. Barnes DM: c-erbB-2 amplification in mammary carcinoma. J Cell Biochem *17G:* 132-138, 1993.
384. Symmans WF, Liu J, Knowles DM, Inghirami G: Breast cancer heterogeneity: evaluation of clonality in primary and metastatic lesions. Hum Pathol *26:* 210-216, 1995.
385. Hiesiger EM, Hayes RL, Pierz DM, Budzilovich GN: Prognostic relevance of epidermal growth factor receptor (EGF-R) and c-neu/erbB2 expression in glioblastomas (GBMs). J Neurooncol *16:* 93-104, 1993.
386. Yu D, Wang SS, Dulski KM, Tsai CM, Nicolson GL, Hung MC: c-erbB-2/neu overexpression enhances metastatic potential of human lung cancer cells by induction of metastasis-associated properties. Cancer Res *54:* 3260-3266, 1994.
387. Tsugawa K, Fushida S, Yonemura Y: Amplification of the c-erbB-2 gene in gastric carcinoma: correlation with survival. Oncology *50:* 418-425, 1993.
388. Swanson PE, Frierson HF JR, Wick MR: c-erbB-2 (HER2/neu) oncopeptide immunoreactivity in localized, high grade transitional cell carcinoma of the bladder. Mod Pathol *5:* 531-536, 1992.
389. Kuhn EJ, Kurnot RA, Sesterhenn IA, Chang EH, Moul JW: Expression of the c-erbB-2 (HER2/neu) oncoprotein in human prostatic carcinoma. J Urol *150:* 1427-1433, 1993.
390. Ross JS, Nazeer T, Church K, Amato C, Figge H, Rifkin MD, Fisher HA: Contribution of HER-2/neu oncogene expression to tumor grade and DNA content analysis in the prediction of prostatic carcinoma metastasis. Cancer *72:* 3020-3028, 1993.
391. Berchuck A, Rodriguez G, Kinney RB, Soper JT, Dodge RK, Clarke-Pearson DL, Bast RC Jr: Overexpression of HER-2/neu in endometrial cancer is associated with advanced stage disease. Am J Obstet Gynecol *164:* 15-21, 1991.
392. Hetzel DJ, Wilson TO, Keeney GL, Roche PC, Cha SS, Podartz KC: HER-2/neu expression: a major prognostic factor in endometrial cancer. Gynecol Oncol *47:* 179-185, 1992.
393. Reinartz JJ, George E, Lindgren BR, Niehans GA: Expression of p53, transforming growth factor β, epidermal growth factor receptor, and c-erbB-2 in endometrial carcinoma and correlation with survival and known predictors of survival. Hum Pathol *25:* 1075-1083, 1994.

394. Pisani AL, Barbuto DA, Chen D, Ramos L, Lagasse LD, Karlan BY: HER2-neu, p53, and DNA analyses as prognostic factors for survival in endometrial carcinoma. Obstet Gynecol *85:* 729-734, 1995.
395. Saffari B, Jones LA, El-Naggar A, Felix JC, George J, Press MF: Amplification and overexpression of HER-2/neu (c-erbB2) in endometrial cancers: correlation with overall survival. Cancer Res *55:* 5693-5698, 1995.
396. Ro J, El-Naggar A, Ro JY, Blick M, Fraschini F, Fritsche H, Hortobagyi G: c-erbB-2 amplification in node-negative breast cancer. Cancer Res *49:* 6941-6944, 1989.
397. Walker RA, Gullick WJ, Varley JM: An evaluation of immunoreactivity for c-erbB-2 protein as a marker of poor short-term prognosis in breast cancer. Br J Cancer *60:* 426-429, 1989.
398. Wright C, Angus B, Nicholson S, Sainsbury RC, Cairns J, Gullick WJ, Kelly P, Harris AL, Horne CHW: Expression of c-erbB-2 oncoprotein: a prognostic indicator in human breast cancer. Cancer Res *49:* 2087-2090, 1989.
399. Borg A, Baldetorp B, Ferno M, Killander D, Olsson H, Ryden S, Sigurdsson H: erbB-2 amplification in breast cancer with a high rate of proliferation. Oncogene *60:* 137-143, 1991.
400. Gullick WJ, Love SB, Wright C, Barnes DM, Gusterson B, Harris AL, Altman DG: c-erbB-2 protein overexpression in breast cancer is a risk factor in patients with involved and uninvolved lymph nodes. Br J Cancer *63:* 434-438, 1991.
401. Kallioniemi O-P, Holli K, Visakorpi T, Koivula T, Helin HH, Isola JJ: Association of c-erbB-2 oncogene overexpression with high rate of cell proliferation, increased risk for visceral metastasis and poor long-term survival in breast cancer. Int J Cancer *49:* 650-655, 1991.
402. Lovekin C, Ellis IO, Locker A, Robertson JF, Bell J, Nicholson R, Gullick WJ, Elston CW, Blamey RW: c-erbB-2 oncoprotein expression in primary and advanced breast cancer. Br J Cancer *63:* 439-443, 1991.
403. O'Reilly SM, Barnes DM, Camplejohn RS, Bartkova J, Gregory WM, Richards MA: The relationship between c-erbB-2 expression, S-phase fraction and prognosis in breast cancer. Br J Cancer *63:* 444-446, 1991.
404. Paterson MC, Dietrich KD, Danyluk J, Paterson AH, Lees AW, Jamil N, Hanson J, Jenkins H, Krause BE, McBlain WA, Slamon DJ, Fourney RM: Correlation between c-erbB-2 amplification and risk of early relapse in node-negative breast cancer. Cancer Res *51:* 556-567, 1991.
405. Rilke F, Colnaghi MI, Cascinelli N, Andreola S, Baldini MT, Bufalino R, Della Porta G, Menard S, Pierotti MA, Testori A: Prognostic significance of HER-2/neu expression in breast cancer and its relationship to other prognostic factors. Int J Cancer *49:* 44-49, 1991.
406. Winstanley J, Cooke T, Murray GD, Platt-Higgins A, George WD, Holt S, Myskov M, Spedding A, Barraclough BR, Rudland PS: The long term prognostic significance of c-erbB-2 in primary breast cancer. Br J Cancer *63:* 447-450, 1991.
407. Campani D, Sarnelli R, Fontanini G, Martini L, Cecchetti D, De Luca F, Squartini F: Receptor status, proliferating activity, and c-erbB2 oncoprotein. An immunocytochemical evaluation in breast cancer. Ann NY Acad Sci *698:* 167-173, 1993.
408. Press MF, Pike MC, Chazin VR, Hung G, Udove JA, Markowicz M, Danyluk J, Godolphin W, Sliwkowski M, Akita R, Brandeis J, Paterson MC, Slamon DJ: HER-2/neu expression in node-negative breast cancers: direct tissue quantitation by computerized image analysis and association of overexpression with increased risk of recurrent disease. Cancer Res *53:* 4960-4970, 1993.

409. Tsuchiya A, Katagata N, Kimijima I, Abe R: Immunohistochemical overexpression of c-erbB-2 in the prognosis of breast cancer. Surg Today *23:* 885-890, 1993.
410. Horiguchi J, Iino Y, Takei H, Yokoe T, Ishida T, Morishita Y: Immunohistochemical study on the expression of c-erbB-2 oncoprotein in breast cancer. Oncology *51:* 47-51, 1994.
411. Muss HB, Thor AD, Berry DA, Kute T, Liu ET, Koerner F, Cirrincione CT, Budman DR, Wood WC, Barcos M: c-erbB-2 expression and response to adjuvant therapy in women with node-positive early breast cancer. N Engl J Med *330:* 1260-1266, 1994.
412. Pechoux C, Chardonnet Y, Noel P: Immunohistochemical studies on c-erbB-2 oncoprotein expression in paraffin embedded tissues in invasive and non-invasive human breast lesions. Anticancer Res *14:* 1343-1360, 1994.
413. Zschiesche W, Schonborn I, Minguillon C, Spitzer E: Significance of immunohistochemical c-erbB-2 product localization pattern for prognosis of primary human breast cancer. Cancer Lett *81:* 89-94, 1994.
414. Szöllösi J, Balázs M, Feuerstein BG, Benz CC, Waldman FM: ERBB-2 (HER2/neu) gene copy number, p185HER-2 overexpression, and intratumor heterogeneity in human breast cancer. Cancer Res *55:* 5400-5407, 1995.
415. Ali IU, Campbell G, Lidereau R, Callahan R: Lack of evidence for the prognostic significance of c-erbB-2 amplification in human breast carcinoma. Oncogene Res *3:* 139-146, 1988.
416. Barnes DM, Lammie GA, Millis RR, Gullick WL, Allen DS, Altman DG: An immunohistochemical evaluation of c-erbB-2 expression in human breast carcinoma. Br J Cancer *58:* 448-452, 1988.
417. Gusterson BA, Machin LG, Gullick WJ, Gibbs NM, Powles TJ, Elliott C, Ashley S, Monaghan P, Harrison S: c-erbB-2 expression in benign and malignant breast disease. Br J Cancer *58:* 453-457, 1988.
418. Van De Vijver MJ, Peterse JL, Mooi WJ, Wisman P, Lomans J, Dalesio O, Nusse R: neu-protein overexpression in breast cancer. Association with comedo-type ductal carcinoma *in situ* and limited prognostic value in stage II breast cancer. N Engl J Med *319:* 1239-1245, 1988.
419. Zhou D-J, Ahuja H, Cline MJ: Proto-oncogene abnormalities in human breast cancer: c-erbB-2 amplification does not correlate with recurrence of disease. Oncogene *4:* 105-108, 1989.
420. Kury F, Sliutz G, Schemper M, Reiner G, Reiner A, Jakesz R, Wrba F, Zeillinger R, Knogler W, Huber J, Holzner H, Spona J: HER-2 oncogene amplification and overall survival of breast carcinoma patients. Eur J Cancer *26:* 946-949, 1990.
421. Richner J, Gerber HA, Locher GW, Goldhirsch A, Gelber RD, Gullick WJ, Berger MS, Groner B, Hynes NE: c-erbB-2 protein expression in node negative breast cancer. Ann Oncol *1:* 263-268, 1990.
422. Clark GM, McGuire WL: Follow-up study of HER-2/neu amplification in primary breast cancer. Cancer Res *51:* 944-948, 1991.
423. Press MF, Hung G, Godolphin W, Slamon DJ: Sensitivity of HER-2/neu antibodies in archival tissue samples: potential source of error in immunohistochemical studies of oncogene expression. Cancer Res *54:* 2771-2777, 1994.
424. Lemoine NR, Barnes DM, Hollywood DP, Hughes CM, Smith P, Dublin E, Prigent SA, Gullick WJ, Hurst HC: Expression of erbB3 gene product in breast cancer. Br J Cancer *66:* 1116-1121, 1992.
425. Poller DN, Spendlove I, Baker C, Church R, Ellis IO, Plowman GD, Mayer RJ: Production and characterisation of a polyclonal antibody to the c-erbB3 protein:

Examination of c-erbB3 protein expression in adenocarcinomas. J Pathol *168:* 275-280, 1992.
426. Sanidas EE, Filipe MI, Linehan J, Lemoine NR, Gullick WJ, Rajkumar T, Levison DA: Expression of the c-erbB3 gene product in gastric cancer. Int J Cancer *54:* 935-940, 1993.
427. Rajkumar T, Gooden CSR, Lemoine NR, Gullick WJ: Expression of the c-erbB3 protein in gastrointestinal tract tumours determined by monoclonal antibody RTJ1. J Pathol *170:* 271-278, 1993.
428. Lemoine NR, Lobresco M, Leung H, Barton C, Hughes CM, Prigent SA, Gullick WJ, Kloppel G: The erbB3 gene in human pancreatic cancer. J Pathol *168:* 269-273, 1992.
429. Arteaga CL: ErbB-targeted therapeutic approaches in human cancer. Exp Cell Res *284:* 122-130, 2003.
430. Gill S, Thomas RR, Goldberg RM: New targeted therapies in gastrointestinal cancers. Curr Treat Options Oncol *4:* 393-403, 2003.
431. Janmaat ML, Giaccone G: The epidermal growth factor receptor pathway and its inhibition as anticancer therapy. Drugs Today (Barc) *39 Suppl C:* 61-80, 2003.
432. Batinac T, Gruber F, Lipozencic J, Zamolo-Koncar G, Stasic A, Brajac I: Protein p53--structure, function, and possible therapeutic implications. Acta Dermatovenerol Croat *11:* 225-230, 2003.
433. Demonacos C, La Thangue NB: Drug discovery and the p53 family. Prog Cell Cycle Res *5:* 375-382, 2003.
434. Melino G, Lu X, Gasco M, Crook T, Knight RA: Functional regulation of p73 and p63: development and cancer. Trends Biochem Sci *28:* 663-670, 2003.
435. Alarcon RM, Rupnow BA, Graeber TG, Knox SJ, Giaccia AJ: Modulation of c-Myc activity and apoptosis *in vivo*. Cancer Res *56:* 4315-4319, 1996.
436. Marshall CJ: Tumor suppressor genes. Cell *64:* 313-326, 1991.
437. Bodey B, Bodey B Jr, Siegel SE: Tumor Suppressor Genes in Childhood Malignancies. A Review. Int J Pediatric Hematol/Oncol *6:* 47-64, 1998.
438. Lane DP, Crawford LV: T-antigen is bound to host protein in SV40-transformed cells. Nature *278:* 261-263, 1979.
439. Sarnow P, Ho YS, Williams J, Levine AJ: Adenovirus E1b-58kd tumor antigen and SV40 large tumor antigen are physically associated with the same 54 kd cellular protein in transformed cells. Cell *28:* 387-394, 1982.
440. Werness BA, Levine AJ, Howley PM: Association of human papillomavirus types 16 and 18 E6 proteins with p53. Science *248:* 76-79, 1990.
441. Bartek J, Bartkova J, Vojtesek B, Staskova Z, Lukas J, Rejthar A, Kovarik J, Midgley CA, Gannon JV, Lane DP: Aberrant expression of the p53 oncoprotein is a common feature of a wide spectrum of human malignancies. Oncogene *6:* 1699-1703, 1991.
442. Carbon De Fromentel C, Soussi T: TP53 tumor suppressor gene: a model for investigating human mutagenesis. Genes Chromosomes Cancer *4:* 1-15, 1992.
443. Greenblatt MS, Bennett WP, Hollstein M, Harris CC: Mutations in the p53 tumor suppressor genes: clues to cancer etiology and molecular pathogenesis. Cancer Res *54:* 4855-4878, 1994.
444. Hollstein M, Sidransky D, Vogelstein B, Harris C: p53 mutations in human cancers. Science *253:* 252-254, 1991.
445. Malkin D, Li FP, Strong LC, Fraumeni JF JR, Nelson CE, Kim DH, Kassel J, Gryka MA, Bischoff FZ, Tainsky MA: Germ line p53 mutations in a familial syndrome of breast cancer, sarcomas, and other neoplasms. Science *250:* 1233-1238, 1990.

446. Srivastava S, Zou ZQ, Pirollo K Blattner W, Chang EH: Germ-line transmission of a mutated p53 gene in cancer-prone family with Li-Fraumeni syndrome. Nature *348:* 747-749, 1990.
447. Gaidano G, Ballerini P, Gong JZ, Inghirami G, Neri A, Newcomb EW, Magrath IT, Knowles DM, Dalla-Favera R: p53 mutations in human lymphoid malignancies: Association with Burkitt lymphoma and chronic lymphocytic leukemia. Proc Natl Acad Sci USA *88:* 5413-5417, 1991.
448. Chen P, Iavarone A, Fick J, Edwards M, Prados M, Israel MA: Constitutional p53 mutations associated with brain tumors in young adults. Genet Cytogenet *82:* 106-115, 1995.
449. Kyritsis AP, Xu R, Bondy ML, Levin VA, Bruner JM: Correlation of p53 immunoreactivity and sequencing in patients with glioma. Molec Carcinogen *15:* 1-4, 1996.
450. Cho MY, Jung SH, Kim TS: p53 protein overexpression in astrocytic neoplasms. Yonsei Med J *36:* 521-526, 1995.
451. Ellison DW, Steart PV, Bateman AC, Pickering RM, Palmer JD, Weller RO: Prognostic indicators in a range of astrocytic tumours: an immunohistochemical study with Ki-67 and p53 antibodies. J Neurol Neurosurg Psych *59:* 413-419, 1995.
452. Kordek R, Biernat W, Alwasiak J, Maculewicz R, Yanagihara R, Liberski PP: p53 protein and epidermal growth factor receptor expression in human astrocytomas. J Neuro-Oncol *26:* 11-16, 1995.
453. Sarkar C, Ralte AM, Sharma MC, Mehta VS: Recurrent astrocytic tumours--a study of p53 immunoreactivity and malignant progression. Br J Neurosurg *16:* 335-342, 2002.
454. Lee CS, Pirdas A, Lee MW: p53 in cutaneous melanoma: immunoreactivity and correlation with prognosis. Australasian J Dermatol *36:* 192-195, 1995.
455. Bergman R, Shemer A, Levy R, Friedman-Birnbaum R, Trau H, Lichtig C: Immunohistochemical study of p53 protein expression in Spitz nevus as compared with other melanocytic lesions. Amer J Dermatopathol *17:* 547-550, 1995.
456. Sparrow LE, English DR, Heenan PJ, Dawkins HJ, Taran J: Prognostic significance of p53 over-expression in thin melanomas. Melanoma Res *5:* 387-392, 1995.
457. Weiss J, Heine M, Arden KC, Korner B, Pilch H, Herbst RA, Jung EG: Mutation and expression of TP53 in malignant melanomas. Recent Results Cancer Res *139:* 137-154, 1995.
458. Takahashi T, Nau MM, Chiba T, Birrer MJ, Rosenberg RK, Vinocour M, Levitt M, Pass H, Gazdar AF, Minna JD: p53: a frequent target for genetic abnormalities in lung cancer. Science *246:* 491-494, 1989.
459. Miller CW, Simon K, Aslo A, Kok Y, Yokota J, Buys CHCM, Terada M, Koeffler HP: p53 mutations in human lung tumors. Cancer Res *52:* 1695-1698, 1992.
460. Husgafvel-Pursiainen K, Ridanpaa M, Anttila S, Vainio H: p53 and ras gene mutations in lung cancer: implications for smoking and occupational exposures. J Occupat Environ Med *37:* 68-76, 1995.
461. Kawajiri K, Eguchi H, Nakachi K, Sekiya T, Yamamoto M: Association of CYP1A1 germ line polymorphisms with mutations of the p53 gene in lung cancer. Cancer Res *56:* 72-76, 1996.
462. Boers JE, Ten Velde GP, Thunnissen FB: p53 in squamous metaplasia: a marker for risk of respiratory tract carcinoma. Amer J Resp Crit Care Med *153:* 411-416, 1996.
463. Kondo K, Tsuzuki H, Sasa M, Sumimoto M, Uyama T, MondeN Y: The dose-response relationship between the frequency of p53 mutations and tobacco consumption in lung cancer patients. J Surg Oncol *61:* 20-26, 1996.

464. Tsai CM, Chang KT, Wu LH, Chen JY, Gazdar AF, Mitsudomi T, Chen MH, Pering RP: Correlations between intrinsic chemoresistance and HER-2/neu gene expression, p53 gene mutations, and cell proliferation characteristics in non-small cell lung cancer cell lines. Cancer Res *56:* 206-209, 1996.
465. Goldblum JR, Bartos RE, Carr KA, Frank TS: Hepatitis B and alterations of the p53 tumor suppressor gene in hepatocellular carcinoma. Am J Surg Pathol *17:* 1244-1251, 1993.
466. Ozturk M: p53 mutation in hepatocellular carcinoma after aflatoxin exposure. Lancet *338:* 1356-1359, 1991.
467. Hsu HC, Tseng HJ, Lai PL, Lee PH, Peng SY: Expression of p53 gene in 184 unifocal hepatocellular carcinomas: association with tumor growth and invasiveness. Cancer Res *53:* 4691-4694, 1993.
468. Tabor E: Tumor suppressor genes, growth factor genes, and oncogenes in hepatitis B virus-associated hepatocellular carcinoma. J Med Virol *42:* 357-365, 1994.
469. Campbell IG, Eccles DM, Dunn B, Davis M, Leake V: p53 polymorphism in ovarian and breast cancer. Lancet *347:* 393-394, 1996.
470. Horne GM, Anderson JJ, Tiniakos DG, McIntosh GG, Thomas MD, Angus B, Henry JA, Lennard TW, Horne CH: p53 protein as a prognostic indicator in breast carcinoma: a comparison of four antibodies for immunohistochemistry. Brit J Cancer *73:* 29-35, 1996.
471. Callahan R: p53 mutations, another breast cancer prognostic factor. J Natl Cancer Inst *84:* 826-827, 1992.
472. Harris AL: p53 expression in human breast cancer. Adv Cancer Res *59:* 69-88, 1992.
473. Eeles RA, Bartkova J, Lane DP, Bartek J: The role of TP53 in breast cancer development. Cancer Surv *18:* 57-75, 1993.
474. Elledge RM, Allred DC: The p53 tumor suppressor gene in breast cancer. Breast Cancer Res Treat *32:* 39-47, 1994.
475. Karameris AM, Worthy E, Gorgoulis VG, Quezado M, Anastassiades OT: p53 gene alterations in special types of breast carcinoma: a molecular and immunohistochemical study in archival material. J Pathol *176:* 361-372, 1995.
476. Ozbun MA, Butel JS: Tumor suppressor p53 mutations and breast cancer: a critical analysis. Adv Cancer Res *66:* 71-141, 1995.
477. Kovach JS, Hartmann A, Blaszyk H, Cunningham J, Schaid D, Sommer SS: Mutation detection by highly sensitive methods indicates that p53 gene mutations in breast cancer can have important prognostic value. Proc Natl Acad Sci USA *93:* 1093-1096, 1996.
478. Liu B, Sun D, Xia W, Hung MC, Yu D: Cross-reactivity of C219 anti-p170(mdr-1) antibody with p185(c-erbB2) in breast cancer cells: cautions on evaluating p170(mdr-1). J Natl Cancer Inst *89:* 1524-1529, 1997.
479. Nigro JM, Baker SJ, Preisinger AC, Jessup JM, Hostetter R, Clearly K, Bigner SH, Davidson N, Baylin S, Devilee P, Glover T, Collins FS, Weston A, Modali R, Harris CC, Vogelstein B: Mutations in the p53 gene occur in diverse human tumour types. Nature *342:* 705-708, 1989.
480. Lane DP: p53, guardian of the genome. Nature *358:* 15-16, 1992.
481. Kern S, Kinzler KW, Bruskin A, Jarosz D, Friedman P, Prives C, Vogelstein B: Identification of p53 as a sequence-specific DNA-binding protein. Science *252:* 1708-1711, 1991.
482. Bargonetti J, Friedman PN, Kern SE, Vogelstein B, Prives C: Wild-type but not mutant p53 immunopurified proteins bind to sequences adjacent to the SV40 origin of replication. Cell *65:* 1083-1091, 1991.
483. Marx J: New link found between p53 and DNA repair. Science *266:* 1321-1322, 1994.

484. Bischoff JR, Friedman PN, Marshak DR, Prives C, Beach D: Human p53 is phosphorylated by p60-cdc2 and cyclin B-cdc2. Proc Natl Acad Sci USA *87:* 4766-4770, 1990.
485. Meek DW, Simon S, Kikkawa U, Eckhart W: The p53 tumor suppressor protein is phosphorylated ar serine 389 by casein kinase II. EMBO J *9:* 3253-3260, 1990.
486. Moll UM, Ostermeyer AG, Haladay R, Winkfield B, Frazier M, Zambetti G: Cytoplasmic sequestration of wild-type p53 protein impairs the G1 checkpoint after DNA damage. Mol Cell Biol *16:* 1126-1137, 1996.
487. Nikolaev AY, Li M, Puskas N, Qin J, Gu W: PARC: a cytoplasm anchor for p53. Cell *112:* 1-2, 2003.
488. Douc-Rasy S, Benard J: A new view on p53 protein cytoplasmic sequestration. Bull Cancer *90:* 380-382, 2003.
489. Nikolaev AY, Gu W: PARC: a potential target for cancer therapy. Cell Cycle *2:* 169-171, 2003.
490. Fridman JS, Lowe SW: Control of apoptosis by p53. Oncogene *22:* 9030-9040, 2003.
491. SuN Y, Nakamura K, Wendel E, Colburn NH: Progression toward tumor cell phenotype is enhanced by overexpression of a mutant p53 tumor suppressor gene isolated from nasopharyngeal carcinoma. Proc Natl Acad Sci USA *90:* 2827-2831, 1993.
492. Bodey B, Gröger AM, Bodey B Jr, Siegel SE, Kaiser HE: Immunocytochemical detection of p53 protein overexpression in primary human osteosarcomas. Anticancer Res *17:* 493-498, 1997.
493. Eeles RA, Warren W, Knee G, Bartek J, Averill D, Stratton MR, Blake PR, Tait DM, Lane DP, Easton DF: Constitutional mutation in exon 8 of the p53 gene in a patient with multiple primary tumours: molecular and immunohistochemical findings. Oncogene *8:* 1269-1276, 1993.
494. Nose H, Imazeki F, Ohto M, Omata M: p53 gene mutations and 17p allelic deletions in hepatocellular carcinoma from Japan. Cancer *72:* 355-360, 1993.
495. Renault B, van den Broek M, Fodde R, Wijnen J, Pellegata NS, Amadori D, Khan PM, Ranzani GN: Base transitions are the most frequent genetic changes at p53 in gastric cancer. Cancer Res *53:* 2614-2617, 1993.
496. Iggo R, Gatter K, Bartek J, Lane D, Harris AL: Increased expression of mutant forms of p53 oncogene in primary lung cancer. Lancet *335:* 675-679, 1990.
497. John J, Frech M, Wittinghofer A: Biochemical properties of Ha-ras encoded p21 mutants and mechanism of the autophosphorylation. J Biol Chem *263:* 11792-11799, 1988.
498. Mazur M, Glickman BW: Sequence specificity of mutations induced by benzo[a]pyrene-7,8-diol-9,10-epoxide at endogenous aprt gene in CHO cells. Somat Cell Mol Genet *14:* 393-400, 1988.
499. Stüzbecher HW, Chumakov P, Welch WJ, Jenkins JR: Mutant p53 proteins bind hsp 72.73 cellular heat shock-related proteins in SV40-transformed monkey cells. Oncogene *1:* 201-211, 1987.
500. Finlay CA, Hinds PW, Tan T-H, Eliyahu D, Oren M, Levine AJ: Activating mutations for transformation by p53 produce a gene product that forms an hsc70-p53 complex with an altered half-life. Mol Cell Biol *8:* 531-539, 1988.
501. von Deimling A, Louis DN, Wiestler OD: Molecular pathways in the formation of gliomas. Glia *15:* 328-338, 1995.
502. von Deimling A, Bender B, Jahnke R, Waha A, Kraus J, Albrecht S, Welenreuther R, Faßbender F, Nagel J, Menon AG, Louis DN, Lenartz DD, Schramm J, Wiestler OD:

Loci associated with malignant progression in astrocytomas: A candidate on chromosome 19q. Cancer Res *54:* 1397-1401, 1994.
503. Fults D, Petronio J, Noblett BD, Pedone CA: Chromosome 11p15 deletions in human malignant astrocytomas and primitive neuroectodermal tumors. Genomics *14:* 799-801, 1992.
504. James CD, Carlblom E, Dumanski JP, Hansen M, Nordenskjold M, Collins VP, Cavenee WK: Clonal genomic alterations in glioma malignancy stages. Cancer Res *48:* 5546-5551, 1988.
505. James CD, He J, Carlblom E, Nordenskjold M, Cavenee WK, Collins VP: Chromosome 9 deletion mapping reveals interferon α and interferon β-1 gene deletions in human glial tumors. Cancer Res *51:* 1684-1688, 1991.
506. Olopade OI, Buchhagen DL, Malik K, Sherman J, Nobori T, Bader S, Nau MM, Gazdar AF, Minna JD, Diaz MO: Homozygous loss of the interferon genes defines the critical region on 9p that is deleted in lung cancers. Cancer Res *53:* 2410-2415, 1993.
507. Ransom DT, Ritland SR, Kimmel DW, Moertel CA, Dahl RJ, Scheithauer BW, Kelly PJ, Jenkins BR: Cytogenetic and loss of heterozygosity studies in ependymoma, pilocytic astrocytoma and oligodendrogliomas. Genes Chromosom Cancer *5:* 348-356, 1992.
508. Venter DJ, Bevan KL, Ludwig RL, Riley TEW, Jat PS, Thomas DGT, Noble MD: Retinoblastoma gene deletions in human glioblastomas. Oncogene *6:* 445-448, 1991.
509. von Deimling A, Eibl RH, Ohgaki H, Louis DN, von Ammon K, Petersen I, Kleihues P, Chung RY, Wiestler OD, Seizinger BR: p53 mutations are associated with 17p allelic loss in grade II and grade III astrocytoma. Cancer Res *52:* 2987-2990, 1992.
510. Liu T, Yan H, Kuismanen S, Percesepe A, Bisgaard ML, Pedroni M, Benatti P, Kinzler KW, Vogelstein B, Ponz de Leon M, Peltomaki P, Lindblom A: The role of hPMS1 and hPMS2 in predisposing to colorectal cancer. Cancer Res *61:* 7798-7802, 2001.
511. Trojan J, Zeuzem S, Randolph A, Hemmerle C, Brieger A, Raedle J, Plotz G, Jiricny J, Marra G: Functional Analysis of hMLH1 Variants and HNPCC-Related Mutations Using a Human Expression System. Gastroenterology *122:* 211-219, 2002.
512. Deng G, Chen A, Pong E, Kim YS: Methylation in hMLH1 promoter interferes with its binding to transcription factor CBF and inhibits gene expression. Oncogene *20:* 7120-7127, 2001.
513. Muller-Koch Y, Kopp R, Lohse P, Baretton G, Stoetzer A, Aust D, Daum J, Kerker B, Gross M, Dietmeier W, Holinski-Feder E: Sixteen rare sequence variants of the hMLH1 and hMSH2 genes found in a cohort of 254 suspected HNPCC (hereditary non-polyposis colorectal cancer) patients: mutations or polymorphisms? Eur J Med Res *6:* 473-482, 2001.
514. Shin KH, Shin JH, Kim JH, Park JG: Mutational Analysis of Promoters of Mismatch Repair Genes hMSH2 and hMLH1 in Hereditary Nonpolyposis Colorectal Cancer and Early Onset Colorectal Cancer Patients: Identification of Three Novel Germ-line Mutations in Promoter of the hMSH2 Gene. Cancer Res *62:* 38-42, 2002.
515. Hussein MR, Roggero E, Sudilovsky EC, Tuthill RJ, Wood GS, Sudilovsky O: Alterations of mismatch repair protein expression in benign melanocytic nevi, melanocytic dysplastic nevi, and cutaneous malignant melanomas. Am J Dermatopathol *23:* 308-314, 2001.
516. Yeh CC, Lee C, Dahiya R: DNA mismatch repair enzyme activity and gene expression in prostate cancer. Biochem Biophys Res Commun *285:* 409-413, 2001.
517. Chung TK, Cheung TH, Wang VW, Yu MY, Wong YF: Microsatellite instability, expression of hMSH2 and hMLH1 and HPV infection in cervical cancer and their clinico-pathological association. Gynecol Obstet Invest *52:* 98-103, 2001.

518. Peiro G, Diebold J, Mayr D, Baretton GB, Kimmig R, Schmidt M, Lohrs U: Prognostic relevance of hMLH1, hMSH2, and BAX protein expression in endometrial carcinoma. Mod Pathol *14:* 777-783, 2001.
519. Aubry MC, Halling KC, Myers JL, Tazelaar HD, Yang P, Thibodeau SN: DNA mismatch repair genes hMLH1, hMSH2, and hMSH6 are not inactivated in bronchioloalveolar carcinomas of the lung. Cancer *92:* 2898-2901, 2001.
520. Wang L, Bani-Hani A, Montoya DP, Roche PC, Thibodeau SN, Burgart LJ, Roberts LR: hMLH1 and hMSH2 expression in human hepatocellular carcinoma. Int J Oncol *19:* 567-570, 2001.
521. Derradji H, Baatout S: Apoptosis: a mechanism of cell suicide. In Vivo *17:* 185-192, 2003.
522. Xerri L, Devilard E, Ayello C, Brousset P, Reed JC, Emile JF, Hassoun J, Parmentier S, Birg F: Cysteine protease CPP32, but not Ich1-L, is expressed in germinal center B cells and their neoplastic counterparts. Hum Pathol *28:* 912-921, 1997.
523. Srinivasan A, Roth KA, Sayers RO, Shindler KS, Wong AM, Fritz LC, Tomaselli KJ: *In situ* immunodetection of activated caspase-3 in apoptotic neurons in the developing nervous system. Cell Death Differ *5:* 1004-1016, 1998.
524. Kopper L: Apoptozis es a daganatok. Magyar Onkologia *47:* 123-131, 2003.
525. Ockner RK: Apoptosis and liver diseases: recent concepts of mechanism and significance. J Gastroenterol Hepatol *16:* 248-260, 2001.
526. Kerr JF, Wyllie AH, Currie AR: Apoptosis: a basic biological phenomenon with wide-ranging implications in tissue kinetics. Br J Cancer *26:* 239-257, 1972.
527. Kuan CY, Roth KA, Flavell RA, Rakic P: Mechanisms of programmed cell death in the developing brain. Trends Neurosci *23:* 291-297, 2000.
528. Arends MJ, Wyllie AH: Apoptosis: Mechanisms and roles in pathology. Int Rev Exp Pathol *32:* 223-254, 1991.
529. Patel T, Gores GJ, Kaufmann SH: The role of proteases during apoptosis. *FAS*EB J *10:* 587-597, 1996.
530. Ellis RE, Jacobson DM, Horvitz HR: Genes required for the engulfment of cell corpses during programmed cell death in *Caenorhabditis elegans*. Genetics *129:* 79-94, 1991.
531. Yuan J, Shaham S, Ledoux S, Ellis HM, Horvitz HR: The *C. elegans* death gene ced-3 encodes a protein similar to mammalian interleukin-1 β-converting enzyme. Cell *75:* 641-652, 1993.
532. Hengartner MO Horvitz HR: *C. elegans* cell survival gene ced-9 encodes a functional homologue of the mammalian protooncogene bcl-2. Cell *76:* 665-676, 1994.
533. Wang S, El-Deiry WS: TRAIL and apoptosis induction by TNF-family death receptors. Oncogene *22:* 8628-8633, 2003.
534. Mischak R: Assessment of caspase activity: synthetic substrates and inhibitors. Bioconcepts, *9.2*: 1-20, 2003.
535. Thornberry NA, Rano TA, Peterson EP, Rasper DM, Timkey T, Garcia-Calvo M, Houtzager VM, Nordstrom PA, Roy S, Vaillancourt JP, Chapman KT, Nicholson DW: A combinatorial approach defines specificities of members of the caspase family and granzyme B. J Biol Chem *272:* 17907-17911, 1997.
536. Chang HY, Yang X: Proteases for cell suicide: Function and regulation of caspases. Mol Biol Rev *64:* 821-846, 2000.
537. Earnshaw WC, Martins LM, Kaufmann SH: Mammalian caspases: structure, activation, substrates and functions during apoptosis. Ann Rev Biochem *68:* 383-424, 1999.
538. Hengartner MO: The biochemistry of apoptosis. Nature *407:* 769-776, 2000.
539. Kohler C, Orrenius S, Zhivotvsky B: Evaluation of caspase activity in apoptotic cells. J Immunol Methods *265:* 97-110, 2002.

540. Ravagnan L, Roumier T, Kroemer G: Mitochondria, the killer organelles and their weapons. J Cell Physiol *192:* 131-137, 2002.
541. Liu X, Kim CN, Yang J, Jemmerson R, Wang X: Induction of apoptotic program in cell-free extracts: requirement for dATP and cytochrome c. Cell *86:* 147-157, 1996.
542. Jia L, Patwari Y, Kelsey SM, Srinivasula SM, Agrawal SG, Alnemri ES, Newland AC: Role of Smac in human leukaemic cell apoptosis and proliferation. Oncogene *22:* 1589-1599, 2003.
543. Stennicke HR, Salvesen GS: Biochemical characteristics of caspases-3, -6, -7 and -8. J Biol Chem *272:* 25719-25723, 1997.
544. Faleiro L, Kobayashi R, Fearnhead H, Lazebnik Y: Multiple species of CPP32 and Mch2 are the major active caspases present in apoptotic cells. EMBO J *16:* 2271-2281, 1997.
545. Hirata H, Takahashi A, Kobayashi S, Yonehara S, Sawai H, Okazaki T, Yamamoto K, Sasada M: Caspases are activated in a branched protease cascade and control distinct downstream processes in *Fas*-induced apoptosis. J Exp Med *187:* 587-600, 1998.
546. Srinivasula SM, Ahmad M, MacFarlane M, Luo Z, Huang Z, Fernandes-Alnemri T, Alnemri ES: Generation of constitutively active recombinant caspases-3 and -6 by rearrangement of their subunits. J Biol Chem *273:* 10107-10111, 1998.
547. Kang JJ, Schaber MD, Srinivasula SM, Alnemri ES, Litwack G, Hall DJ, Bjornsti MA: Cascades of mammalian caspase activation in the yeast *Saccharomyces cerevisiae*. J Biol Chem *274:* 3189-3198, 1999.
548. Fernandes-Alnemri T, Litwack G, Alnemri ES: Mch 2, a new member of the apoptotic Ced-3/Ice cysteine protease gene family. Cancer Res *55:* 2737-2742, 1995.
549. Fernandes-Alnemri T, Armstrong RC, Krebs J, Srinivasula SH, Wang L, Bullrich F, Fritz LC, Trapani JA, Tomaselli KJ, Litwack G, Alnemri ES: *In vitro* activation of CPP32 and Mch3 by Mch4, a novel human apoptotic cysteine protease containing two FADD-like domains. Proc Natl Acad Sci USA *93:* 7464-7469, 1996.
550. Slee EA, Harte MT, Kluck RM, Wolf BB, Casiano CA, Newmeyer DD, Wang HG, Reed JC, Nicholson DW, Alnemri ES, Green DR, Martin SJ: Ordering the cytochrome c-initiated caspase cascade: hierarchical activation of caspases-2, -3, -6, -7, -8 and -10 in a caspase-9-dependent manner. J Cell Biol *144:* 281-292, 1999.
551. Srinivasula SM, Ahmad M, Fernandes-Alnemri T, Litwack G, Alnemri ES: Molecular ordering of the *Fas*-apoptotic pathway: the *Fas*/APO-1 protease Mch5 is a CrmA-inhibitable protease that activates multiple Ced-3/ICE-like cysteine proteases. Proc Natl Acad Sci USA *93:* 14486-14491, 1996.
552. Peter ME, Kischkel FC, Scheuerpflug CG, Medema JP, Debatin KM, Krammer PH: Resistance of cultured peripheral T cells towards activation-induced cell death involves a lack of recruitment of FLICE (MACH/caspase 8) to the CD95 death-inducing signaling complex. Eur J Immunol *27:* 1207-1212, 1997.
553. Perera LP, Waldmann TA: Activation of human monocytes induces differential resistance to apoptosis with rapid down regulation of caspase-8/FLICE. Proc Natl Acad Sci USA *95:* 14308-14313, 1998.
554. Wang CY, Mayo MW, Korneluk RG, Goeddel DV, Baldwin AS Jr: NF-kappaB antiapoptosis: induction of TRAF1 and TRAF2 and c-IAP1 and c-IAP2 to suppress caspase-8 activation. Science *281:* 1680-1683, 1998.
555. Skulachev VP: Cytochrome c in the apoptotic and antioxidant cascades. FEBS Lett *423:* 275-280, 1998.
556. Li P, Nijhawan D, Budihardjo I, Srinivasula SM, Ahmad M, Alnemri ES, Wang X: Cytochrome c and dATP-dependent formation of Apaf-1/caspase-9 complex initiates an apoptotic protease cascade. Cell *91*: 479-489, 1997.

557. Hakem R, Hakem A, Duncan GS, Henderson JT, Woo M, Soengas MS, Elia A, de la Pompa JL, Kagi D, Khoo W, Potter J, Yoshida R, Kaufman SA, Lowe SW, Penninger JM, Mak TW: Differential requirement for caspase 9 in apoptotic pathways *in vivo*. Cell *94*: 339-352, 1998.
558. Srinivasula SM, Ahmad M, Fernandes-Alnemri T, Alnemri ES: Autoactivation of procaspase-9 by Apaf-1-mediated oligomerization. Mol Cell *1:* 949-957, 1998.
559. Nakagawara A, Nakamura Y, Ikeda H, Hiwasa T, Kuida K, Su MS, Zhao H, Cnaan A, Sakiyama S: High levels of expression and nuclear localization of interleukin-1 β converting enzyme (ICE) and CPP32 in favorable human neuroblastomas. Cancer Res. *57:* 4578-4584, 1997.
560. Ray SK, Patel SJ, Welsh CT, Wilford GG, Hogan EL, Banik NL: Molecular evidence of apoptotic death in malignant brain tumors including glioblastoma multiforme: upregulation of calpain and caspase-3. J Neurosci Res *69:* 197-206, 2002.
561. Trauth BC, Klas C, Peters AMJ, Matzku S, Moller P, Falk W, Debatin K-M, Krammer PH: Monoclonal antibody-mediated tumor regression by induction of apoptosis. Science *245:* 301-305, 1989.
562. Itoh N, Yonehara S, Ishii A, Yonehara M, Mizushima S, Sameshima M, Hase A, Seto Y, Nagata S: The polypeptide encoded by the cDNA for human cell surface antigen *Fas* can mediate apoptosis. Cell *66:* 233-243, 1991.
563. Shi Y, Glynn JM, Guilbert LJ, Cotter TG, Bissonnette RP, Green DR: Role for c-myc in activation-induced apoptotic cell death in T cell hybridomas. Science *257:* 212-214, 1992.
564. Evan GI, Wyllie AH, Gilbert CS, Littlewood TD, Land H, Brooks M, Waters CM, Penn LZ, Hancock DC: Induction of apoptosis in fibroblasts by c-myc protein. Cell *69:* 119-128, 1992.
565. Korsmeyer SJ: Bcl-2: a repressor of lymphocyte death. Immunol Today *13:* 285-288, 1992.
566. Itoh N, Nagata S: A novel protein domain required for apoptosis. Mutational analysis of human *Fas* antigen. J Biol Chem *268:* 10932-10937, 1993.
567. Hoffman B, Liebermann DA: Molecular controls of apoptosis: differentiation/growth arrest primary response genes, proto-oncogenes, and tumor suppressor genes as positive and negative modulators. Oncogene *9:* 1807-1812, 1994.
568. Ehl S, Hoffmann-Rohrer U, Nagata S, Hengartner H, Zinkernagel R: Different susceptibility of cytotoxic T cells to CD95 (*Fas*/Apo-1) ligand-mediated cell death after activation *in vitro* versus *in vivo*. J Immunol *156:* 2357-2360, 1996.
569. Bodey B, Bodey B Jr, Siegel SE, Kaiser HE: *Fas* (APO-1, CD95) receptor expression and new options of immunotherapy in childhood medulloblastomas. Anticancer Res *19:* 3293-3314, 1999.
570. Bodey B, Bodey B Jr, Siegel SE, Kaiser HE: *Fas* (APO-1, CD95) receptor expression in childhood astrocytomas. Is it a marker of the major apoptotic pathway or a signaling receptor for immune escape of neoplastic cells? In Vivo *13:* 357-373, 1999.
571. Durkop H, Latza U, Hummel M, Eitelbach F, Seed B, Stein H: Molecular cloning and expression of a new member of the nerve growth factor receptor family that is characteristic for Hodgkin's disease. Cell *68:* 421-427, 1992.
572. Calderhead DM, Buhlmann JE, van den Eertwegh AJM, Claassen E, Noelle RJ, Fell HP: Cloning of mouse Ox40: a T cell activation marker that may mediate T-B cell interactions. J Immunol *151:* 5261-5271, 1993.
573. Smith CA, Farrah T, Goodwin RG: The TNF receptor superfamily of cellular and viral proteins: activation, costimulation, and death. Cell *76:* 959-962, 1994.

574. Nagata S: *Fas* and *Fas* ligand: a death factor and its receptor. Adv Immunol *57:* 129-144, 1994.
575. Ogasawara J, Suda T, Nagata S: Selective apoptosis of $CD4^+CD8^+$ thymocytes by the anti-*Fas* antibody. J Exp Med *181:* 485-491, 1995.
576. Alderson MR, Armitage RJ, Maraskovsky E, Tough TW, Roux E, Schooley K, Ramsdell F, Lynch DH: *Fas* transduces activation signals in normal human T lymphocytes. J Exp Med *178:* 2231-2235, 1993.
577. Yonehara S, Ishii A, Yonehara M: A cell-killing monoclonal antibody (anti-*Fas*) to a cell surface antigen co-downregulated with the receptor of tumor necrosis factor. J Exp Med *169:* 1747-1756, 1989.
578. Ogasawara J, Watanabe-Fukunaga R, Adachi M, Matsuzawa A, Kasugai T, Kitamura Y, Itoh N, Suda T, Nagata S: Lethal effect of the anti-*Fas* antibody in mice. Nature *364:* 806-809, 1993.
579. Suda T, Takahashi T, Golstein P, Nagata S: Molecular cloning and expression of the *Fas* ligand: a novel member of the tumor necrosis factor family. Cell *75:* 1169-1178, 1993.
580. Suda T, Nagata S: Purification and characterization of the *Fas* ligand that induces apoptosis. J Exp Med *179:* 873-878, 1994.
581. Nagata S, Golstein P: The *Fas* death factor. Science *267:* 1449-1456, 1995.
582. Schulze-Osthoff K: The *Fas*/APO-1 receptor and its deadly ligand. Trends Cell Biol *4:* 421-426, 1995.
583. Cohen PL, Eisenberg RA: Lpr and gld: single gene models of systemic autoimmunity and lymphoproliferative disease. Annu Rev Immunol *9:* 243-269, 1991.
584. Watanabe-Fukunaga R, Brannan CI, Copeland NG, Jenkins NA, Nagata S: Lymphoproliferation disorder in mice explained by defects in *Fas* antigen that mediates apoptosis. Nature *356:* 314-317, 1992.
585. Takahashi T, Tanaka M, Brannan CI, Jenkins NA, Copeland NG, Suda T, Nagata S: Generalized lymphoproliferative disease in mice, caused by a point mutation in the *Fas* ligand. Cell *76:* 969-976, 1994.
586. Zhou T, Bluethmann H, Eldridge J, Berry K, Mountz JD: Abnormal thymocyte development and production of autoreactive T cells in TCR transgenic autoimmune mice. J Immunol *147:* 466-474, 1991.
587. Zhou T, Mountz JD, Edwards III CK, Berry K, Bluethmann H: Defective maintenance of T cell tolerance to a superantigen in MRL-lpr/lpr mouse. J Exp Med *176:* 1063-1072, 1992.
588. Zhou T, Bluethmann H, Eldridge J, Berry K, Mountz JD: Origin of $CD4^-CD8^-B220^+$ T cells in MRL-lpr/lpr mice. Clues from a T cell receptor β transgenic mouse. J Immunol *150:* 3651-3667, 1993.
589. Owen-Schaub LB, Yonehara S, Crump WL III, Grimm E: DNA fragmentation and cell death is selectively triggered in activated human lymphocytes by *Fas* antigen engagement. Cell Immunol *140:* 197-205, 1992.
590. Klas C, Debatin K-M, Jonker RR, Krammer PH: Activation interferes with the APO-1 pathway in mature human T cells. Int Immunol *5:* 625-630, 1993.
591. Su X, Zhou T, Wu J, Jope R, Mountz JD: Dephosphorylation of a 65kD protein associated with signaling for *Fas*-mediated apoptosis. (Abstract) *FAS*EB J *8:* A218, 1994.
592. Mountz JD, Zhou T, Wu J, Wang W, Su X, Cheng J: Regulation of apoptosis in immune cells. J Clin Immunol *15:* 1-16, 1995.

593. Tachibana O, Nakazawa H, Lampe J, Watanabe K, Kleihues P, Ohgaki H: Expression of *Fas*/APO-1 during the progression of astrocytomas. Cancer Res *55:* 5528-5530, 1995.
594. Tachibana O, Lampe J, Kleihues P, Ohgaki H: Preferential expression of *Fas*/APO1 (CD95) and apoptotic cell death in perinecrotic cells of glioblastoma multiforme. Acta Neuropathol *92:* 431-434, 1996.
595. Dietrich P-Y, Walker PR, Saas P, de Tribolet N: Immunobiology of gliomas: new perspectives for therapy. *In:* Challenges and opportunities in pediatric oncology (Holmes FF, Kepes JJ, Vats TS, Schuler D, Nyary I, eds). New York, Ann NY Acad Sci *824:*124-140, 1997.
596. Cohen JJ: Apoptosis. Immunol Today *14:* 126-130, 1993.
597. Strater J, Wellisch I, Riedl S, Walczak H, Koretz K, Tandara A, Krammer PH, Moller P: CD95 (APO-1/*Fas*)-mediated apoptosis in colon epithelial cells: a possible role in ulcerative colitis. Gastroenterology *113:* 160-167, 1997.
598. Reyher von U, Strater J, Kittstein W, Gschwendt M, Krammer PH, Moller P: Colon carcinoma cells use different mechanisms to escape CD95-mediated apoptosis. Cancer Res *58:* 526-534, 1998.
599. Weller M, Schuster M, Pietsch T, Schabet M: CD95 ligand-induced apoptosis of human medulloblastoma cells. Cancer Lett *128:* 121-126, 1998.
600. Schiffer D, Cavalla P, Chio A, Giordana MT, Marino S, Mauro A, Migheli A: Tumor cell proliferation and apoptosis in medulloblastoma. Acta Neuropathol *87:* 362-370, 1994.
601. Schubert TE, Cervos-Navarro J: The histopathological and clinical relevance of apoptotic cell death in medulloblastomas. J Neuropathol Exp Neurol *57:* 10-15, 1998.
602. Reed JC: Bcl-2 family proteins: regulators of apoptosis and chemoresistance in hematologic malignancies. Semin Hematol *34(4 Suppl 5):* 9-19, 1997.
603. Basu A, Haldar S: Microtubule-damaging drugs triggered bcl2 phosphorylation-requirement of phosphorylation on both serine-70 and serine-87 residues of bcl2 protein. Int J Oncol *13:* 659-664, 1998.
604. Hockenbery DM: The bcl-2 oncogene and apoptosis. Semin Immunol *4:* 413-420, 1992.
605. Vile GF, Tyrrell RM: Oxidative stress resulting from ultraviolet A irradiation of human skin fibroblasts leads to a heme oxygenase-dependent increase in ferritin. J Biol Chem *268:* 14678-14681, 1993.
606. Haldar S, Jena N, Croce CM: Inactivation of bcl-2 by phosphorylation. Proc Natl Acad Sci USA *92:* 4507-4511, 1995.
607. Haldar S, Chintapalli J, Croce CM: Taxol-induced bcl-2 phosphorylation and death of prostate cancer cells. Cancer Res *56:* 1253-1255, 1996.
608. Blagosklonny MV, Schulte T, Nguyen P, Trepel J, Neckers LM: Taxol-induced apoptosis and phosphorylation of bcl-2 protein involves c-Raf-1 signal transduction pathway. Cancer Res *56:* 1851-1854, 1996.
609. Vaux DL, Cory S, Adams JM: Bcl-2 gene promotes hematopoietic cell survival and cooperates with c-myc to immortalize pre-B cells. Nature *355:* 440-442, 1988.
610. Hockenbery DM, Nunez G, Milliman C, Schreiber RD, Korsmeyer SJ: Bcl-2 is an inner mitochondrial membrane protein that blocks programmed cell death. Nature *348:* 334-336, 1990.
611. Hockenbery DM, Oltvai ZN, Yin X-M, Milliman CL, Korsmeyer SJ: Bcl-2 functions in an antioxidant pathway to prevent apoptosis. Cell *75:* 241-251, 1993.
612. Kaufmann SH: Induction of endonucleolytic DNA cleavage in human acute myelogenous leukemia cells by etoposide, camptothecin, and other cytotoxic anticancer drugs: a cautionary note. Cancer Res *49:* 5870-5878, 1989.

613. Martin SJ, Lennon SV, Bonham AM, Cotter TG: Induction of apoptosis (programmed cell death) in human leukemic HL-60 cells by inhibition of RNA or protein synthesis. J Immunol *145:* 1859-1867, 1990.
614. Barry MA, Behnke CA, Eastman A: Activation of programmed cell death (apoptosis) by cisplatin, other anticancer drugs, toxins and hyperthermia. Biochem Pharmacol *40:* 2353-2362, 1990.
615. Perotti M, Toddei F, Mirabelli F, Vairetti M, Bellomo G, McConkey DJ, Orrenius S: Calcium-dependent DNA fragmentation in human synovial cells exposed to cold shock. FEBS Lett *259:* 331-334, 1990.
616. Kruman II, Matylevich NP, Beletsky IP, Afanasyev VN, Umansky SR: Apoptosis of murine BW 5147 thymoma cells induced by dexamethasone and gamma-irradiation. J Cell Physiol *148:* 267-273, 1991.
617. Martin SJ, Cotter TG: Ultraviolet B irradiation of human leukaemia HL-60 cells *in vitro* induces apoptosis. Int J Radiat Biol *59:* 1001-1016, 1991.
618. Del Bino G, Lassota P, Darzynkiewicz Z: The S-phase cytotoxicity of camptothecin. Exp Cell Res *193:* 27-35, 1991.
619. Del Bino G, Darzynkiewicz Z: Camptothecin, teniposide, or 4'-(9-acridinylamino)-3-methanesulfon-m-anisidide, but not mitoxantrone or doxorubicin, induces degradation of nuclear DNA in the S phase of HL-60 cells. Cancer Res *51:* 1165-1169, 1991.
620. Bertrand R, Sarang M, Jenkin J, Kerrigan D, Pommier Y: Differential induction of secondary DNA fragmentation by topoisomerase II inhibitors in human tumor cell lines with amplified c-myc expression. Cancer Res *51:* 6280-6285, 1991.
621. O'Connor PM, Wassermann K, Sarang M, Magrath I, Bohr VA, Kohn KW: Relationship between DNA cross-links, cell cycle, and apoptosis in Burkitt's lymphoma cell lines differing in sensitivity to nitrogen mustard. Cancer Res *51:* 6550-6557, 1991.
622. Hara A, Hirose Y, Yoshimi N, Tanaka T, Mori H: Expression of Bax and bcl-2 proteins, regulators of programmed cell death, in human brain tumors. Neurol Res *19:* 623-628, 1997.
623. Yew DT, Wang HH, Zheng DR: Apoptosis in astrocytomas with different grades of malignancy. Acta Neurochir *140:* 341-347, 1998.
624. Schiffer D, Cavalla P, Migheli A, Chio A, Giordana MT, Marino S, Attanasis A: Apoptosis and cell proliferation in human neuroepithelial tumours. Neurosci Lett *195:* 81-84, 1995.
625. Schiffer D, Cavalla P, Migheli A, Giordana MT, Chiado-Piat L: Bcl-2 distribution in neuroepithelial tumours: an immunohistochemical study. J Neurooncol *27:* 101-109, 1996.
626. Gratas C, Tohma Y, Van Meir EG, Klein M, Tenan M, Ishii N, Tachibana O, Kleihues P, Ohgaki H: *Fas* ligand expression in glioblastoma cell lines and primary astrocytic brain tumors. Brain Pathol *7:* 863-869, 1997.
627. Tohma Y, Gratas C, Van Meir EG, Desbaillets I, Tenan M, Tachibana O, Kleihues P, Ohgaki H: Necrogenesis and *Fas*/APO-1 (CD95) expression in primary (*de novo*) and secondary glioblastomas. J Neuropathol Exp Neurol *57:* 239-245, 1998.
628. Krajewski S, Krajewska M, Ehrmann J, Sikorska M, Lach B, Chatten J, Reed JC: Immunohistochemical analysis of Bcl-2, Bcl-X, Mcl-1, and Bax in tumors of central and peripheral nervous system origin. Am J Pathol *150:* 805-814, 1997.
629. Altieri DC, Marchisio PC, Marchisio C: Survivin apoptosis: an interloper between cell death and cell proliferation in cancer. Lab Invest *79:* 1327-1333, 1999.
630. Ambrosini G, Adida C, Altieri DC: A novel anti-apoptosis gene, survivin, expressed in cancer and lymphoma. Nat Med *3:* 917-921, 1997.

631. Li F, Ambrosini G, Chu EY, Plescia J, Tognin S, Marchisio PC, Altieri DC: Control of apoptosis and mitotic spindle checkpoint by survivin. Nature *396:* 580-584, 1998.
632. Hoffman WH, Biade S, Zilfou JT, Chen J, Murphy M: Transcriptional repression of the anti-apoptotic survivin gene by wild type p53. J Biol Chem *277:* 3247-3257, 2002.
633. Li F, Ackermann EJ, Bennett CF, Rothermel AL, Plescia J, Tognin S, Villa A, Marchisio PC, Altieri DC: Pleiotropic cell-division defects and apoptosis induced by interference with survivin function. Nat Cell Biol *1:* 461-466, 1999.
634. Deveraux QL, Reed JC: IAP family proteins--suppressors of apoptosis. Genes Dev *13:* 239-252, 1999.
635. Chantalat L, Skoufias DA, Kleman JP, Jung B, Dideberg O, Margolis RL: Crystal structure of human survivin reveals a bow tie-shaped dimer with two unusual α-helical extensions. Mol Cell *6:* 183-189, 2000.
636. Miller LK: An exegesis of IAPs: salvation and surprises from BIR motifs. Trends Cell Biol *9:* 323-328, 1999.
637. Sun C, Cai M, Gunasekera AH, Meadows RP, Wang H, Chen J, Zhang H, Wu W, Xu N, Ng SC, Fesik SW: NMR structure and mutagenesis of the inhibitor-of-apoptosis protein XIAP. Nature *401:* 818-822, 1999.
638. Shi Y: Survivin structure: crystal unclear. Nat Struct Biol *7:* 620-623, 2000.
639. O'Connor DS, Grossman D, Plescia J, Li F, Zhang H, Villa A, Tognin S, Marchisio PC, Altieri DC: Regulation of apoptosis at cell division by p34cdc2 phosphorylation of survivin. Proc Natl Acad Sci USA *97:* 13103-13107, 2000.
640. Ambrosini G, Adida C, Sirugo G, Altieri DC: Induction of apoptosis and inhibition of cell proliferation by survivin gene targeting. J Biol Chem *273:* 11177-11182, 1998.
641. Conway EM, Pollefeyt S, Steiner-Mosonyi M, Luo W, Devriese A, Lupu F, Bono F, Leducq N, Dol F, Schaeffer P, Collen D, Herbert JM: Deficiency of survivin in transgenic mice exacerbates *Fas*-induced apoptosis *via* mitochondrial pathways. Gastroenterology *123:* 619-631, 2002.
642. Suzuki A, Hayashida M, Ito T, Kawano H, Nakano T, Miura M, Akahane K, Shiraki K: Survivin initiates cell cycle entry by the competitive interaction with Cdk4/p16(INK4a) and Cdk2/cyclin E complex activation. Oncogene *19:* 3225-3234, 2000.
643. Shin S, Sung BJ, Cho YS, Kim HJ, Ha NC, Hwang JI, Chung CW, Jung YK, Oh BH: An anti-apoptotic protein human survivin is a direct inhibitor of caspase-3 and -7. Biochemistry *40:* 1117-1123, 2001.
644. Kobayashi K, Hatano M, Otaki M, Ogasawara T, Tokuhisa T: Expression of a murine homologue of the inhibitor of apoptosis protein is related to cell proliferation. Proc Natl Acad Sci USA *96:* 1457-1462, 1999.
645. Katoh M, Wilmotte R, Belkouch MC, de Tribolet N, Pizzolato G, Dietrich PY: Survivin in brain tumors: an attractive target for immunotherapy. J Neurooncol *64:* 71-76, 2003.
646. Kajiwara Y, Yamasaki F, Hama S, Yahara K, Yoshioka H, Sugiyama K, Arita K, Kurisu K: Expression of survivin in astrocytic tumors: correlation with malignant grade and prognosis. Cancer *97:* 1077-1083, 2003.
647. Tsujimoto Y, Cossman J, Jaffe E, Croce CM: Involvement of the bcl-2 gene in human follicular lymphoma. Science *228:* 1440-1443, 1985.
648. Adida C, Crotty PL, McGrath J, Berrebi D, Diebold J, Altieri DC: Developmentally regulated expression of the novel cancer anti-apoptosis gene survivin in human and mouse differentiation. Am J Pathol *152:* 43-49, 1998.
649. Lu CD, Altieri DC, Tanigawa N: Expression of a novel anti-apoptosis gene, survivin, correlated with tumor cell apoptosis and p53 accumulation in gastric carcinomas. Cancer Res *58:* 1808-1812, 1998.

650. Rabbani SA: Metalloproteases and urokinase in angiogenesis and tumor progression. In Vivo *12:* 135-142, 1998.
651. Czubayko F, Liadet-Coopman EDE, Aigner A, Tuveson AT, Berchem GJ, Wellstein A: A secreted FGF-binding protein can serve as the angiogenic switch in human cancer. Nature Med *3:* 1137-1140, 1997.
652. Rak J, Kerbel RS: bFGF and tumor angiogenesis--back in the limelight? Nature Med *3:* 1083-1084, 1997.
653. Czubayko F, Smith RV, Chung HC, Wellstein A: Tumor growth and angiogenesis induced by a secreted binding protein for fibroblast growth factors. J Biol Chem *269:* 28243-28248, 1994.
654. Morrison RS, Giordano S, Yamaguchi F, Hendrickson S, Berger MS, Palczewski K: Basic fibroblast growth factor expression is required for clonogenic growth of human glioma cells. J Neurosci Res *34:* 502-509, 1993.
655. Redekop GJ, Naus CC: Transfection of bFGF sense and antisense cDNA resulting in modification of malignant glioma growth. J Neurosurg *82:* 83-90, 1995.
656. Stiles JD, Ostrow PT, Balos LL, Greenberg SJ, Plunkett R, Grand W, Heffner RR Jr: Correlation of endothelin-1 and transforming growth factor β 1 with malignancy and vascularity in human gliomas. J Neuropathol Exp Neurol *56:* 435-439, 1997.
657. Gougos A, Letarte M: Primary structure of endoglin, an RGD-containing glycoprotein of human endothelial cells. J Biol Chem *265:* 8361-8364, 1990.
658. Lopez-Casillas F, Cheifetz S, Doody J, Andres JL, Lane WS, Massague J: Structure and expression of the membrane proteoglycan β-glycan, a component of the TGF-β receptor system. Cell *67:* 785-795, 1991.
659. Altomonte M, Montagner R, Fonsatti E, Colizzi F, Cattarossi I, Brasoveanu LI, Nicotra MR, Cattelan A, Natali PG, Maio M: Expression and structural features of endoglin (CD105), a transforming growth factor β1 and β3 binding protein, in human melanoma. Br J Cancer *74:* 1586-1591, 1996.
660. Burrows FJ, Derbyshire EJ, Tazzari PL, Amlot P, Gazdzar AF, King SW, Letarte M, Vitetta ES, Thorpe PE: Up-regulation of endoglin on vascular endothelial cells in human solid tumors: Implications for diagnosis and therapy. Clin Cancer Res *1:* 1623-1634, 1995.
661. Kumar P, Wang JM, Bernabeu C: CD 105 and angiogenesis. J Pathol *178:* 363-366, 1996.
662. Fernandez-Ruiz E, St-Jacques S, Bellon T, Letarte M, Bernabeu C: Assignment of the human endoglin gene (END) to 9q34→qter. Cytogenet Cell Genet *64:* 204-207, 1993.
663. Wang JM, Kumar S, van Agthoven A, Kumar P, Pye D, Hunter RD: Irradiation induces up-regulation of E9 protein (CD105) in human vascular endothelial cells. Int J Cancer *62:* 791-796, 1995.
664. Yamada N, Kato M, Yamashita H, Nister M, Miyazono K, Heldin CH, Funa K: Enhanced expression of transforming growth factor-β and its type-I and type-II receptors in human glioblastoma. Int J Cancer *62:* 386-392, 1995.
665. Henriksen R, Gobl A, Wilander E, Oberg K, Miyazono K, Funa K: Expression and prognostic significance of TGF-β isotypes, latent TGF-β 1 binding protein, TGF-β type I and type II receptors, and endoglin in normal ovary and ovarian neoplasms. Lab Invest *73:* 213-220, 1995.
666. Szekanecz Z, Haines GK, Harlow LA, Shah MR, Fong TW, Fu R, Lin SJ, Rayan G, Koch AE: Increased synovial expression of transforming growth factor (TGF)β receptor endoglin and TGF-β1 in rheumatoid arthritis: possible interactions in the pathogenesis of the disease. Clin Immunol Immunopathol *76:* 187-194, 1995.

667. Zhang H, Shaw AR, Mak A, Letarte M: Endoglin is a component of the transforming growth factor (TGF)β receptor complex of human pre-B leukemic cells. J Immunol *156:* 564-573, 1996.
668. Griffioen AW, Damen CA, Blijham GH, Groenewegen G: Endoglin/CD 105 may not be an optimal tumor endothelial treatment target. Breast Cancer Res Treat *39:* 239-242, 1996.
669. Shovlin CL, Scott J: Inherited diseases of the vasculature. Annu Rev Physiol *58:* 483-507, 1996.
670. Pichuantes S, Vera S, Bourdeau A, Pece N, Kumar S, Wayner EA, Letarte M: Mapping epitopes to distinct regions of the extracellular domain of endoglin using bacterially expressed recombinant fragments. Tissue Antigens *50:* 265-276, 1997.
671. Zagzag D: Angiogenic growth factors in neural embryogenesis and neoplasia. Am J Pathol *146:* 293-309, 1995.
672. Bouck N: Angiogenesis: a mechanism by which oncogenes and tumor suppressor genes regulate tumorigenesis. Cancer Treat Res *63:* 359-371, 1992.
673. Takano S, Yoshii Y, Kondo S, Suzuki H, Maruno T, Shirai S, Nose T: Concentration of vascular endothelial growth factor in the serum and tumor tissue of brain tumor patients. Cancer Res *56:* 2185-2190, 1996.
674. Cheung N, Wong MP, Yuen ST, Leung SY, Chung LP: Tissue-specific expression pattern of vascular endothelial growth factor isoforms in the malignant transformation of lung and colon. Hum Pathol *29:* 910-914, 1998.
675. Sharkey AM, Charnock-Jones DS, Boocock CA, Brown KD, Smith SK: Expression of mRNA for vascular endothelial growth factor in human placenta. J Reprod Fertil *99:* 609-615, 1993.
676. Aase K, Lymboussaki A, Kaipainen A, Olofsson B, Alitalo K, Eriksson U: Localization of VEGF-B in the mouse embryo suggests a paracrine role of the growth factor in the developing vasculature. Dev Dyn *215:* 12-25, 1999.
677. Samoto K, Ikezaki K, Ono M, Shono T, Kohno K, Kuwano M, Fukui M: Expression of vascular endothelial growth factor and its possible relation with neovascularization in human brain tumors. Cancer Res *55:* 1189-1193, 1995.
678. Berkman RA, Merrill MJ, Reinhold WC, Monacci WT, Saxena A, Clark WC, Robertson JT, Ali IU, Oldfield EH: Expression of the vascular permeability factor /vascular endothelial growth factor gene in central nervous system neoplasms. J Clin Invest *91:* 153-159, 1993.
679. Morii K, Tanaka R, Washiyama K, Kumanishi T, Kuwano R: Expression of vascular endothelial growth factor in capillary hemangioblastoma. Biochem Biophys Res Commun *194:* 749-755, 1993.
680. Wizigmann-Voos S, Breier G, Risau W, Plate KH: Up-regulation of vascular endothelial growth factor and its receptors in von Hippel-Lindau disease-associated and sporadic hemangioblastomas. Cancer Res *55:* 1358-1364, 1995.
681. Plate KH, Breier G, Weich HA, Mennel HD, Risau W: Vascular endothelial growth factor and glioma angiogenesis: coordinate induction of VEGF receptors, distribution of VEGF protein and possible *in vivo* regulatory mechanisms. Int J Cancer *59:* 520-529, 1994.
682. Godard S, Getz G, Delorenzi M, Farmer P, Kobayashi H, Desbaillets I, Nozaki M, Diserens AC, Hamou MF, Dietrich PY, Regli L, Janzer RC, Bucher P, Stupp R, de Tribolet N, Domany E, Hegi ME: Classification of human astrocytic gliomas on the basis of gene expression: a correlated group of genes with angiogenic activity emerges as a strong predictor of subtypes. Cancer Res *63:* 6613-6625, 2003.

683. Breier G, Albrecht U, Sterrer S, Risau W: Expression of vascular endothelial growth factor during embryonic angiogenesis and endothelial cell differentiation. Development *114:* 521-532, 1992.

684. Risau W: Embryonic angiogenesis factors. Pharmacol Ther *51:* 371-376, 1991.

685. Risau W: Molecular biology of blood-brain barrier ontogenesis and function. Acta Neurochir (Wien) [Suppl] *60:* 109-112, 1994.

686. Shim JW, Koh YC, Ahn HK, Park YE, Hwang DY, Chi JG: Expression of bFGF and VEGF in brain astrocytoma. J Korean Med Sci *11:* 149-157, 1996.

687. Melnyk O, Shuman MA, Kim KJ: Vascular endothelial growth factor promotes tumor dissemination by a mechanism distinct from its effect on primary tumor growth. Cancer Res *56:* 921-924, 1996.

688. Machein MR, Kullmer J, Fiebich BL, Plate KH, Warnke PC: Vascular endothelial growth factor expression, vascular volume, and, capillary permeability in human brain tumors. Neurosurgery *44:* 732-740, 1999. discussion 740-741.

689. Liotta LA, Abe S, Robey PG, Martin GR: Preferential digestion of basement membrane collagen by an enzyme derived from a metastatic murine tumor. Proc Natl Acad Sci USA *76:* 2268-2272, 1979.

690. Seltzer JL, Adams SA, Grant GA, Eisen AZ: Purification and properties of a gelatin-specific neutral protease from human skin. J Biol Chem *256:* 4662-4668, 1981.

691. Seltzer JL, Eisen AZ, Bauer EA, Morris NP, Glanville RW, Burgeson RE: Cleavage of type VII collagen by interstitial collagenase and type IV collagenase (gelatinase) derived from human skin. J Biol Chem *264:* 3822-3826, 1989.

692. Seltzer JL, Akers KT, Weingarten H, Grant GA, McCourt DW, Eisen AZ: Cleavage specificity of human skin type IV collagenase (gelatinase). Identification of cleavage sites in type I gelatin, with confirmation using synthetic peptides. J Biol Chem *265:* 20409-20413, 1990.

693. Gadher SJ, Schmid TM, Heck LW, Woolley DE: Cleavage of collagen type X by human synovial collagenase and neutrophil elastase. Matrix *9:* 109-115, 1989.

694. Welgus HG, Fliszar CJ, Seltzer JL, Schmid TM, Jeffrey JJ: Differential susceptibility of type X collagen to cleavage by two mammalian interstitial collagenases and 72-kDa type IV collagenase. J Biol Chem *265:* 13521-13527, 1990.

695. Senior RM, Griffin GL, Fliszar CJ, Shapiro SD, Goldberg GI, Welgus HG: Human 92- and 72-kilodalton type IV collagenases are elastases. J Biol Chem *266:* 7870-7875, 1991.

696. Hibbs MS, Hoidal JR, Kang AH: Expression of a metalloproteinase that degrades native type V collagen and denatured collagens by cultured human alveolar macrophages. J Clin Invest *80:* 1644-1650, 1987.

697. Niyibizi C, Chan R, Wu JJ, Eyre D: A 92 kDa gelatinase (MMP-9) cleavage site in native type V collagen. Biochem Biophys Res Commun *202:* 328-333, 1994.

698. Pourmotabbed T: Relation between substrate specificity and domain structure of 92-kDa type IV collagenase. Ann NY Acad Sci *732:* 372-374, 1994.

699. Pourmotabbed T, Solomon TL, Hasty KA, Mainardi CL: Characteristics of 92 kDa type IV collagenase/gelatinase produced by granulocytic leukemia cells: structure, expression of cDNA in E. coli and enzymic properties. Biochim Biophys Acta *1204:* 97-107, 1994.

700. Fosang AJ, Neame PJ, Last K, Hardingham TE, Murphy G, Hamilton JA: The interglobular domain of cartilage aggrecan is cleaved by PUMP, gelatinases, and cathepsin B. J Biol Chem *267:* 19470-19474, 1992.

701. Nguyen Q, Murphy G, Hughes CE, Mort JS, Roughley PJ: Matrix metalloproteinases cleave at two distinct sites on human cartilage link protein. Biochem J *295:* 595-598, 1993.
702. Okada Y, Nagase H, Harris ED Jr: A metalloproteinase from human rheumatoid synovial fibroblasts that digests connective tissue matrix components. Purification and characterization. J Biol Chem *261:* 14245-14255, 1986.
703. Muller D, Quantin B, Gesnel MC, Millon-Collard R, Abecassis J, Breathnach R: The collagenase gene family in humans consists of at least four members. Biochem J *253:* 187-192, 1988.
704. Flannery CR, Lark MW, Sandy JD: Identification of a stromelysin cleavage site within the interglobular domain of human aggrecan. Evidence for proteolysis at this site *in vivo* in human articular cartilage. J Biol Chem *267:* 1008-1014, 1992.
705. Nguyen Q, Murphy G, Roughley PJ, Mort JS: Degradation of proteoglycan aggregate by a cartilage mettaloproteinase. Evidence for the involvement of stromelysin in the degradation of link protein heterogeneity *in situ.* Biochem J *259:* 61-67, 1989.
706. Wu JJ, Lark MW, Chun LE, Eyre DR: Sites of stromelysin cleavage in collagen types II, IX, X, and XI of cartilage. J Biol Chem *266:* 5625-5628, 1991.
707. Mott JD, Khalifah RG, Nagase H, Shield CF 3rd, Hudson JK, Hudson BG: Nonenzymatic glycation of type IV collagen and matrix metalloproteinase susceptibility. Kidney Int *52:* 1302-1312, 1997.
708. Wilhelm SM, Shao ZH, Housley TJ, Seperack PK, Baumann AP, Gunja-Smith Z, Woessner JF Jr: Matrix metalloproteinase-3 (stromelysin-1). Identification as the cartilage acid metalloprotease and effect of pH on catalytic properties and calcium affinity. J Biol Chem *268:* 21906-21913, 1993.
709. Enghild JJ, Salvesen G, Brew K, Nagase H: Interaction of human rheumatoid synovial collagenase (matrix metalloproteinase 1) and stromelysin (matrix metalloproteinase 3) with human α_2-macroglobulin and chicken ovostatin. Binding kinetics and identification of matrix metalloproteinase cleavage sites. J Biol Chem *264:* 8779-8785, 1989.
710. Mast AE, Enghild JJ, Nagase H, Suzuki K, Pizzo SV, Salvesen G: Kinetics and physiologic relevance of the inactivation of α 1-proteinase inhibitor, α 1-antichymotrypsin, and antithrombin III by matrix metalloproteinases-1 (tissue collagenase), -2 (72-kDa gelatinase/type IV collagenase), and -3 (stromelysin). J Biol Chem *266:* 15810-15816, 1991.
711. Harrison R, Teahan J, Stein R: A semicontinuous, high-performance liquid chromatography-based assay for stromelysin. Annals Biochem *180:* 100-113, 1989.
712. Fowlkes JL, Enghild JJ, Susuki K, Nagase H: Matrix metalloproteinasees degrade insulin-like growth factor-binding protein-1 in dermal fibroblast cultures. J Biol Chem *269:* 25742-25746, 1994.
713. Mayer U, Mann K, Timpl R, Murphy G: Sites of nidogen cleavage by proteases involved in tissue homeostasis and remodeling. Eur J Biochem *217:* 877-884, 1993.
714. Sasaki T, Gohring W, Mann K, Maurer P, Hohenester E, Knauper V, Murphy G, Timpl R: Limited cleavage of extracellular matrix protein BM-40 by matrix metalloproteinases increases its affinity for collagens. J Biol Chem *272:* 9237-9243, 1997.
715. Sasaki T, Mann K, Murphy G, Chu ML, Timpl: Different susceptibilities of fibulin-1 and fibulin-2 to cleavage by matrix metalloproteinases and other tissue proteases. Eur Biochem *240:* 427-434, 1996.

716. Bini A, Itoh Y, Kudryk BJ, Nagase H: Degradation of cross-linked fibrin by matrix metalloproteinase 3 (stromelysin 1): Hydrolysis of the γGly^{404}-Ala^{405} peptide bond. Biochem *35:* 13056-13063, 1996.
717. Imai K, Shikata H, Okada Y: Degradation of vitronectin by matrix metalloproteinases-1, -2, -3, -7, and -9. FEBS Lett *369:* 249-251, 1995.
718. Knauper V, Wilhelm SM, Seperack PK, DeClerck YA, Langley KE, Osthues A, Tschesche H: Direct activation of human neutrophil procollagenase by recombinant stromelysin. Biochem J *295:* 581-586, 1993.
719. Knauper V, Lopez-Otin C, Smith B, Knight G, Murphy G: Biochemical characterization of human collagenase-3. J Biol Chem *235:* 187-191, 1996.
720. Nagase H, Enghild JJ, Suzuki K, Salvesen G: Stepwise activation mechanisms of the precursor of matrix metalloproteinase 3 (stromelysin) by proteinases and 4-aminophenyl)mercuric acetate. Biochemistry. *29:* 5783-5789, 1990.
721. Ogata Y, Enghild JJ, Nagase H: Matrix metalloproteinase 3 (stromelysin) activates the precursor for the human matrix metalloproteinase 9. J Biol Chem. *267:* 3581-3584, 1992.
722. Suzuki K, Enghild JJ, Morodomi T, Salvesen G, Nagase H: Mechanisms of activation of tissue procollagenase by matrix metalloproteinase 3 (stromelysin). Biochemistry *29:* 10261-10270, 1990.
723. Freije JM, Diez-Itza I, Balbin M, Sanchez LM, Blasco R, Tolivia J, Lopez-Otin C: Molecular cloning and expression of collagenase-3, a novel human matrix metalloproteinase produced by breast carcinomas. J Biol Chem *269:* 16766-16773, 1994.
724. Knauper V, Will H, Lopez-Otin C, Smith B, Atkinson SJ, Stanton H, Hembry RM, Murphy G: Cellular mechanisms for human procollagenase-3 (MMP-13) activation. Evidence that MT1-MMP (MMP-14) and gelatinase a (MMP-2) are able to generate active enzyme. J Biol Chem *271:* 17124-17131, 1996.
725. Billinghurst RC, Dahlberg L, Ionescu M, Reiner A, Bourne R, Rorabeck C, Mitchell P, Hambor J, Diekmann O, Tschesche H, Chen J, Van Wart H, Poole AR: Enhanced cleavage of type II collagen by collagenases in osteoarthritic articular cartilage. J Clin Invest *99:* 1534-1545, 1997.
726. Mitchell PG, Magna HA, Reeves LM, Lopresti-Morrow LL, Yocum SA, Rosner PJ, Geoghegan KF, Hambor JE: Cloning, expression, and type II collagenolytic activity of matrix metalloproteinase-13 from human osteoarthritic cartilage. J Clin Invest *97:* 761-768, 1996.
727. Lampert K, Machein U, Machein MR, Conca W, Peter HH, Volk B: Expression of matrix metalloproteinases and their tissue inhibitors in human brain tumors. Am J Pathol *153:* 429-437, 1998.
728. Bodey B, Bodey B Jr, Siegel SE, Kaiser HE: Matrix metalloproteinase expression in childhood medulloblastomas/primitive neuroectodermal tumors. In Vivo *14:* 667-673, 2000.
729. Bodey B, Bodey B Jr, Siegel SE, Kaiser HE: Matrix metalloproteinase expression in childhood astrocytomas. Anticancer Res *20:* 3287-3292, 2000.
730. Rooprai HK, McCormick D: Proteases and their inhibitors in human brain tumours: a review. Anticancer Res *17:* 4151-4162, 1997.
731. Rooprai HK, Van Meter T, Rucklidge GJ, Hudson L, Everall IP, Pilkington GJ: Comparative analysis of matrix metalloproteinases by immunocytochemistry, immunohistochemistry and zymography in human primary brain tumours. Int J Oncol *13:* 1153-1157, 1998.

732. Nakagawa T, Kubota T, Kabuto M, Sato K, Kawano H, Hayakawa T, Okada Y: Production of matrix metalloproteinases and tissue inhibitor of metalloproteinases-1 by human brain tumors. J Neurosurg *81:* 69-77, 1994.
733. Nakano A, Tani E, Miyazaki K, Yamamoto Y, Furuyama J: Matrix metalloproteinases and tissue inhibitors of metalloproteinases in human gliomas. J Neurosurg *83:* 298-307, 1995.
734. Nakagawa T, Kubota T, Kabuto M, Fujimoto N, Okada Y: Secretion of matrix metalloproteinase-2 (72 kD gelatinase/type IV collagenase = gelatinase A) by malignant human glioma cell lines: implications for the growth and cellular invasion of the extracellular matrix. J Neurooncol *28:* 13-24, 1996.
735. Sawaya RE, Yamamoto M, Gokaslan ZL, Wang SW, Mohanam S, Fuller GN, McCutcheon IE, Stetler-Stevenson WG, Nicolson GL, Rao JS: Expression and localization of 72 kDa type IV collagenase (MMP-2) in human malignant gliomas *in vivo*. Clin Exp Metastasis *14:* 35-42, 1996.
736. Rao JS, Yamamoto M, Mohaman S, Gokaslan ZL, Fuller GN, Stetler-Stevenson WG, Rao VH, Liotta LA, Nicolson GL, Sawaya RE: Expression and localization of 92 kDa type IV collagenase/gelatinase B (MMP-9) in human gliomas. Clin Exp Metastasis *14:* 12-18, 1996.
737. Vince GH, Wagner S, Pietsch T, Klein R, Goldbrunner RH, Roosen K, Tonn JC: Heterogeneous regional expression patterns of matrix metalloproteinases in human malignant gliomas. Int J Dev Neurosci *17:* 437-445, 1999.
738. Janckila AJ, Yam LT, Li C-Y: Immunoalkaline phosphatase cytochemistry. Amer J Clin Pathol *84:* 476-480, 1985.
739. Zeltzer PM, Bodey B, Marlin A, Kemshead J: Immunophenotype profile of childhood medulloblastomas and supratentorial primitive neuroectodermal tumors using 16 monoclonal antibodies. Cancer *66:* 273-283, 1990.
740. Yam LT, Janckila AJ, Epremian BE, Li C-Y: Diagnostic significance of levamisole-resistant alkaline phosphatase in cytochemistry and immunocytochemistry. Amer J Clin Pathol *91:* 31-36, 1989.
741. Strasburger CJ, Amir-Zaltsman Y, Kohen F: The avidin-biotin reaction as an universal amplification system in immunoassays. Prog Clin Biol Res *285:* 79-100, 1988.
742. Wilchek M, Bayer EA: Introduction to avidin-biotin technology. Methods Enzymol *184:* 5-13, 1990.
743. Duhamel RC, Whitehead JS: Prevention of nonspecific binding of avidin. Methods Enzymol *184:* 201-207, 1990.
744. Diamandis EP, Christopoulos TK: The biotin-(strept)avidin system: principles and applications in biotechnology. Clin Chem *37:* 625-636, 1991.
745. Bodey B: Cancer-Testis antigens: promising targets for antigen directed anti-neoplastic immunotherapy. EOBT *2:* 577-584, 2002.
746. Bodey B: Genetically engineered antibodies for direct anti-neoplastic treatment and neoplastic cells directed delivery of various therapeutic agents. EOBT *1:* 603-617, 2001.
747. Ma Z, Khatlani TS, Li L Sasaki K, Okuda M, Inokuma H, Onishi T: Molecular cloning and expression analysis of feline melanoma antigen (MAGE) obtained from a lymphoma cell line. Vet Immunol Immunopathol *83:* 241-252, 2001.
748. Weiser TS, Ohnmacht GA, Guo ZS, Fischette MR, Chen GA, Hong JA, Nguyen DM, Schrump DS: Induction of MAGE-3 expression in lung and esophageal cancer cells. Ann Thorac Surg *71:* 295-301; discussion 301-302, 2001.

749. Chomez P, De Backer O, Bertrand M, De Plaen E, Boon T, Lucas S: An overview of the MAGE gene family with the identification of all human members of the family. Cancer Res *61:* 5544-5551, 2001.
750. Ohman Forslund K, Nordqvist K: The melanoma antigen genes--any clues to their functions in normal tissues? Exp Cell Res *265:* 185-194, 2001.
751. van der Bruggen P, Traversari C, Chomez P, Lurquin C, De Plaen E, Van den Eynde B, Knuth A, Boon T: A gene encoding an antigen recognized by cytolytic T lymphocytes on a human melanoma. Science *254:* 1643-1647, 1991.
752. De Plaen E, Arden K, Traversari C, Gaforio JJ, Szikora JP, De Smet C, Brasseur F, van der Bruggen P, Lethe B, Lurquin C, *et al.*: Structure, chromosomal localization, and expression of 12 genes of the MAGE family. Immunogenetics *40:* 360-369, 1994.
753. Rogner UC, Wilke K, Steck E, Korn B, Poutska A: The melanoma antigen (MAGE) family is clustered in the chromosomal band Xq28. Genomics *29:* 725-731, 1995.
754. Lurquin C, De Smet C, Brasseur F, Muscatelli F, Martelange V, De Plaen E, Brasseur R, Monaco AP, Boon T: Two members of th human MAGEB gene family loacted in Xp21.3 are expressed in tumors of various histological origins. Genomics *46:* 397-408, 1997.
755. Lucas S, De Smet C, Arden KC, Viars CS, Lethe B, Lurquin C, Boon T: Identification of a new MAGE gene with tumor-specific expression by representational difference analysis. Cancer Res *58:* 743-752, 1998.
756. Pold M, Zhou J, Chen GL, Hall JM, Vescio RA, Berenson JR: Identification of a new, unorthodox member of the MAGE gene family. Genomics *59:* 161-167, 1999.
757. Aubry F, Satie AP, Rioux-Leclercq N, Rajpert-De Meyts E, Spagnoli GC, Chomez P, De Backer O, Jegou B, Samson M: MAGE-A4, a germ cell specific marker, is expressed differentially in testicular tumors. Cancer *92:* 2778-2785, 2001.
758. Gillespie AM, Coleman RE: The potential of melanoma antigen expression in cancer therapy. Cancer Treat Rev *25:* 219-227, 1999.
759. Chen Y-T, Old LJ: Cancer-testis antigens: targets for cancer immunotherapy. Cancer J from Scientific American *5:* 16-17, 1999.
760. Marchand M, van Baren N, Weynants P, Brichard V, Dreno B, Tessier MH, Rankin E, Parmiani G, Arienti F, Humblet Y, Bourlond A, Vanwijck R, Lienard D, Beauduin M, Dietrich PY, Russo V, Kerger J, Masucci G, Jager E, De Greve J, Atzpodien J, Brasseur F, Coulie PG, van der Bruggen P, Boon T: Tumor regressions observed in patients with metastatic melanoma treated with an antigenic peptide encoded by gene MAGE-3 and presented by HLA-A1. Int J Cancer *80:* 219-230, 1999.
761. Boon T, Old LJ: Tumor antigens. Curr Opin Immunol *9:* 681-683, 1997.
762. Sahin U, Koslowski M, Tureci O, Eberle T, Zwick C, Romeike B, Moringlane JR, Schwechheimer K, Feiden W, Pfreundschuh M: Expression of cancer testis genes in human brain tumors. Clin Cancer Res *6:* 3916-3922, 2000.

II. ANTI-NEOPLASTIC BIOLOGICAL THERAPIES

ANTINEOPLASTIC BIOLOGICAL THERAPIES

"There is at bottom only one genuine scientific treatment for all diseases, and that is to stimulate the phagocytes."
(George Bernard Shaw - The Doctor's Dilemma, 1909)

Classical anti-neoplastic therapeutic modalities such as surgery, radiation, and chemotherapy not only fail to cure the great majority of malignant tumors (including AA, GBM, and MED/PNET) but their employment often leads to severe and debilitating side effects, with the latter two even being associated with severe neoplasm-related morbidity. Childhood PNETs and GBM are among the most malignant brain tumors in children (1-8). Despite advances in current diagnostic procedures and multimodality treatment, the five year survival rate in these children is about 50% (1-3). MED/PNETs are traditionally classified as WHO grade IV brain tumors. The majority of high grade glial tumors are incurable with the currently employed classical therapeutic modalities (9), which may well be the direct result of the biological cell variability of these tumors' microenvironment, *e.g.* multiple stem cell lines, intrinsic and acquired multidrug resistance. Today no therapeutic regimen can reliably cure MED/PNET. Recent years have seen major advances in the elucidation of the biology of malignant childhood brain tumors, which in turn have led to the current development of innovative therapeutic strategies (10,11). The question confronting us at the beginning of twenty-first century is whether we should continue to use and investigate chemotherapy or whether the time has come for experimental treatments. Assessment of more than 20 years of chemotherapy trials is discouraging despite a few areas of modest success. Only patients with specific histology (oligodendroglioma, anaplastic ASTR) and good prognostic factors (young age, good performance status) may benefit from chemotherapy, with a possible reversal of neurological dysfunction. However, the real impact on survival is small (anaplastic ASTR) or undefined (oligodendroglioma). Furthermore, it is unfortunately obvious that the outcome of GBM patients is not significantly modified by chemotherapy. We agree with the majority of clinical oncologists that the time has come to explore the potential of novel biological therapies in GBM patients. This could also be proposed for AA and oligodendroglioma patients after the failure of chemotherapy.

As we stated above, the majority of PNETs/MEDs and malignant glial tumors are incurable with the current classical therapeutic modalities, including surgical resection, radiotherapy, and chemotherapy (10,11). This may well be the direct result of the biological variability of these tumors, *e.g.* multiple stem cell lines, intrinsic and acquired multidrug resistance, heterogenic IP, cellular IP dedifferentiation, and activity of several tumor escape mechanisms, including the presence of *FasR-FasL* apoptotic pathway.

It has been well established that GBM is the most common and extremely malignant of brain tumors. It is highly suitable for antigen epitope targeted molecular therapy as part of the immunological anti-neoplastic therapy. GBM overexpress oncogenes, such as the mutated EGFRvIII and PDGFR, with other multiple mutations and deletions of suppressor genes, such as PTEN and TP53 being identified (12,13).

The comparative aspects of neoplastic transformation in the brain have been explored and may allow for the better observation of novel therapies of xenotransplanted malignancies in animals, as they may relate to humans. It still remains clear that in all neoplastic diseases, a combination of therapeutic modalities with adjuvant therapies that target different aspects of the malignant neoplastic growth should be used in an "individualized cocktail" to ensure the best possible results.

A recent article examined the new treatments currently being investigated for malignant glioma. The authors stated that it is well known that immunotherapy is theoretically very attractive because it offers the potential for high glial tumor cell directed and specific cytotoxicity (14). There are increasing reports demonstrating that systemic immunotherapy employing dendritic cells in tumor vaccines are capable of inducing a host's antiglioma response. Therefore, dendritic cell based immunotherapy could be a new treatment modality for patients with glioma. Dendritic cell based immunotherapy strategies appear promising as an approach to successfully induce an anti-tumor immune response and increase survival in patients with glioma. The development of methods for manipulating dendritic cells for the purpose of vaccination will enhance the clinical usefulness of these cells for biotherapy for malignant glioma.

What kinds of immunotherapies are being developed in the time span of one century? We are suggesting that the never-ending development of new strategies, treatments and more sophisticated forms of application could free mankind from the devastating neoplastic "epidemy".

Chapter 3

EXPERIMENTAL THERAPIES IN BRAIN TUMORS

Several experimental studies have shown that genetically engineered thymidine kinase (tk)-defective herpes simplex virus type 1 (HSV-1) can effectively and selectively destroy gliomas in animal models (15). The consequences of viral infection and tumor regression must be fully characterized before this kind of therapy can be applied in human clinical trials. Jia and co-workers injected immunocompetent rats harboring 9L gliosarcomas intratumorally with a tk-defective HSV-1, KOS-SB, at titers that previously have been demonstrated to cause tumor regression. In animals surviving 3 months or longer following viral treatment, there was no evidence of persistent infection or inflammation in neither peritumoral brain tissue nor in remote systemic organs studied with routine histological and immunocytochemical analyses. Polymerase chain reaction using primers specific for HSV-1 detected HSV-1 DNA in peritumoral tissue only in animals sacrificed within 3 months of viral injection. There was no evidence of HSV-1 DNA in systemic tissues at any time after treatment. The authors concluded that stereotactic intratumoral injection of tk-deficient HSV can be attempted for the treatment of brain tumors without risk of systemic infection or significant toxicity to normal brain tissue or remote proliferating tissues.

Cell motility within the central nervous system (CNS) "neurophil" may be largely restricted, yet local infiltration by glioma cells is commonly observed. Glioma cells remodel nervous tissue and may assemble extracellular matrix in order to migrate (16). The rat C6 glioma cell line was examined for laminin expression and response *in vitro* and following engraftment into rat spinal cord. C6 cell cultures expressed laminin-2, and C6 cells attached equally well to substrates of purified laminin-1 and laminin-2 and laminin-2-enriched C6 conditioned medium. In contrast, C6 cell migration was substantially greater on laminin-2 and C6-derived substrata than on laminin-1. Glioma cell attachment to laminin-1 and -2 was largely inhibited by antibody to the laminin receptor LBP110 and by an IKVAV peptide, but not by YIGSR or control peptides. IKVAV peptide and anti-LBP110 antibodies also inhibited glioma cell invasion through the synthetic basement membrane. Anti-β1 integrin antibody selectively inhibited cell migration and invasion on laminin-2 substrata without affecting percent cell attachment. These findings suggest C6 cell migration and invasion are promoted by autocrine release of laminin-2 and involve LPB110 and β1 integrin laminin receptors. A possible role for laminin-2 in CNS infiltration *in vivo* was also examined following glioma engraftment into rat spinal cord. Engrafted C6 tumors share many histologic features with invasive human glioma. Engrafted glioma cells expressed laminin, LBP110 and β1 integrin antigens, indicating the molecular mechanisms of C6 motility observed in culture may be quite similar to glioma invasion *in vivo*. NMR and corroborative immunocytochemistry provided precise means to monitor tumor progression following glioma engraftment into rat spinal cord. Advantages of this glioma model as they relate to the assessment of anti-adhesive therapies *in vivo* are discussed further by the authors.

As a step towards the development of novel and improved therapies for MED/PNET, the ability of a neuroattenuated HSV-1 ICP34.5 mutant (variant-1716) to replicate within and destroy an authentic MED cell line [Med 283 (D283)] was assessed employing immunocytochemistry, *in situ* hybridization, and viral titrations (17). *In vitro* studies have demonstrated that variant-1716 replicates in and destroys monolayers of D283 cells with kinetics similar to wild-type strain 17^{+}. When D283 tumor-bearing animals were treated with variant-1716 injected directly into the tumor, there was a statistically significant increase in survival ($p < 0.02$) compared to mock-treated tumor-bearing mice. Additionally, several novel scientific facts emerged from this study. Most importantly, the authors detected focal acute viral replication within murine brain cells, a finding not previously reported with HSV-1 ICP-34.5 variants. Further, the brains of tumor-bearing animals treated with variant-1716 demonstrated persistent viral replication within tumors for several weeks. Variant-1716 thus causes a statistically significant increase in survival of experimental animals that had been injected with a MED cell line. Further analysis of the replication of HSV-1 ICP34.5 mutants within the mammalian CNS is necessary to assess potential long-term toxicity.

These are just a few of the experimental advances being made in the treatment of brain tumors. These advances, however, are built upon a long history of research that began exploring immunobiological treatment options approximately a century ago.

REFERENCES

1. Stahli C, Staehelin T, Miggiano V, Schmidt J, Haring P: High frequencies of antigen-specific hybridomas: dependence on immunization parameters and prediction by spleen cell analysis. J Immunol Methods *32:* 297-304, 1980.
2. Laurent JP, Cheek WR: A new staging method versus TNM staging in children with posterior fossa primitive neuroectodermal tumor (medulloblastoma). Childs Nerv Syst *2:* 238-241, 1986.
3. Collins VP, Loeffler RK, Tivey H: Observations on growth rates of human tumors. Amer J Radiol *76:* 988, 1956.
4. Quest DO, Brisman R, Antunes JL, Housepian EM: Period of risk for recurrence of medulloblastoma. J Neurosurg *48:* 159-163, 1978.
5. Bodey B, Bodey B Jr, Siegel SE, Kaiser HE: Immunophenotypical (IP) analysis and immunobiology of childhood primary brain tumors. Anticancer Res *19:* 2973-2992, 1999.
6. Harisiadis L, Chang CH: Medulloblastoma in children: a correlation between staging and results of treatment. Int J Radiat Oncol Biol Phys *2:* 833-841, 1977.
7. Bruce DA, Schut L: Cerebellar medulloblastoma, sarcoma and hemangioblastoma. p. 375, In: American Association of Neurological Surgeons, J Neurosurgery, Grune & Stratton, 1982.
8. Bodey B, Bodey B Jr, Siegel SE, Kaiser HE: *Fas* (APO-1, CD95) receptor expression and new options of immunotherapy in childhood medulloblastomas. Anticancer Res *19:* 3293-3314, 1999.
9. Kaye AH, Laidlaw JD: Chemotherapy of gliomas. Curr Opin Neurol Neurosurg *5:* 526-533, 1992.
10. Jendrossek V, Belka C, Bamberg M: Novel chemotherapeutic agents for the treatment of glioblastoma multiforme. Expert Opin Investig Drugs *12:* 1899-1924, 2003.
11. Novotny L, Szekeres T: Cancer therapy: new targets for chemotherapy. Hematology *8:* 129-137, 2003.
12. Kleihues P, Ohgaki H: Primary and secondary glioblastomas: from concept to clinical diagnosis. Neurooncol *1:* 44-51, 1999.
13. Kuan CT, Wikstrand CJ, Bigner DD: EGF mutant receptor vIII as a molecular target in cancer therapy. Endocr Relat Cancer *8:* 83-96, 2001.
14. Yamanaka R, Homma J, Yajima N, Tsuchiya N, Tanaka R: Dendritic cell therapy for malignant glioma. No To Shinkei *55:* 771-780, 2003.
15. Jia WW, Tan J, Redekop GJ, Goldie JH: Toxicity studies in thymidine kinase-deficient herpes simplex virus therapy for malignant astrocytoma. J Neurosurg *85:* 662-666, 1996.
16. Muir D, Johnson J, Rojiani M, Inglis BA, Rojiani A, Maria BL: Assessment of laminin-mediated glioma invasion *in vitro* and by glioma tumors engrafted within rat spinal cord. J Neurooncol *30:* 199-211, 1996.
17. Lasner TM, Kesari S, Brown SM, Lee VM, Fraser NW, Trojanowski JQ: Therapy of a murine model of pediatric brain tumors using a herpes simplex virus type-1 ICP34.5 mutant and demonstration of viral replication within the CNS. J Neuropathol Exp Neurol *55:* 1259-1269, 1996.

Chapter 4

BIOLOGIC ANTI-NEOPLASTIC THERAPIES

"The world fears a new experience more than it fears anything. Because a new experience displaces so many old experiences."

D. H. Lawrence, English author (1885-1930)

1. INTRODUCTION

By the year 2010, with the everyday numerical growth of the retired population, neoplastic disease may be the leading cause of death in the United States of America. During the last few decades, one in five deaths from all causes in the USA was from neoplasms (1-6). Despite the great advances being made in cancer prevention and therapeutic strategies, about 472,000 Americans will succumb to their malignant illness annually (7). Decreases in incidence and mortality of the Hodgkin's disease, cervical cancer, and different clinico-pathological forms of stomach cancer are typical of improvements through life-style and eating habit changes. However, negative changes have occurred as well, including lung cancer, which was rare at the beginning of this century, now being the current leader of cancer related death.

Despite basic and dynamic advances and developments in neoplastic chemotherapy, partly successful therapies with new clinical trials, 62% of all five-year cancer survivors are provided by surgery alone (8). Specifically, cryosurgery destructs the malignant tissue *in situ* employing subzero temperatures (9). In 1893, Charles Tripler, a physician in New York City, introduced the use of liquid air in the treatment of skin malignancies. The cryosurgical method for elimination of hepatic tumors is a rutin today.

Unfortunately the chemotherapy, after much enthusiasm during its beginning, has not only reached a plateau, but has now started to decline in its use and enthusiasm. A true evaluation of the current chemotherapy and future expectation of the necessary changes in the aims of anti-neoplastic therapy was given by Dr. Jako, the former President of American Society for Laser Medicine and Surgery (8): "The biological therapy in 15 years its contribution as a single or combined modality will add to the arsenal, as much as chemotherapy has over the past 20 years." In their critical article in "Science", Pastan and Fitzgerald (10) went even further:

> "Because chemotherapy is not a cure for the common types of cancer in adults, new therapies must be developed. One approach is to target a cytotoxic agent to the cancer cell. To accomplish this, the cytotoxic agent is attached to an antibody or a growth factor that preferentially binds to cancer cells. The targets for this type of therapy can be growth factor receptors, differentiation antigens, or other less characterized cell surface antigens."

"Immunotherapy" represents a new, fourth modality of anti-neoplastic, biologic therapy. It is intended to employ and activate the host's own immune system in the elimination of neoplastic growth. The major questions of anti-neoplastic immunotherapy has not changed, only our knowledge in how to answer these questions has been expanded.

Historically the concept of an immunobiological treatment of neoplasms dates back at least to Ehrlich, who 80 years ago predicted the scientific development of neoplastic cell directed, specific to antigen epitopes antibodies; his "magic bullets" that target drugs and toxins to neoplastic cells and thereby assure their destruction during a specific "immunological" treatment of human neoplasms. The earliest pure empirical lead came a little more than a century ago, from Coley's (11) observation of complete regression of a head and neck sarcoma after the patient suffered a local streptococcal infection. These rare results suggested that a treatment with bacterial toxins may be successful and Coley in 1893 (12) was able to submit the first results of his pioneer work, obtained after nonspecific treatment of sarcomas, with bacterial toxins. Eighty-five years later, Sinkovics (13) stated in his review article about clinical immunotherapy:

> "Immunologic intervention has become an accepted investigational modality for treatment of human cancer. Immunotherapy... has progressed along three major avenues: 1) use of animal model systems; 2) empirical use

> of immunotherapy in man, without investigation of its mechanism of action; and 3) use of immunotherapy with monitoring by various assays to establish the mechanism of action."

Today, we are able to monitor, and partly understand how immunologic techniques and principles can be *Fas*hioned into powerful therapies. We have found out that cellular immunomodulation and anti-neoplastic treatment employ the newly discovered biological response modifiers (BRM) and MoABs in clinical trials to add this modality to the therapy of human malignancies. The extremely rapid advances in the discoveries in the disciplines of (1) immunobiology of T lymphocyte maturation and differentiation in the field of cellular immunology; and (2) molecular biology, with gene detection and expression in different solid neoplasms, and the methods of genetic engineering are the two major disciplines resulting in the development of a more effective immunotherapy in the last two decades.

The x-ray crystallographic structure of Pseudomonas exotoxin (PE) was the model for the synthesis of genetically engineered recombinant (made in *E. coli*) toxins. Replacing the I domain of PE with the transforming growth factor-alpha (TGF-α) resulted in a chimeric toxin TGF-α-PE40, extremely cytotoxic to cells that contain epithelial growth factor (EGF) receptors. Clinical trials have begun in which TGF-α-PE40 is installed into the urinary bladder for treatment of bladder cancer. Further, recombinant toxins have now been engineered by combining IL-2, IL-4, IL-6, and acidic fibroblast growth factors (FGF) with PE40 and diphteria toxin (DT) with IL-2 and melanocyte-stimulating hormone (10, 14, 15).

Growth factors exhibit a leading role in the regulation of normal cell proliferation and differentiation by initiating a well balanced cascade of intracytoplasmic biochemical reactions after they bind to their own cell surface located glycoprotein receptors (16, 17). Interest in growth factors escalated when their receptors was found to be present on a variety of tumors and that tumors are able to synthesize and secrete these factors (18). Thus, neoplastic cell growth is an autocrine system, independent of an exogenous source of growth factors (19). Other tumors are not only capable of secreting growth factors, but under certain microenvironmental circumstances, some of these peptides actually inhibit growth. In addition at a certain pathobiological moment, the growth factor's level or chemical structure appears to be abnormal (20, 21) The inhibition of the interaction between growth factors and their specific receptors, utilization of negative peptides, changing the signal transduction pathways or altering their increased or decreased cell surface receptor expression are all possibilities for a new and specific cell biological treatment and could result in neoplastically transformed cell death.

Early in the 1990s, single chain MoABs were also combined with PE40, including antibodies against growth factors, and the human transferrin receptor, OVB3 and B3 (10, 22). The B3 MoAB, which binds to a carbohydrate antigen present on the cell surface of many carcinomas in combination with PE40, is an engineered toxin that causes full regression of transplanted human tumors in nude mice.

Deeper analyses of neoplasm bearing host humoral immune response against his or her own tumor concluded that for enhanced response, immunizations should include also glycoproteins, peptides, glycolipids or anti-idiotype vaccines and some synthetic carbohydrate determinants (23).

Some of the future tendencies of immunotherapies in our opinion will be immune-effector cell oriented. We agree with John Holcenberg MD, vice president of the biotech firm Immunex, who said in an interview in 1990: "In the next ten years, the biotech firms will figure out how to make cells increase or decrease in numbers and activity." Experimental data from the Johns Hopkins University (Baltimore) provided a rationale for the use of lymphokine gene-transfected autologous tumor cells as a modality for cellular immunotherapy of cancer (24). In the first few years of the twenty-first century, immunotherapy (as part of biological therapy) comprises such things as active specific immunotherapy ("cancer vaccines"), nonspecific immunostimulation with cytokines, and the inhibition of suppressor influences exerted or elicited by the tumor (25). Just as cancer chemotherapy began with the employment of single agents and evolved into clinical trial combination therapy, so immunotherapeutic agents have been combined with each other and with chemotherapy. The alkylating agent cyclophosphamide (Cytoxan; CYC) has been used for many years to inhibit tumor-derived suppressor influences in rodents and has been exploited for the same use in humans. Combinations of CYC and cancer vaccines such as autologous tumor cells, Melacine, large multivalent immunogen (LMI), and Theratope have been tested with some success in humans for more than a decade. In this use, the CYC is a biological response modifier rather than an anti-tumor agent and is intended to inhibit suppressor influences. CYC and low- to moderate-dose IL-2 has also been a useful regimen in treating human melanomas. IL-2 is itself a useful component of combination immunotherapy, such as with melanoma peptide vaccines or with interferon α-2b, (IFN-α), as a dual combination or part of a biochemotherapy regimen. Several different combinations of drugs and biological agents have been used as biochemotherapy for melanoma, but although there are higher response (regression) rates, the long-range survival benefits have been marginal, not justifying the severe toxicity. Combinations of 5-fluorouracil

(5-FU) and IFN-α or levamisole have had efficacy in colon, head, and neck cancers, but here the biological agents have been biochemical modulators and not used as immunotherapy. Although experience with combinations of monoclonal antibodies and chemotherapy has been limited, it appears that trastuzumab (Herceptin) potentiates anti-tumor therapy in breast cancer but also increases the cardiotoxicity of those regimens.

We are sure that the scientific efforts of the last few decades will develop a combined anti-neoplastic immunotherapeutic modality after surgery or laser ray surgery, which will include the employment of tumor associated antigens (TAAs)-directed immunoactive, tumor infiltrating cytotoxic lymphocytes (CTL) as main effector cells of the cellular immune response and specific MoABs delivering chemical, toxic, growth factor or biological response modifier substances to the same tumor specific or associated antigenic epitope.

2. ACTIVE NONSPECIFIC IMMUNOMODULATION OF THE NATURAL IMMUNITY

The main goal of active but nonspecific immunomodulation is to achieve systematic, immune defense enhancement by active manipulation of the host immune system. The term "nonspecific" refers to the absence of neoplasm-type specificity, the absence of the effect of tumor associated antigens (TAAs) on the host immune system. These nonspecific stimulations have employed different adjuvants, viable mycobacterial organisms such as bacilli *Calmette-Guerin* (BCG), viable but inactivated organisms such as OK-432, penicillin inactivated, streptococcal bacteria, and methanol killed microorganisms such as *Corynebacterium parvum*.

The questions of whether neoplastically transformed cells are strongly or weakly antigenic or whether cancer bearing patients can respond to neoplastic antigens are not without significance in nonspecific immunotherapy. The nonspecific immunotherapy and macrophage and other antigen presenting cell activation, natural killer cell and MHC non restricted T lymphocyte actions have been very effective in cancer patients. The new era of nonspecific immunomodulation started again after the late 1950's, with numerous clinical studies having attempted a BCG vaccination. In Japan, historically controlled manipulations showed prolongation of cancer remission time and the patient's survival (26, 27).

The active nonspecific immunomodulators act mainly *via* the major antigen presenting cellular elements of the reticuloendothelial system (DCs, Langerhans cells, interdigitating cells, *etc*). The second element of their anti-tumor effectiveness is presumed to be indirect and complex. For example, the stimulation of macrophages and the other cells of RES by BCG administration is followed by the exposure of cells to endotoxin or lipopolysaccharide which results in the release of tumor necrosis factor (28). Another important part of the anti-tumor mechanism of active nonspecific immunotherapy is the induction of cytokines that may have direct or indirect anti-tumor activity.

At the present time, these mechanisms are well understood. For example, BCG administration increased the antibody formation in experimental animals (29), accelerated homograft rejection (30), increased resistance to infection (31), and enhanced resistance to transplanted tumors (32).

In the late 1960's, Mathe and co-workers (33) designed an experimental BCG or BCG mixed with irradiated tumor cells nonspecific immunotherapy for patients with acute lymphocytic leukemia after they received chemotherapy. The remission duration time and the survival time were significantly longer in patients who received non-specific immunotherapy. This study was responsible for approximately 15 years of very active clinical oncology research on the use of BCG and other nonspecific immunomodulators in the treatment of malignant tumors.

In their famous experiment, Morton and co-workers (34) injected BCG directly in the cutaneous lesions of malignant melanoma, which resulted in regression of the injected metastases. In addition, if the patient was immunocompetent, the treatment also eliminated some non-injected lesions.

In 1976 Morales, Eidinger and Bruce (35) first reported the use of intravesical *Calmette-Guerin bacillus* (BCG) (Frappier strain) or extractable residue of BCG in the treatment of superficial bladder tumors. BCG eliminated the visible tumors and prevented tumor recurrences in 8/9 patients. Since their initial report, randomized studies have shown that intravesical BCG is superior to endoscopic resection alone or intravesical chemotherapy (36-38). The exact mechanism of BCG anti-tumor activity remains unclear, but it seems that adequate *in situ* contact between the tumor cells and BCG [5 x 108 to 2 x 109 colony forming units] is necessary, because employing only systematic administration of BCG failed to reduce bladder tumors (39). The BCG doses were: 5 x 107 viable units intradermally and 120 mg dissolved in 50 ml saline intravesically. Brosman (36) conducted a randomized study with 49 patients comparing intravesical BCG with intravesical chemotherapy. 27 of the patients received BCG and the remaining 22 had thiotepa. No intadermal injections of BCG were given. BCG was administered for a total of 24 months: weekly for the first six weeks, every two weeks for three months, and monthly thereafter. No tumor recurrences were seen in the BCG-treated patients, whereas 9/19 (40%) of patients who completed the thiotepa therapy had recurrent tumors within the two years.

BCG, as was mentioned above, has a significant stimulatory effect on the cells of reticulo-endothelial system (40-44) and this enhancement has been associated with the inhibition of tumor growth in preclinical animal models (45). 175 patients involved in studies of active nonspecific immunotherapy, using BCG, (between 1982-1984) demonstrated augmentation of NK activity, monocyte mediated antibody dependent cytotoxicity (ADCC), *in vitro* lymphocyte proliferation, and delayed-type hypersensitivity responses after single or multiple doses of BCG-MER. During the nonspecific activation of the immune system, the immunoregulatory suppressor cells were also activated. By the administration of active nonspecific immunotherapeutic agents, such as BCG, the suppressor cell activity was reduced, and as a result, the cellular and humoral immune responses augmented.

Synthetic oligodeoxynucleotides containing CG motifs (CpG ODN) have potent immunostimulatory properties, and have potential as immunotherapeutic agents in human neoplastic disease (46). Preclinical, animal models suggest CpG ODN can activate a variety of immune effector cells, such as natural killer (NK) cells, and also enhance the efficacy of tumor immunization when used as immune adjuvants or to directly activate DCs, the major antigen-presenting cells. CpG ODN is also capable of altering the expression of a number of antigens by malignant B-cells, including those targeted by MoABs and those involved in communication with T lymphocytes. The ability of CpG ODN to activate the immune effector cells that participate in antibody-dependent cellular cytotoxicity (ADCC), upregulate target antigen, and perhaps induce development of an active immune response, suggest these agents may be capable of enhancing the efficacy of anti-neoplastic MoAB therapy. Such enhanced efficacy has been demonstrated in preclinical animal models and is now undergoing evaluation in clinical trials.

Recently employed nonspecific immunomodulators are in three categories:

1) those that require a healthy, functional immune system;
2) those with maximum effects on a suppressed immune system; and
3) those that act on normal and immunosuppressed systems.

3. THYMIC HORMONES

Several experimental (preclinical) and clinical studies used thymic hormones as nonspecific immunostimulants. The thymic extracts belong to the group of natural and synthetic immunomodulators:

Thymic hormones are: thymosin fraction 5 (TF5), thymosin α-1 (TA1), thymosin α-7 (TA7), thymosin α-11 (TA11), thymosins β-3 to β-11 (TB3-TB11), TP1, TP5, thymic humoral factor (THF), thymulin or factuer thymique serique (FTS), homeostatic thymus hormone (HTH), lymphocyte stimulating hormone, thymus polypeptide fraction, thymostimulin, thymosterines, and thymopoietin (47-49).

The main function of thymic extracts is to mediate the intra- and extra-thymic T lymphocyte differentiation at several stages and to influence the terminal T cell proliferation. TA1 enhances the production of lymphokines, such as IL-2, and the expression of IL-2Rs and the production of α and gamma interferon and nuclear influx of phosphoprotein. Thymosin β-4 induces in immunosuppressed mice and humans *in vivo* and *in vitro* TdT expression in T lymphocyte precursors of bone marrow, which have already been immigrated into the thymic cortex as non-differentiated thymocytes. It also stimulates the release of lutenizing hormone releasing factor (LH-RH) (50) and also has a corticotropin-releasing activity (51).

A second messenger mechanism can be proposed with three effective pathways of thymosin-mediated activation of lymphocytes at the cellular level:

1) Activation of guanyl cyclase and elevation of cyclic guanosine monophosphate (c-GMP);
2) Altered influx of calcium (Ca^{++}) followed by binding to Calmodulin;
3) Changes in the arachidonic acid metabolism. AA leave the cytoplasm and are followed by the generation of prostaglandins and leukotrienes (49).

Preclinical experimental data support the notion of a hormonal regulatory role of the thymus *via* thymopoietin or TA-1 on natural killer (NK) cytotoxic activity, lymphokine production and IL-2R expression (52-54). Similar functions were reported for TP5 (55) and for FTS (56). One of the most important effects of TA1, TF5 and TB4 is the early stimulation of lymphokines. Stimulated lymphokines include the migration inhibiting factor, lymphotoxin, IL-2 and α- and gamma interferons.

There has been an increasing interest into the possibility that immunoactive factors produced by effective cells of the immune system may also participate in the controlling of the release of ACTH, β-endorphin and other peptides derived from the corticotrope precursor, proopiomelanocortin (POMC). The thymus may act as a key source for immunologic mediators of hypothalamic pituitary-adrenal axis, through the release of hormones and cytokines produced by the cells of the thymic reticulo-epithelial network.

3.1 Use of thymic extracts in clinical trials

Thymosin α-1 was produced by conventional chemical synthesis and its chemical composition is identical to the polypeptide found in blood and mammalian tissues; it does not have local or systemic toxicity. Extensive experience with thymosin fraction 5 demonstrated presence of a dominant immunoactive peptide responsible for numerous cellular immune restorative properties of this thymic extract (57). Clinical benefit obtained with the thymic peptides is directly related to the elimination of several deficiencies and associated syndromes of the cellular immunology. The immune modulation with thymic extracts never resulted in the generation of a hyperimmune state in the patients and clinical improvement has been usually detectable by one to four months after the therapy. No adverse effects have developed during a long-term treatment with thymic extracts.

In a randomized, double-blind, placebo controlled trial in non-small-cell lung cancer patients treated by radiation therapy, thymosin α-1 (TA1) significantly increased the time of survival. The goal of immunomodulation therapy is to restore the number of circulating immunocompetent T lymphocytes in the peripherial blood and the increase in the T lymphocyte precursors. Any therapeutic advantages accrue from these results. It was postulated that after irradiation, a well functioning cellular immune system will exert an inhibiting effect on the growth of the residual tumor. The administration of thymosin α-1 began immediately after the last session of irradiation.

Thymosin Fraction 5 and thymosin α-1 were used in Phase I and II studies in 1987 in the Veteran Administration Medical Center in La Jolla, CA (58). During the study, 12 patients with advanced colon cancer were treated thrice weekly with thymosin fraction 5 at a dose of 120 mg/m^2 and 10 patients with non-small cell lung cancer received thymosin α-1 at 1.2 mg/m^2 also thrice weekly. There were no tumor responses observed in any of these patients and neither was immune system augmentation obtained. These negative results probably resulted from the early observations immediately after the experimental immunomodulation. These non-TAA specific involved host defense mechanisms were poorly understood as the field of nonspecific immunotherapy was developed in preclinical animal models and in cancer patients.

In completed phase II and III trials of Radiation Therapy Oncology Group (RTOG), patients with non-small-cell lung carcinoma who displayed tumor regression or stabilization after irradiation were randomized to receive thymosin α-1 or placebo twice weekly until tumor recurrence or for a maximum of a year. Among stage 3 tumor patients, those who recurred or died during the one year (minimal time of effective therapy was 26 weeks) treatment with thymosin α-1 lived significantly (50%) longer than those who received only the placebo.

3.2 Thymic hormones as adjuvants to virus vaccines

Influenza is still a major cause of death among the elderly population and in children under two years of age. During each epidemic year, there are more than 20,000 influenza related deaths in the United States alone. Children and high risk group adults usually receive immunization. However, the anti-influenza vaccines often are not effective in these patients. Therefore, adjuvants were employed to enhance the number of successful immunizations. Phase II clinical trials designed to detect the ability of TA1 to increase the responsiveness to anti-viral vaccines in elderly patients have been completed at the Cornell Medical Center (59). The results showed that 6/7 patients developed high levels of anti-influenza antibodies. A second trial was performed at the University of Wisconsin Veterans' Administration Medical Center on 91 veterans over 64 years of age. 68% of those who received TA1 with the influenza vaccine were effectively immunized, compared with only 46% receiving placebo with the vaccine. The study demonstrated that the efficacy of the influenza vaccine is increased by 50% if TA1 was used as adjuvant. Similar clinical trial was conducted in 23 patients with renal failure who were receiving chronic dialysis at the University of Maryland with hepatitis vaccine (60). Seven of eleven patients, who had previously failed to respond to the vaccine, developed protective antibody levels if the hepatitis vaccine was administered together with TA1 as adjuvant. Future developments in this area are predicted.

A number of other agents have been or are being tested as members of the "nonspecific immunomodulators."

3.2.1 Bacterial Products

Extracts from some Streptococcus (OK-432 preparation from Japan, potent immunostimulator, activating macrophages and T lymphocytes and increasing the production of TNF and several other cytokines in both animal and human experiments (61) and Staphylococcus increase the proliferation of suppressor cells); Bordetella pertussis vaccines (62) - in cases of lymphoma and leukemia and inducing an immunosuppressive effect in the host mice, also in experiments with AKR mice); CGP 31362, a synthetic bacterial cell wall analogue and activator of macrophages was used in a preclinical mouse experimental model for the treatment of metastatic renal cell carcinoma (63). The tumoricidal immunomodulator was delivered in its multi-lamellar phospholipid liposomes-encapsulated form (MLV-CGP 31362) and was able to activate the alveolar macrophages *in situ* (64, 65). Therapy with repeated i.v. injections significantly reduced the number of spontaneous lung metastases. Optimal therapeutic effectivity was achieved by combining the use of i.v. MLV-CGP 31362 with the s.c. injections of recombinant murine gamma interferon (66). Several lines of evidence indicate that *in situ* activation of macrophages was responsible for the therapeutic effects of MLV-CGP 31362.

3.2.2 Immunostimulation with Oncolysates

The oncolytic product of neoplasm tissue destruction after infection with oncolytic viruses (67-70). Koprowski and co-workers (71) first noted that animals recovering from ascites tumors after oncolytic virus infection were resistant to rechallenging with the same tumors. Lindenmann and Klein (72) demonstrated that cell-free lysates of tumor cells infected by the influenza virus could confer stronger protection against rechallenge than uninfected tumor cell lysates. Adenoviruses have been critical in the development of the molecular approaches to brain tumors (73). They have been engineered to function as vectors for delivering therapeutic genes in gene therapy strategies and as direct cytotoxic agents in oncolytic viral therapies. Today, the employment of adenoviruses in brain tumor immunotherapy is at the phase III clinical trials of adenovirus-mediated p53 gene therapy and the application of two conditionally replicative adenoviruses (CRAds) ONYX-015 and Delta 24.

The smallpox vaccine (vaccinia virus) was the only one able to duplicate the lytic action of the influenza on the Erhlich ascites cells. The immune mechanisms operative in the vaccinia virus augmentation regimen is the induction of virus specific T helper lymphocytes, and the potential significance of both cellular and humoral effector mechanisms (74, 75).

Csatary, a Hungarian born physician, recognized the anti-neoplastic cellular effect of the Newcastle Disease Virus (NDV) infection and in 1968, with the Ethical Commity's permission at the Jefferson Memorial Hospital in Alexandria, VA, he began clinically treating stage IV, metastatic, terminal cancer patients with NDV (MTH-68/H). Results were published (76).

In 1988-90, several phase II clinical trials were carried out in Budapest Hungary, under the direct supervision of Dr. Eckhardt, Director of the Hungarian National Institute of Cancer (77).

MTH-68/H product contains the live, naturally attenuated, mesogenic Hertfordshire (H) strain of NDV, (classified as avian paramyxovirus-1). MTH-68/H is an oncolytic virus selectively replicating in various tumor cells leaving healthy cells intact, having direct and indirect antineoplastic efficacy. Its presumed antineoplastic activity includes cell lysis, induction of proapoptotic mechanism and also stimulates specific as well as aspecific anti-tumor immune mechanism in the host organism. It is a highly purified product, manufactured according to the human GMP requirements and administered intravenously.

In 2004, Csatary, myself, and others published a paper describing case histories of high-grade glioma patients who after exhausting conventional therapies received

MTH-68/H treatment IV was employed as a form of anti-neoplastic therapy. Five to nine years of survival was achieved (78). In the case of GBM, such a long survival rate has never before been reported. Against all odds, each patient resumed a lifestyle that resembled their previous daily routines and now enjoy a high quality of life.

Preclinical, animal brain tumor models and *in vitro* experiments in cell cultures have demonstrated the feasibility of RV mediated gene transduction and the killing of glioma cells by toxicity generating transgenes. Phase I and II clinical studies in patients with recurrent malignant glioma have shown a favorable safety profile and some efficacy of RV mediated gene therapy. On the other hand, a prospective randomized phase III clinical study of RV gene therapy in primary malignant glioma failed to demonstrate significant extension of the progression-free or overall survival times in RV treated patients. The failure of this RV gene therapy study may be due to the low brain tumor cell transduction rate observed *in vivo*. The biological effects of the treatment may also heavily depend on the choice of transgene/prodrug system and on the vector delivery methods. Retrovirus clinical trials in malignant glioma have nevertheless produced a substantial amount of data and have contributed toward the identification of serious shortcomings of the non-replicating virus vector gene therapy strategy. Novel types of therapeutic virus vector systems are currently being designed and new clinical protocols are being created based on the lessons learned from the RV gene therapy trials in patients with malignant brain tumors.

3.2.3 Vitamins and Vitamin Analogs

Vitamin A analogs (retinal, retinol, retinyl acetate, all-trans retinoic acid), vitamin C tilorone, L-fucose. The all-trans retinoic acid induces "terminal differentiation" and complete remission of the malignant cells of acute promyelocytic leukemia (APL). As a method of treatment, the aim of retinoic acid therapy is in connection with the differentiation level of the cancer cell. The neoplastic cell differentiation is blocked at some primitive stage, and if we are successful in overcoming or reversing this blockage, the neoplastically transformed cells will once again follow a normal cell growth pattern. All-trans retinoic acid *in vitro* induces leukemic cells from patients with acute promyelocytic leukemia (M3) to differentiate into mature granulocytes, which express the CD15 antigen and are capable of respiratory burst function (79).

3.2.4 Chemical Immunomodulators

Sulfur containing Levamisole, diethyldithiocarbamates; cyanoaziridines - Azimexon, ciamexon, imexon; tellurium based - AS101; Taxol, which is derived from the Pacific yew tree (a study on women with advanced epithelial ovarian cancer was done after they failed three chemotherapy regimens); tellurium chloride, carbocyclic nucleoside (Cyclopentenyl-cytosine) (80, 81); and synthetic polynucleotides or interferon inducers - Ampligen, poly A-U, maleic anhydride, poly I-C; Pyan copolymers (MVE-2).

3.2.5 Regulators of Normal Physiological Functions

A progesterone-like drug - lynestrenol, PGE-2 synthesis inhibitors like Indomethacin, H-2 receptor antagonists like Cimetidine and suppressor cell regulation by cytotoxic chemotherapy - cyclophosphamide;

3.2.6 Glucans

Yeast glucan, a component of the cell wall of Saccharomyces cerevisiae, stimulates the cells of reticulo-endothelial system (RES), as well as enhancing both the humoral and cellular-mediated immunity. The glucans also potentiate anti-tumor immunity. Other glucans used in the clinic are fungal derivates like Lentinan, Pachymaran, and Scizophyllan.

4. INTERFERONS: BASIC AND PRECLINICAL STUDIES

Interferons (IFNs) are cytokines, a family of natural proteins produced by specific cell types in response to virus, antigens, mitogens and double-stranded ribonucleic acid (RNA). Viral infections result in the production of interferon (IFN), discovered in 1957 (82). IFNs are active, low molecular weight proteins capable of preventing viral replication. Interferons have very high anti-viral specific activity. Their anti-viral effect can be detected at low concentrations as 10(-12) to 10(-13). These proteins are capable of inducing an anti-viral state and interfering with the viral challenge of target cells. The IFNs are protective against both RNA and DNA viruses (83).

IFNs had their beginning with the discovery by Hoskins (84): the attenuated neurotropic variant of yellow fever protected monkeys against an otherwise fatal simultaneous infection with pantropic yellow fever virus.

The biochemical mechanism of anti-viral and anti-microbial activity of IFNs is based on inducing one of the following enzymatic systems: 1) activation of 2',5' oligoadenylate (2',5' A) synthetase followed by endogenous ribonuclease activation that catalyzes the cleavage of both viral and host cellular RNA, and could be broad-spectrum against nearly all viruses but very narrow-spectrum with regard to host species in which the interferon was secreted and in which it could be demonstrated to show activity (85); 2) a 2',5' phosphodiesterase that catalyzes RNA degradation and thus further inhibiting new protein synthesis, but also degrading 2',5' synthetase (5, 6); 3) a protein kinase that phosphorylates proteins P1 and elongation initiation factor EIF2α, which in turn inhibits the binding of transfer RNA to the surface of ribosomes. The last inhibition results in a discrimination between host cell and viral mRNA (86, 87); 4) indolamine 2,3 dioxygenase decreasing the intracellular tryptophan level (88).

Interferons also mediate a wide range of other biologic responses, probably by initiating the production of a series of new proteins: anti-proliferative effects, cell transformation suppression, direct cytotoxic effects, immunomodulation (intensification of the host's immunity) and regulation of lipid metabolism.

The mechanism by which gene expression is modified by interferons requires protein binding to its receptor pertains. Deactivation of oncogenes (89), gene activation and tumor cell differentiation induction (90, 91), interference with growth factors or their receptors, and direct inhibition of tumor-induced angioneogenesis (92) also belonged to interferons.

Interferons, and substances capable of inducing endogenous IFN synthesis, have been demonstrated to have direct and indirect cytotoxicity against numerous animal tumors (93-98).

The detection of ribonucleic acid (RNA)-dependent desoxyribonucleic acid (DNA) polymerase in RNA oncoviruses, as well as in human leukemic and other tumor cells, has provided additional information about the "virus infection aetiology" theory of carcinogenesis. Virus infections should be thought of as the second most significant risk factor for cancer development in humans. Human cancers associated with papillomaviruses (PV), hepatitis B virus, Epstein-Barr virus, and T lymphocyte leukemia-lymphoma virus infections are responsible for about 15% of the neoplasm incidence. Cases of uteral cervix cancer and hepatocellular carcinoma represent 80% of virus-linked malignancies. JC and BK virus genomic DNA is present in a fraction of tumor cells in gliomas and insulinomas. PVs are a group of eptheliotropic viruses that cause benign epithelial proliferations. The cottontail rabbit papillomavirus (CRPV) induces growth of papillomas that often convert into squamous cell carcinomas. More than 60 pathogenic humans PVs (HPVs) genotypes have been isolated (99). A large number of HPV infect the anogenital tract, with anogenital cancers representing about 10% of all cancers occurring around the globe.

Interferons are produced when viruses, viral RNA, or other inducers interact with a number of viable cells. The precise intracellular stimulus is unknown. Nonviral substances capable of inducing IF synthesis are classified into four groups: 1) microorganism (bacteria, protozoa, chlamydia, rickettsiae) and some viruses (Herpes simplex) (100) with the requirement of intracellular localization for replication; 2) microbial extracts, such as endotoxin and plant mitogens like PHA and pokeweed extract; 3) natural and synthetic polymers. The natural substances are double-stranded RNAs extracted from plants or normal liver or detected in fungal extracts such as statolon and helenine. Synthetic substances are also double-stranded RNAs, *e.g.* polyriboinosinic-polyribocytidylic acid (poly I:C), and the polycarboxylate, pyran copolymer, and polysulfate families of plastics. The best analyzed IF-inducer is poly ICLC, a hydrophilic complex formed between poly-l-lysine, carboxymethylcellulose, and the polyribonucleotide composed of inosinic and cytidylic acids. Poly ICLC interacts with a cellular gene to release repressors from the control sites of interferon gene transcription. Diethylaminoethyl dextran compound is capable of stabilizing and increasing the cellular uptake of single-stranded RNAs, which are poor IF inducing agents; and 4) chemically defined, purified low molecular weight substances like kanamycin, cycloheximide, tilorone hydrochloride, *etc*. It has been demonstrated in both animal preclinical models and human experiments that the drug flavone-8-acetic acid has potent IF-inducing properties. They appear to act in paracrine capacity in the immune system with a wide range of effects on the endocrine system (101).

There are basically three interferon systems separated by standard chemical procedures and classified according to their source: α (from leukocytes), β (from fibroblasts), and gamma (from antigen- or mitogen activated T lymphocytes and natural killer - NK cells) (101, 102). In fact, the α and β types appear to be produced by virtually every cell in the human body. There are marked homologies between α and β interferons: their genes cluster on chromosome 9 and they both express a cell surface receptor encoded on chromosome 21.

Interferons α and β are closely related (type I interferon). In contrast, the gamma gene was defined on chromosome 12 and shows no homology with other interferons. The typical surface receptor is encoded by chromosome 6 and is different from that present on α or β interferons (103).

Sinkovics (89) submitted a simplified, useful nomenclature of human IFNs:

1) IFN-α = natural leukocyte IFN = HU-nIFN-α;
2) IFN-α-N1 = Namalva IFN = Lymphoblastoid IFN-α = Hu-IFN-α Ly;
3) IFN-α2a = recombinant IFN-A = Hu-rIFN-α2a (AA at position 23 and 34; lys; his);
4) IFN-α2b = Hu-rIFN-α2b (AA at position 23 and 34; arg; his);
5) IFN-α2c = Hu-rIFN-2c (AA at position 23 and 34 arg; arg);
6) IFN-β = Hu-nIFNβ, Hu-rIFNβ; and
7) IFN-γ = Hu-nIFN- γ; Hu-rIFN- γ.

Interferons were purified by precipitation and gel exclusion chromatography from the sources mentioned above.

The advent of gene cloning has allowed the complete structure of interferons to be understood (102-104). A-IFN was the first to be cloned. Gene cloning identified at least 15 α-forms, two subtypes of β (1 and 2), and only one gamma species, each with subtle amino acid differences (105). IFNs produced by recombinant DNA technology in *E. coli* are not glycosylated but this lack of glycosylation did not alter their biochemical efficacy.

Interferon induced changes in human cell gene expression:

1) Increased expression:
   HLA-A,B,C (in various cells of human body increased expression of MHC Class I and Class II antigens); Tubulin (in lymphoblasts); β2-microglobulin (in various cell types); Metallothionein (in neuroblastoma); 2'5' oligoadenylate synthetase (in various cells all over the body); 6-16 and 9-27 (in neuroblastoma); HLA-DR (in myeloid leukemia).

2) Decreased expression of oncogenes:
c-fos (in 3T3 fibroblasts); c-myc and c-fgr (in Burkitts lymphoma - Daudi); c-Ha-ras and c-src (in bladder carcinoma).

The major identified immunological regulations of INFs include: 1) induction of expression of class I and II MHC antigens on the surface of potential target tumor cells; 2) activation of a natural immunity, including of NK, K and probably lymphokine-activated killer (LAK) cell responses (inducing an early cytotoxicity and antibody-dependent cell-mediated cytotoxicity (106-113); 3) intensification of antigen specific (TAA), $CD8^+$, MHC restricted T cell mediated cytotoxicity; 4) stimulation of helper, $CD4^+$ lymphocyte function; 5) macrophage activation and induction of much Fc receptor expression and; 6) suppression (with early application into the immunologic response) or activation (applied permanently, but using low dosage) of antibody responses (89-114). The optimum dose for immune stimulation is unclear. Lower doses of α-IFN produced higher NK activity than larger doses did (115).

4.1 Lymphokine activated killer (LAK) cell system

The study of Damle and Doyle (116) with α, β, and γ IFNs was undertaken to determine their regulatory effect in the induction of the effector phase of LAK cell function. Neither endogenously induced nor exogenously added IFNs had any effect on the LAK cell subpopulation, which is in contrast to their augmenting effects on NK cells and CTL.

Interferons can alter effector cell function, antibacterial activity and cause multiple changes in the surface phenotype by altering the cell membrane and increasing the expression of Class I and II MHC molecules (human leukocyte antigens-HLAs) and β-2 microglobulin on both tumor and normal cells (117-119). Interferon activates NK cell migration in tissues and can enhance ADCC mediated by killer (K) cells. Other immune altering activities common to all interferons include activation of B cells and augmentation of other lymphokines, such as interleukin-2 (IL-2) and tumor necrosis factor (TNF). Gamma INF appears to have the widest immune stimulation among interferons. It is the most potent activator of Fc receptor expression on monocytes/macrophages. All MHC Class II antigens (HLA-DR, HLA-DC, HLA-SB) (120, 121) are under the regulatory function of INFs. Interferons are also capable of inhibiting angioneogenesis induced by tumor humoral factors (122). Other cellular changes are achieved by alterations in tubulin production, as mentioned above, which is so important in the normal architecture of the cytoskeleton.

Clinical oncologists have long noted that a small number of cancer patients will enter a spontaneous remission following a viral infection. Interferons have been found to be useful in the therapy of numerous infections and immunological disorders and they actually represent the treatment of choice for some malignant tumors (118, 123-125). Phase I-III trials have been conducted in more than 5,000 patients suffering from different neoplasms and various virus infections. Phase I and Phase II clinical trials were initiated in 1981. These trials were designed to ascertain the maximal tolerated doses (MTD) and dose-limiting toxicities (DLT) in various schedules and administration routes, as well as direct cytotoxic effects, regulation of immunomodulation and cell differentiation and synergy with irradiation therapy or with chemotherapeutic agents.

4.2 Clinical responses with interferons

Interferon (human, leukocyte) was employed for the first time for anti-neoplastic immunotherapy in advanced stage nodular, poorly differentiated lymphomas, with complete failure to demonstrate any tumor regression for all three observed histocytic type lymphomas (126). A study with the same IFN for anti-neoplastic treatment used the partially purified (about 0.1%) human leukocyte INF in 17 advanced breast cancer patients, 10 patients with multiple myeloma and 11 with malignant lymphomas (127), with significant regressions only in multiple myeloma patients who showed a disappearance of Bence-Jones protein from urine. Similar results were reported by Mellstedt and co-workers (128) with other studies having been conducted in other types of advanced malignancies.

There were several other preparations, but none recombinant interferon preparations, considered for immunotherapy and studied clinically in detail. Only four preparations with virtually the same biologically active ingredient, leukocyte or recombinant α interferon, remained in clinical trial: 1) Hu-IFN α (Le), produced by the Helsinki Red Cross; 2) Hu-IFN α (Ly), from Welcome; 3) Hu-rIFN α A, from Hoffmann-La Roche; and 4) Hu-rIFN α2, from Schering-Plough.

"Both β- and γ- interferons are only just starting Phase II trials", wrote Goldstein and Laszlo in 1988 (20), but two years later in 1990, the first successful therapy of the hairy cell leukemia with a type I interferon, called "β-ser", a potent inhibitor of cell proliferation (129) had been described. "β-ser" was administered three times weekly as an intravenous injection of 90 million units per dose. All of the response rates mentioned in the literature were defined as either "complete" or "partial". A complete response (CR) meant total disappearance of all neoplastic tissue following the therapy and a partial response (PR) meant greater than 50% shrinkage of all measurable tumor masses. The significance of the results not meeting the standard 50% regression was under controversy (130).

Between 1980 and 1982, two other devastating cancers were found to be sensitive to α-IFN treatment, such as renal cell carcinoma (131), but showing disappointing results with other, conventional therapies and Kaposi's sarcoma, which is associated with acquired immunodeficiency syndrome (AIDS) (132).

Since 1982, the best results have been reported in hematologic malignancies, most notably in hairy cell leukemia, a rare B cell disorder for which no conventional therapy existed and with reported CR rate up to 100%, defined as clearing of all leukemic cells from the bone marrow also including greater than 90% PRs, defined as disappearance of hairy cells and increase of hematologic parameters (133-136). Samuels and co-workers (137) demonstrated the direct biochemical effect of the recombinant α-IFN on hairy leukemia cells in terms of the induction of the synthesis of specific proteins. It was also shown that α- and γ-IFNs interact with separate receptors on hairy cells, termed type I for α-IFN and type II for γ-IFN (138, 139). According to Niederle and co-workers (140), γ-IFN was found ineffective in the therapy of hairy cell leukemia when compared to α-IFN. Exposure of hairy cells to α-IFN resulted in the induction of the synthesis of several proteins, ranging from 58-139 kD, with the most prominent being 80 kD. The most prominent protein induced by γ-IFN was 62 kD. Two other proteins of approximately 75 and 120 kD were also induced by both α- and γ-IFNs. There are six other malignant diseases where interferon as a single agent may have a role to play: 1) multiple myeloma (141-144); 2) non-Hodgkin's lymphoma, renal cell carcinoma (145-151), melanoma (152-154), Kaposi's sarcoma (132, 155, 156) and juvenile laryngeal papilloma. Complete responses, often of considerable duration, have been observed in these diseases. Others have also demonstrated efficacy of recombinant α-IFN against cutaneous T-cell lymphoma (mycosis fungoides, Sezary syndrome) (157) and in endocrine pancreatic tumors (158). Clinical studies after 1988 also included β and γ IFNs, and have thus further established the place of IFNs among the immunotherapeutic agents of human cancer. γ-IFN increases the expression of HLA-DR antigens on tumor cells and they become vulnerable to cytotoxic lymphocytes (159-161). Recombinant γ-IFN was tested against human renal cell carcinoma that had been transplanted into a preclinical nude mice model (162). Treatment with r-γ-IFN inhibited tumor growth in 13/17 mice. Rinehart and co-workers (163) undertook a Phase I trial of rgamma-IFN in thirteen patients with metastatic renal cell carcinoma (no history of previous therapy). γ-IFN was given by a 4 h i.v. infusion and a second dosing was applied in 10 minutes i.v. infusion twice weekly. The applied r-IFN-γ had very limited objective biologic effects *in vivo* (lymphocyte subsets, proliferation rates or 2',5'-oligonucleotide synthetase were not altered). Clinical beneficiary effect after γ-IFN treatment also was not noticed. It was reported that a phase I study of the effect of the combined use of rIL-2 and α-interferon-2a in patients of renal cell cancer, colorectal cancer, melanoma and B cell malignancy (165). The defined maximum tolerated doses (MTD) were: 5 x 10^6 U/m^2 for the rIL-2, recombinant α and α-IFN 6 x 10^6 U/m^2. Thirty-four patients were evaluated for clinical response, resulting in 4/18 (22%) renal cell carcinoma patients having a partial response. No response was noticed in patients with melanoma, lymphoma, or colorectal cancer.

4.3 Toxicity of interferons

A variety of side effects and toxicities have been recognized since the start of the INF era in human cancer treatment (135, 166). In 1987, Rinehart and co-workers described several side effects after using rIFN β-serine and r-IFN-gamma in Phase I and II trials on metastatic renal cancer patients. Both IFs caused the usual *in vivo* toxicity (*i.e.* fever, chills, and malaise) and rIFN β-serine exhibited marked immunomodulatory and anti-tumor effects *in vitro* and *in vivo*.

A dose of 3 million units/day of partially purified α-IFN demonstrated its efficacy in patients with hairy cell leukemia, lymphomas, breast cancer, renal cell carcinoma and multiple myeloma (135, 142, 166). The maximum tolerated dose in most patients is of the order of 30 million units/m^2 surface area/week. Therefore, IFNs use requires careful clinical monitoring.

The pyrexia and myalgia are not dose-related side effects; their one week duration has been reported (102). The other side effects (*i.e.* nausea and vomiting, headache, and lethargy) are dose-related and have an average of 2-3 weeks duration. Central nervous system toxicity, anorexia and weight lose are the most serious, dose-dependent side effects, with a duration time of approximately one month.

4.4 Conclusions

1) Endogenous interferons are produced in small amounts and probably work in a paracrine *Fas*hion. They may play a role in the host's defense against viral and bacterial infection, in protecting against neoplastic transformation, and in the permanent process of hemopoiesis.
2) Interferon-α has shown significant anticancer effect against a wide range of human neoplasms, especially in the treatment of hematological disorders. In the light of current data of clinical trials it

seems that for future immunotherapy the α and β IFNs will be used more successfully.

3) Interferons also potentiate *via* direct or indirect mechanisms that respond to other agents. For example, synergestic interaction between IFN-α and IFN-γ and between interferons and cytokines (IL-1, IL-2, tumor necrosis factor).

4) The immunologically active γ-IFN has special immuno-biological effects on tumor cells, especially on the expression of different receptors on cell surfaces.

5) The successful use of the members of cytokine regulation is likely to be coupled with the early phase in the course of cytokine irregulatory disease.

6) The capacity of viruses and neoplastic processes to interfere with the IFN system is thought to represent a "virus-against-host" or "cancer-against-host" defense mechanism (167). Four resistance factors have been identified: 1) release of free IFN- α/β type 1 receptors into the circulation that, at appropriate concentrations, capture and inactivate IFNs; 2) a new IFN inhibitory protein has been isolated and its chemical structure is under study; 3) prostaglandin E2, which is produced by certain tumor cells, inhibits IFN production; and 4) high levels of cAMP phosphodiesterases present, for example in certain tumor cells, reduces cAMP, an important second messenger in IFN synthesis. Studies are under way to reverse these inhibitory effects and to increase endogenous interferon production.

5. TUMOR NECROSIS FACTORS

5.1 Basic and preclinical characterization

Tumor Necrosis Factor alpha (TNF-α) is a tumorcidal cytokine, produced and released from endotoxin influenced macrophages, which induce necrosis formation in tumors. TNF, in combination with interleukin-1 and IL-6, is able to mediate many aspects of the acute stage of inflammation. Lymphotoxin (LT), a tumorlytic protein produced by activated lymphocytes, is 30% homologous with TNF and appears to attack the same receptor, and has therefore, also been called TNF-β. Cachectin, the protein of chronic wasting disorders and sepsis, also secreted by macrophages, was identified in the early 1980s (168). Based on sequence analysis, it became clear that TNF and cachectin in fact represent the same molecule (169). Thus, scientists interested in such diverse disorders as wasting syndromes, septic shock, and tumor necrosis have all been introduced to a small group of closely related molecules with diverse effects on cells.

The close relationship between cancer growth and the influence of bacterial infection on it was already mentioned in connection with the pioneer studies of a New York surgeon, Dr. William B. Coley, who observed a hemorrhagic necrosis in a neck sarcoma patient, who had contacted an intercurrent streptococcal infection. He administered erysipelas cultures to ten advanced tumor patients and observed tumor inhibitory effects. Later he used only extracts from Streptococcus pyogenes and Serratia marcesens and he demonstrated that this mixture of bacterial extracts was as effective as the inoculation of bacterial cultures. Shear and co-workers (170) defined biochemically a polysaccharide factor from *S. marcesens* culture filtrates that was able to cause "hemorrhagic necrosis" of a sarcoma tissue transplanted from another animal. The active substance of the extract was purified and identified as lipopolysaccharide (LPS) originated from the bacterial cell wall. O'Malley and co-workers (171) reported that serum of mice in shock could initiate necrotic changes on other tumors transplanted to the animal. Carswell and co-workers (172) similarly observed that a serum factor could elicit hemorrhagic necrosis of transplantable tumors *in vivo* in mice primed with BCG and treated with LPS. The name tumor necrosis factor (TNF) was given to a factor, present in the sera of BCG-primed, endotoxin (LPS)-treated mice, which was able to cause dramatic necrotic changes in a transplantable murine sarcoma (Meth A). Priming with *Corynebacterium parvum* prior to LPS was effective in also inducing TNF production (173). *In vitro* TNF production by isolated macrophages was also soon established. TNF was found to mediate many of effects of endotoxic shock previously ascribed to cachectin.

Aggarwal and co-workers (174) analyzed the molecular weight of mature human TNF isolated from the culture supernatants of HL-60 cells and stimulated with phorbol ester (TPA). After purification to homogeneity, the human TNF-α demonstrated migration with an apparent molecular weight of 17.1 kD, amino acid analysis confirmed a primary peptide structure. The protein was defined to have an isoelectric point (pI) of 5.3, containing an intrachain disulfide bridge between the cysteines in position 67 and 101. The mature human protein was determined to be 157 amino acid residues long. The human precursor propeptide contained 76 additional amino acids attached to the N terminus of the mature protein (in mouse, this number is 79). The mature TNF-α contained seven regularly spaced hydrophobic peaks suggesting the presence of uniform areas that pass through a hydrophilic core region.

Pennica and co-investigators (175) reported the isolation and nucleotide sequences of cloned DNAs for both human tumor necrosis factor (TNF) and lymphotoxin (LT or TNF-β). The study proved that these two immunomodulators have distinct molecules with only 30% structural homology and very similar, if not identical biological activities.

Cytogenetic analyses demonstrated that the genes for human TNF-α and TNF-β are located near the HLA locus on the short arm of chromosome 6, lying in a tandem arrangement in the genome separated by 1100 bases (176, 177). Both genes localize in the 6p23 - 6q12 segment. The TNF-α gene contains three introns, one of which interrupts the sequence encoding the hormone. These genes expressed in *E. coli* encoded the AA sequence of TNF (178). Homology was described in the 5'untranslated region of both the TNF-α and TNF-β genes indicating the possibility of coordinated control of gene expression and that these two genes may have diverged several hundred million years ago (176, 179). Expression of TNF-α or TNF-β belongs to a distinct cell lineage. The TNF molecule is highly conserved across species as well. The mouse and human TNF amino-acid sequences are 79% homologous (180). TNF has two highly active cysteine residues in men, rabbits and mice. Prostaglandin E2 (PGE2) and cyclic nucleotides are crucial mediators of TNF gene regulation. Exogenous PGE2 decreased TNF expression in a dose-dependent *Fas*hion on a transcriptional level (181). PGE2 also increased levels of adenosine 3':5'- cyclic phosphate (cAMP) through receptor-mediated increases in adenyl cyclase activity.

Cellular immunological assays defined that the activated macrophages secrete TNF-α while TNF-β is a product of activated lymphocytes (174, 179, 182). Although TNF appears to have a glycoprotein in nature, there is no evidence for glycosylation of human TNF, which is important for murine IFNs (174, 183). All known TNFs fold into a trimetric structure (184, 185) and as such, the lack of glycosylation does not affect the degree of polymerization. Some reports suggest that the trimer form is required for full biologic activity, others demonstrated that TNF in trimer form has a greater affinity for its receptors (184).

Activated T lymphocytes surfaces express the pluripotent TNF molecule, when signaled with the synergistic combination of a calcium ionophore, ionomycin, and a protein kinase C activator, 12-o-tetradecanoyl phorbol acetate (186). The expression of TNF on activated T cells provides a mechanistic basis for the realization of the effect of TNF in an antigen-specific *Fas*hion. Cell surface radioiodination observations localized the presence of a 26 kD transmembrane protein, a size predicted by TNF cDNA. The induced cell surface expression of TNF can be blocked pretranslationally with cyclosporine and/or methylprednisolone.

Like other cytokines, TNF acts *via* a specific cell surface receptor. The same receptor acts as the receptor for LT, which means the two cytokines are in competitive binding (187). The various TNF-α effects on different cell types are suggestive of molecular specificity at the level of cell membrane receptors. Human TNF receptor has been purified and cloned by several investigators (188). Chensue and co-workers (189) demonstrated intracytoplasmic and cell membrane-associated expression of TNF-α with immunohistochemical antigen detection methods. TNF presence was identified throughout the lymphoid tissue (190), with elevated levels in the serum of patients with lupus erythematosus, Kawasaki disease, systemic histiocytosis, and in B chronic lymphocytic leukemia (191). Tsujimoto and co-workers (193) have detected that 125I-labeled TNF binds to high-affinity cell surface receptors. The number of these high-affinity receptors in different cell lines has shown to range from 0.2 - 10 x 103 sites/cell (193-195), with TNF receptors expressed on both normal and neoplastic cells. Kull and co-workers (193) demonstrated the presence of two polypeptide chains of 75 and 95 kD, both of which participate in the binding of TNF-α to the cell surface. Four cellular polypeptides (138, 90, 75 and 54 kD) have been found, having been isolated from a high-affinity TNF-α receptor complex. The 138 kD protein was detected only in the cell line MCF-7, which is in contrast to the remaining three lower molecular weight proteins that were found in all of the investigated cells. Therefore, it has been suggested that the 138 kD protein is involved either in directing the cytotoxical effect or is responsible for mediating the cellular anti-proliferative effects of TNF-α. Nearly all cells appear to express TNF receptors and their number 100 to 7000 per cell does not predict sensitivity to TNF directed cytolysis (194, 196). The number of TNF receptors is actively regulated. Both type I interferons and gamma-interferon increase the expression of cell surface receptors by two- to threefold in certain carcinoma cell lines through active protein synthesis. Aggarwal and co-authors (197) reported that lectins such as *Concavalin A* may cause a two fold increase in the number of TNF-α cell surface receptors. However, the cytotoxic response of target cells to TNF-α can apparently be abrogated by Con A treatment. The number of TNF receptors is down-regulated by the cytokine IL-1, LPS and phorbol esters (198). Once the ligand binds to the TNF receptor, the complex is internalized and rapidly degraded.

5.2 Physiologic action

As a postreceptor biochemical effector, TNF first arrest cells in the G2 phase (199). Lysosomotropic agents verified the mechanism of endocytosis, suggesting that internalization of TNF into the cells is a requirement and is mediated by active endocytosis (196, 33, 34). Effective TNF modulated cytolysis is on its way (173). However, TNF-α did not affect the rate of progression through the cell cycle. Purified human TNF arrested MCF-7 cells at G1.

The mechanism of cell killing (drug dependent cellular cytotoxicity - DDCC) through toxins involves either:

1) direct perforation (destruction) of plasma membrane units;
2) elimination of cytoplasmic organelles, following active or passive entry of the toxin; and
3) a combination of both mechanisms mentioned above.

The specific activity of purified TNF protein is approximately 10(7) cytolytic units per milligram of protein in the L929 cytolysis assay with actinomycin D. The role of phospholipase activation and lipid metabolism was examined in lymphotoxin (LT), with TNF-mediated *in vitro* destruction of murine L929 cells (202). After LT and TNF were employed, cell destruction began at 4-6 hours and was 99% complete by 30 hours. Cell membrane phospholipids, labeled prior to the experiment at the C2 position with 14C arachidonic acid, were analyzed by two-dimensional thin layer chromatography and quantified over a 30 hour time course after LT or TNF treatment. Radio-labeled arachidonic acid, eicosanoids, and neutral lipids were released into the medium prior to the onset of cell death (4-6 hours) and continued to accumulate linearly throughout the destructive reactions. Heicappell and co-workers (203) demonstrated *in vitro* that a short 30 minute exposure of TNF-α to renal carcinoma cells were able to induce significant cell lysis, when measured 72 hours after. In contrast, Rosenblum and Donato (204) found that exposure of log-phase ME-180 cells to TNF-α for 12 hours had little effect on the cellular proliferation. The optimal anti-proliferative effect can be achieved with at least 24 hours continuous exposure (205).

The standard *in vitro* "biological assay" for TNF activity uses the mouse L-929 fibroblast cell line. Within twenty-four hours after addition, the TNF caused cytolysis in cells pretreated with actinomycin D (172). Without pretreatment, L-929 fibroblasts are virtually resistant to TNF-induced lysis (206, 207). Cytotoxicity of TNF on L929 cells could be blocked by several amiloride analogs, but not by amiloride itself (208). The protection is RNA or protein synthesis independent. Na^+ and H^+ antiporter-negative L-M (TK^-) cells (LAP) could be killed by TNF, showing that the Na^+ and H^+ exchange is not required for TNF mediated cytotoxicity.

Other authors suggest, based upon studies with artificial liposomal preparations, that TNF-α may produce direct plasma membrane damage (209). At a pH below 5.0, TNF-α bound phosphatidylserine-phosphatidylcholine liposomal membranes and caused release of encapsulated dye.

Phosphoesterase inhibitors such as pentoxiphylinne and cAMP analogues also blocked TNF production at the cellular level (210, 211). The serine protease inhibitor p-toluenesulphonyl-L-arginine methyl ester (TAME) inhibits the secretion of TNF from human leukocytes (212). Botulinum D toxin also inhibits secretion of TNF through the adenosine diphosphate ribosylation of a cellular G protein required for protein secretion (196). Phospholipase (LP) inhibitors like quinacrine, hydrocortisone, dexamethasone, and indomethacin are also potent inhibitors of LT and TNF-mediated cell destruction suggesting that selective deacylation of the specific membrane PL by phospholipase activation is an important step in the events that lead up to LT- and TNF-mediated *in vitro* cellular destruction (202). The TNF receptors can be also down-modulated with two hour use of phorbol esters or activators of protein kinase C (213). The phorbol esters induced TNF-α receptor down-modulation, which could be eliminated using the protein kinase C inhibitor H7. This data demonstrates how tightly the TNF binding to activated T cells by the protein kinase C is controlled. The kinase C activation supports an event of direct or indirect phosphorylation of TNF-α, a very important regulator of the internalization or inactivation of TNF-α (214, 215). A polyclonal antibody directed against TNF-α was able to inhibit the cytotoxicity of macrophages (216, 219).

Various sensitive and resistant well-established human malignant cell lines synthesized two proteins of molecular weight 36 and 42kD in response to TNF-α exposure (220). Pretreatment with actinomycin D or cycloheximide inhibited the synthesis of these two polypeptides (p.36 and p.42) while simultaneously augmenting the cytotoxic effect of TNF-α.

We believe that TNF-α has the ability to indirectly inhibit the growth or necrotize solid neoplasms, but this activity appears to be related to host mediated suppression (221). In a study done by Patek and co-authors, tumor lytic activity of natural cytotoxic (NC) cells was compared to the tumor lytic activity of TNF-α against the same cancer target cells (222). NC-resistant tumor cells also showed resistance against TNF-α, supporting the idea that both NC and TNF-α activate the same lytic mechanisms and that TNF may at least be one component of the soluble factors that mediate the lytic action of NC cells. Additionally, a polyclonal antibody against TNF suppresses the lytic function of NC cells. Combinations of TNF and IL-2 or IFNs, including all three together, have resulted in much greater rates of tumor regression when compared to TNF as a single agent (223-225).

In cell culture conditions, TNF-α has anti-growth or cytotoxic effects against numerous normal and neoplastically transformed cell lines: 1) human - ACHN (renal carcinoma), BT-20 (breast carcinoma), BT-475 (breast carcinoma), HS-939T (melanoma), HT-29 (colon carcinoma), HT-1376 (bladder carcinoma), MCF-7 (breast carcinoma), ME-180 (cervical carcinoma), SKBR-3, ZR-75-1 and ZR-75-30 (breast carcinoma), SKCO-1 and WiDr (colon carcinoma), PC10 and SKLU-1 (lung carcinoma), HMV-2 and SKMEL-109 (melanoma), SKOV-4 (ovarian carcinoma), GOTO (neuroblastoma); 2) murine - B6MS2 and B6MS5 (sarcoma), CMS4 and CMS16 (sarcoma), B16 (melanoma), colon 26 (adenocarcinoma), MH134 (hepatoma), L-929 (fibroblast), Meth A and MMT (breast carcinoma), SAC (Moloney virus transformed 3T3), and WEHI-164 (sarcoma) (226, 227). Some normal, mature and fetal cell lines were found to grow *Fas*ter under the stimulating effects of TNF-α: 1) human - U373 (astrocytoma) (228), the same effect was also noted for IL-1, CCD-18Co (normal colon), Detroit 551 (normal fetal skin), FS-4 and HS-27F (foreskin fibroblasts), LL-24 and WI-1003 (normal lung) and WI-38 (normal fetal lung). Growth enhancement activity can be caused by modulation of alternate regulatory biochemical pathways in the cell.

TNF has multiple biological actions on endothelial cells:

1) induces expression of surface antigens that entice leukocyte and lymphocyte adherence, which is probably necessary for diapedesis; TNF-α as well as lymphotoxin are known to be able to induce chemotactic migration in human monocytes and polymorphonuclear leukocytes in low concentrations (below 1 unit/milliliter) (229, 230);
2) increases expression of HLA antigens, both Class I and Class II, together with rIL-1 and γ-IFN;
3) changes the morphological characteristics of endothelial cells from epithelioid to fibroblastoid in synergism with γ-IFN;
4) TNF does not have chemo-attractive activity on large granular lymphocytes or endothelial cells, despite the presence of high-affinity TNF receptors.

5.3 Induction of secondary mediators

Both *in vitro* and *in vivo* TNFs are part of a very complex network of cellular mediators that participate in the transmission of regulatory signals. Ample evidence now suggests that TNF can regulate not only its own reproduction and release by the target cell, but also the synthesis of other lymphokines and cytokines. TNF treatment enhanced its own synthesis and secretion as well as that of interleukin-1 (IL-1) and IL-6 (169). TNF-α together with α- and γ-IFN can induce the production of IL-1 by blood monocytes. Furthermore, cellular response to TNF can also be controlled by regulating the expression of TNF receptors. Treatment of ME-180 cells with γ-IFN increased the number of TNF receptors per cell so that the upregulated cells bind more TNF-α (197, 231). This data partially explains some of the synergistic antiproliferative and antiviral effects observed with combinations of γ-IFN and TNF-α.

Functional and morphological differentiation was induced by TNF-β (lymphotoxin), which was used as a single agent or in combination with retinoic acid on two human myeloid leukemia cell lines (HL-60 and THP-1) (232). The maturation process induced on myeloid cells resulted in an increased production of superoxide anion, IgG Fc receptors and immunophagocytosis.

Since TNF is not able to inhibit virus replication (particularly not VSV replication) in different cell lines, the observed antiviral effects may have been the result of IFN induction by TNF. However, use of neutralizing antibodies to IFN α, β, and γ failed to suppress TNF antiviral activity. Culture supernatants of cells treated with TNF demonstrated no antiviral activity.

In mouse experiments, i.v. and s.c. ways of administration were used (adjacent to the treated tumor skin). TNF-α at doses of 3,000 to 10,000 U/mouse produced dose-related lesions in blood vessels in the sarcoma and the adjacent skin, but not in the skin distant from the sarcoma (233).

5.4 Clinical applications of Tumor Necrosis Factors

Sherwin (234) was able to review the early phase I trials of recombinant TNF (Genetech, rTNF) submitted to the fifth NCl-EORTC Symposium on new drugs in cancer therapy. In his experiments, 46 patients received escalating doses on a twice-weekly schedule for 4 weeks. In other phase I study, reported by Blick and co-workers (225), rTNF was given i.v. and i.m. by bolus injection twice weekly for four weeks to 20 patients with metastatic cancer. Within minutes after i.v. bolus administration, serum levels of rTNF were detectable, with a peak level at about 30 minutes later. There were two responses of the 16 evaluable patients, one being complete regression of a metastatic neck lesion. Feinberg and co-authors (235) investigated a phase I dose escalation in 39 patients and compared 30 minute versus 240 minute infusions at dose levels of 5 to 250 microg/m^2. The MTD was 200 microg/m^2. The authors were unable to notice any anti-tumor response. Given the very short half-life of rTNF after the i.v. bolus administration, several Phase I trials have used the continuous i.v. infusion in patients with advanced cancer (236-239). The rationale for continuous infusion is an attempt to prolong maximal levels of rTNF in the blood and to provide a prolonged and strong exposure of the tumor mass to rTNF. When Phase I trials of rTNF was administered by i.v. in advanced renal cell carcinoma one **complete response** was evident (240). After three repeated administrations of rTNF in doses of 205 mikrogramm/m^2/day for five days, the tumors disappeared. The response only lasted for three months, with the patient subsequently passing away from brain metastases.

In Phase I studies, intratumoral injection of rTNF has been explored in an attempt to ensure very high levels of the cytokine directly within the neoplastic microenvironment. Patients with refractory malignancy, which was accessible for injection, received a single intralesional injection. The MTD was 391 microg/m^2. Of 21 patients treated, there was one complete local response, four partial responses, and four minor local responses (241). Five of 14 patients had tumor shrinkage when injected with rTNF in doses ranging from 25 to 300 microg/m^2 up to three times weekly (242). Almost no anti-tumor efficacy of rTNF was determined after intramuscular administration in Phase I studies (225, 242). Peak serum levels of 0.3 to 0.5 ng/ml at doses of 150 to 175 microg/m^2 were detected two hours after i.v. administration and are substantially lower than levels achieved by similar i.v. doses. The MTD for i.v. rTNF was 150 microg/m^2. The direct cytotoxic effect of rTNF on cancer cells and the frequent use of intraperitoneal (i.p.) administration in preclinical, animal studies have influenced Phase I human studies with i.p. treatment of patients with ovarian and gastrointestinal tumors with advanced intraperitoneal disease. 12/24 patients (50%) with advanced ovarian cancer and malignant ascites demonstrated complete resolution of ascites by ultrasound followed by rTNF treatment with weekly escalating doses. Subcutaneous administration (s.c.) of rTNF has been reported, but significant inflammation, pain, and ulceration at the injection site have limited the dose used during the treatment (243).

Preliminary reports of Phase II trials of i.v. rTNF have been published with similarly disappointing results. A phase II study in 25 patients with advanced malignancies (11 melanoma, 14 renal cancer), TNF was given at an escalating dose 25-50 microg/m^2/decimeters using intramuscular injections (i.m.). There were no responses in 14 patients with advanced colorectal carcinoma, with i.v. rTNF doses of 100-150 microg/m^2 twice daily every other week for four cycles (244). Similar negative results were reported by Schaadt and co-workers (245) in evaluating the non-responses after i.v. rTNF treatment in 15 patients with advanced colorectal carcinoma. Each patient received a dose between 217 to 653 microg/m^2. A phase II trial of TNF in patients with advanced colorectal carcinoma was done in 1990 (246). TNF was administered by a 30 minute intravenous infusion twice daily for five days every other week. No objective responses were detected in 14 previously treated patients. We propose that only future, continued *in vivo* and *in vitro* immunobiochemical observations of important immunomodulators like TNF will provide information on how and when to use this exciting biological response modifier for appropriate cancer therapy.

Preclinical animal models suggest that coupling of TNF with other cytokines may augment the response to TNF. McIntosh and co-workers (247) have demonstrated that the combination of TNF and IL-2 markedly increased the anti-tumor effect in the mouse experimental tumor system. Talmadge and co-authors (248) already defined the benefits of GM-CSF. In fact, it is suggested that careful sequences of these biological agents combined with chemotherapy (cytoxan) may avoid toxicity. The need for participation of the host's lymphocytes for the cure of murine tumors suggests that IL-2 be used. DNA cytometric and histopathological investigations were performed in two mouse tumor models (BP and S180) that differed in sensitivity to TNF (249). TNF administration induced necrosis in both tumors, but only the sarcoma (S180) showed total regression. DNA cytometry revealed an increase of cells in the S-phase in the BR tumor and a loss of aneuploid cell populations in the sarcoma. Mononuclear cell infiltration was observed only in the sarcoma. Are the aneuploid cell populations more sensitive to TNF than the eudiploid? Combinative use of dactinomycin and rTNF in anticancer treatment has been recognized for many years. This mechanism of interaction is realized during the enzyme DNA topoisomerase II, which controls the winding and unwinding of DNA in all cells. Several well-known chemotherapeutic agents (*e.g.* doxorubicin, etoposide, tenopiside) appear to be targets in this enzyme system.

5.5 Toxicity of Tumor Necrosis Factors

The toxicity of rTNF administered by continuous i.v. infusion is similar to that observed after i.v. bolus administration. In a 24-hour infusion protocol, Spriggs and co-workers (237) registered fevers of greater than 38 degrees and chills at doses above 227 microg/m^2/day. The most frequent symptoms related to the therapeutic use of toxicities are: fever, chills, headache, fatigue (constant at doses >150 microg/m^2) (235), sometimes nausea and vomiting, and local pain related to the way of administration and inflammation occurred after subcutaneous or intramuscular injection. Headaches, myalgias, and profound fatigue were observed only at higher doses. Blood hypotension occurred infrequently and was a dose-limiting toxicity only for two patients, but required fluid and vasopressor medication at higher doses. In Feinberg's phase I study (235), granulocyte and platelet counts were decreased with increased triglyceride levels. Mild elevations of hepatic enzymes were seen with continuous infusion and elevations of BUN and creatinine were associated with fluid retention. Effects of rTNF on lipid metabolism have been addressed in numerous clinical trials. An increase of serum triglycerides and a reciprocal decrease in serum cholesterol have been detected. The fluid retention after doses above 454 microg/m^2/day was significant with patients gaining 2 to 4 kg of fluid during one day infusions. Significant neurotoxicity has been observed during continuous infusions while seizures were provocated if pre-existing brain infarcts were present. Weight loss after systematic administration of rTNF has also been demonstrated in animals during preclinical trials, but not in humans to a significant degree. After being combined with IL-2, the toxicity of TNF also increased producing shock and other cardiorespiratory effects (250). It should thus never be forgetten that TNF is a primary mediator of endotoxic shock.

5.6 Conclusion

1) It appears doubtful that rTNF as a single agent will have significant efficacy in the multimodality treatment of human malignancies;
2) Unless the toxicity of rTNF can be significantly decreased by other cytokines or chemotherapeutic agents, or its efficacy modulated in combination with other cytokines at tolerable doses of rTNF, it is unlikely that TNF will play a significant role in future systemic therapy of human cancer.
3) It has now been about two decades since tumor necrosis factor (TNF) was first identified as a protein produced by the immune system that played a major role in suppression of tumor cell proliferation. Extensive research since then has revealed that TNF is a major mediator of inflammation, viral replication, tumor metastasis, transplant rejection, rheumatoid arthritis, and septic shock. As of today, approximately 18 different members of the TNF superfamily have been identified, and most of them have been found to mediate a wide variety of diseases, especially cancer (251). All the cytokines of the TNF superfamily mediate their effects through the activation of the transcription factor NF-kappaB, c-Jun N-terminal kinase, apoptosis, and proliferation. Thus, agents that can either suppress the production of these cytokines or block their action have therapeutic value for a wide variety of diseases.

6. THE DISCOVERY OF INTERLEUKIN-2: BASIC PRINCIPLES

The short period of time between the scientific discovery of IL-2, originally named lymphocyte growth factor, as a biologically important molecule and its successful clinical use represents one of the greatest success stories of modern life science. The T cell growth factor (TCGF), named later interleukin-2 (IL-2) by Morgan, Ruscetti and Gallo (252), is a glycoprotein, produced by mitogen- or alloantigen "activated T lymphocytes" (resting T cells that do not express the necessary number of surface interleukine-2 (IL-2) receptors and do not produce IL-2). IL-2 also allowed the long-term culture of T lymphocytes responding to both antigen and mitogen with the release of cytokines, as well as other effector cell functions, including cytotoxicity (253-255).

The central role of T lymphocytes is the regulation of immune response, especially during tissue destruction, as observed in autoimmune diseases, viral infections and transplant rejection. Further investigations defined that one lymphokine, IL-2, plays a central role in cellular immune regulation and demonstrated that a long-term lymphocyte culture can be controlled in order to remain in active proliferation *in vitro*. In primary thymic cultures from early fetuses, IL-2 has been shown to have an inhibitory effect on thymocyte proliferation (256). Transient expression of IL-2 receptors and responsiveness to IL-2 appear to be a major selective pathway for T lymphocytes during interthymic maturation. The leukemic cell line Jurkat was found to produce high concentrations of human IL-2, and using this cell line, the gene coding for human IL-2 was isolated and expressed in *E. coli* (257, 258). The ability of the T cell growth factor to migrate with other factors has caused effects such as mitogenesis of anti-tumor effector cells, leading to its investigation of its role in cancer immunology.

Human cells contain a single copy of the IL-2 gene, which consist of four exons and three introns in the mid portion on the long arm of chromosome 4 (4q26-28). The cDNA consist of a single open reading frame coding for 153 amino acids, cleaved into 133 amino acids mature molecule, with cysteine residues located at amino acids in positions 58 and 105 and 125 allowing three potential disulfide bond between the cysteines, both essential for the functional activity of IL-2 molecule (259). Site specific mutagenesis of the cysteine with either a serine or alanine substitution at amino acid 125 has allowed active production of this molecule using molecular biological techniques. The molecule is extremely hydrophobic and is stable at pH 2. IL-2 has a molecular weight of approximately 15,420 daltons and can be produced by both helper and suppressor T cells, depending upon the particular stimulus; although in mice there is accumulating evidence that a subset of helper T cells (the TH1 class) produce IL-2 preferentially *in vivo* under physiological conditions. IL-2 can also be produced by other cells such as the reticulo-epithelial (RE) cells in the thymic medulla and astrocytes in the central neuronal system (CNS).

6.1 Physiological role of IL-2

1) autocrine regulation: freshly stimulated (activated) T lymphocytes, LAK cells, NK cells showed presence of mRNA for IL-2R production within one hour of antigenic stimulation, thymocyte mitogenetic effect, induction of interferon gamma production;
2) after the expression of IL-2R, IL-2 has a B cell differentiating effect, including the production of immunoglobulins later (expression of IL-2R and detection of Tac antigen);
3) macrophages (direct increase of cytotoxic ability; IL-2R expression);
4) epidermal Langerhans cells (presence of IL-2R on cultured LC);
5) oligodendroglia cells (stimulation of proliferation and maturation);
6) IL-2 stimulates NK cells to release IL-1;
7) vascular endothelium stimulated by PHA or the mouse anti-human MoAB, anti-CD3 augmented the T cell proliferation, IL-2 synthesis and rapid expression of its receptor (9); and
8) IL-2 serves as a chemotactic factor attracting lymphocytes to sites of inflammation. Administration of IL-2 is followed within four hours by rapid emigration of lymphokine-activated killer (LAK) cell precursors that express the intermediate affinity form of IL-2 receptor.

Interleukin-2 production can be inhibited by T suppressor cells or by immunosuppressive substances such as glucocorticoids (dexamethasone), cyclosporin A and a prostaglandin (PGE2). Numerous other cytokines and allogens can upregulate or induce the IL-2 synthesis, among them being IL-1 and the gamma-interferon (γ-IFN), the mitogens phytohemagglutinin (PHA), phorbol-12 myristate-13 acetate and concavalin A, calcium ionophore A23187 and MoAB anti-CD3 (259). A large advantage of preclinical studies and models with IL-2 is its complete lack of species specificity. Thus, recombinant human IL-2 can be used for *in vivo* immunotherapeutic studies in mice, which is something that cannot be done with immunologic interferon.

IL-2 interacts, through causing proliferation, maturation and differentiation, with T and B lymphocytes, natural killer cells (NK) and thymocytes by binding to specific receptors located on the cell surface. NK cells lack the T cell receptor for antigen bearing Fc receptors, which can be activated in both human and mouse by IL-2 (261) to lyse freshly isolated tumor cells (LAK cell – phenomenon) (262). High affinity receptors mediate the physiologic response of T cells to IL-2 and comprise about 10% of the IL-2 receptors. Another group of receptors containing a 55 kD protein detected with anti-Tac MoAB bind IL-2 with low affinity. A 75 kD IL-2 receptor protein of intermediate affinity has also been identified.

Soluble forms of IL-2R have been found in serum, urine, and malignant effusion or tissue culture supernatant. They were released by strongly activated T cells during *in vitro* mitogen or antigen stimulation (256). "Resting" lymphocytes do not respond to extrinsic IL-2 because this lymphokine requires acquisition of membrane receptors for IL-2. In a very important experiment, resting lymphocytes were activated or re-activated with anti-CD3 MoAB, with this activation process not requiring presence of any accessory, antigen presenting cells (263). When IL-2 stimulates IL-2R positive T cells, they progress in their proliferation from G1 to S phase (264).

Sera of patients with AIDS inhibit IL-2 induced T cell proliferation, but IL-2 was partially able to reconstitute the immune defects caused by the HIV infection. Lymphatic cells incubated with IL-2 develop a capacity to lyse fresh tumor cells (255, 257, 261, 262).

The activation of a non-T and non-B lymphocyte subset, the so-called lymphokine-activated killer (LAK) cell phenomenon, occur both *in vitro* and *in vivo* and represent the basis for the development of adoptive immunotherapeutical protocols against cancer in humans. IL-2 also causes *in vivo* release of other lymphokines and hormones that themselves can mediate anti-tumor effects, often in concert with IL-2. Initial clinical studies used IL-2 derived from the Jurkat cell line, but only small quantities of purified IL-2 could be obtained.

The already mentioned experimental expression of the IL-2 gene in *E. coli* has led to the availability of virtually unlimited amounts of recombinant IL-2 (rIL-2), with most clinical trials using this material.

6.2 Interleukin-2 therapy of nonneoplastic diseases in preclinical animal models

1) Enhances (in dose 200 IU/day i.v.) the activity of adoptively transferred antigen specific immune effector cells against herpes simplex;
2) Administration in dose 8 x 10^4 to 8 x 10^5 IU/kg for 4 days lowers the rates of infection and increases healing of herpes simplex genital lesions;
3) Protects neonatal mice from lethal herpes simplex infection in doses 50-200 IU/day (macrophage regulated with involvement of gamma interferon);
4) Treatment enhances survival (in dose 100 IU/day for 3-5 days) against a lethal infection with *T. gondii*;
5) Administration in doses 1.8 to 7.0 x 10^6 IU/kg i.v. or i.m. protects animals from lethal infection with gram-negative bacteria;

6) Protects and provides a therapeutic effect against fatal *E. coli* sepsis in doses of 5 x 10^5 IU/kg intraperitoneally;
7) Prophylactic and therapeutic effect on respiratory-tract infection with *K. pneumoniae* (in doses 2-20 microg/day SQ x7-14 days);
8) Systematic adjuvant with IL-2 or PEG-IL-2 administration increases the potency of inactivated rabies virus vaccine;
9) Alters genetic nonresponsiveness to malaria-sporozoite peptides (5 x 10^4 IU IL-2 emulsified with the antigen).

The most important finding was that IL-2 clearly enhances the activity of antigen specific adoptively transferred host immune effector cells in the setting of herpes simplex and cytomegalovirus infection (265) and the *in vivo* proliferation and activity of transferred LAK cells (266).

6.3 Experiments using biochemically modified interleukin-2

Numerous attempts were carried out to engineer the IL-2 molecule using either biochemical or molecular biological approaches. Insertion of the IL-2 gene directly into human T cells, as described earlier, is associated with the development of tumors in mice. Employing the gene transfected in tandem combination with vaccinia virus antigens allowed new immunization strategies (267). The recombinant immunogen with the use of vaccinia virus antigens and IL-2 allowed for immune stimulation without the possible complications of an overhelming vaccinia infection. A similar approach was carried out by transfecting the IL-2 gene directly into fibroblasts. Enhanced anti-neoplastic effectivity was noted with its employment in immunotherapy models (268). A similar approach could be carried out to insert the IL-2 gene directly into human neoplastically transformed cells.

Chemically connected to polyethylene glycol, IL-2 was aimed to decrease IL-2 toxicity and to enhance the half-life and efficacy, which would enable a long-term administration of IL-2. Diphteria toxin conjugates have been prepared to be used in immunosuppression and was demonstrated to bind both p55 and p70 subunits of the IL-2 receptor. These conjugates were also capable of lysing neoplastic cells from adult T-cell leukemia patients. A molecular biological construct adding a lysine-rich C-terminal region to the carboxyl terminus allowed derivatization with biotin, without losing the activity and identification of IL-2 reactive cells (269).

6.4 Immunotherapy using recombinant interleukin-2 (rIL-2) as a "single agent"

Although no clinical condition has been identified with total absence of interleukin-2, relative decreases have been detected in several diseases, including patients with certain autoimmune conditions. Further, decreased production and responsiveness to

IL-2 progressed with aging. No differences were detected between the sexes in IL-2 production. MoAB against IL-2R suppresses immune responsiveness (in murine models). Cyclosporin treatment inhibits the generation of IL-2Rs.

Rosenberg and co-workers (270) summarized their enormous experience with use of high-dose IL-2 therapy during the treatment of 652 cancer patients. They administered 1039 courses of high-dose IL-2 for 596 patients with metastatic cancer that either had failed standard, effective therapies or had disease for which no standard effective therapy existed. 56 patients were treated in absence of disease that is able to be evaluated in the adjuvant setting.

The authors used numerous modifications of cancer immunotherapy, forming the following subgroups:

1) IL-2 was administered either alone ("single agent" treatment) - 155 patients;
2) together with LAK cells - 214 patients;
3) with previously *in vitro* expanded in number tumor infiltrating lymphocytes (TIL) -66 patients;

Combined with other cytokines such as:

4) α-IFN - 128 patients;
5) TNF - 38 patients;
6) with monoclonal antibodies -32 patients; and as form of chemoimmunotherapy:
7) with the chemotherapeutic agent cyclophosphamide - 19 patients.

Initial results in subgroups 1 and 2 indicated that cancer regressions could be achieved in 20-35% of patients with selected, advanced metastatic cancers. Although most responses have been noticed in patients with metastatic renal cell cancer, melanoma, colorectal cancer, and non-Hodgkin's lymphoma, many different types of cancer have not been treated in this way yet. The adoptive transfer of TIL combined with IL-2 administration appears to be more effective than the employment of IL-2 as a "single agent". Tumor regressions were as follow: 18 patients achieving a complete response and 10 having not noticed recurrence between the long intervals of 18-52 months. These studies demonstrate that purely immunologic manipulation can induce significant regression of advanced cancers.

A variety of schedules of IL-2 administration have been explored. Most studies have used the bolus administration of IL-2 at doses between 10,000 and 100,000 IU/kg intravenously every 8 hours. Other protocols administered IL-2 in the form of continuous infusion in doses of 1,000,000 to 7,000,000 IU/m^2/day. A lymphopenia occurs after IL-2 treatment, but if a small amount of IL-2 is administered for more than a week, lymphocytosis may occur as well. There was a depletion of LAK precursor cells ($CD2^+$,$CD11^+$,$CD16^+$,$CD3^-$ surface phenotype) from the periphery in minutes after the IL-2 administration was started. In the treatment of human neoplasms in advanced stage, IL-2 has been used alone or in conjunction with the adoptive transfer of expanded *in vitro* in a number of killer cells.

6.5 Toxicity of the rIL-2, administered as a "single agent"

It is quite clear that a major dose limiting toxicity in humans is related to a "vascular (capillary) leak syndrome" (261). This capillary leak syndrome is clearly dose-related, with increased dosage and increased number of days of IL-2 administration being associated with enhanced weight gain and apparent vascular leak. Corticosteroids can stop the syndrome but appear to eliminate the anti-tumor effect of IL-2. Vascular leak syndrome was not detected in nude mice, suggesting the importance of components related to the host. Similar toxicity was observed in rat models, with ascites and pleural effusions, as well as marked tissue infiltration with NK cells and eosinophils. An antibody to the NK 1.1 antigen depletes cells with NK and LAK activity and its use leads to a marked decrease in toxicity. Administration of high dose recombinant IL-2 (rIL-2) can be associated with dose-related and dose limiting toxic side-effects (271, 272). All of these toxicities were resolved within a few days of stopping IL-2 treatment. Numerous side-effects are probably the result of enhanced lymphoid infiltrates in vital organs, lymphocytosis and expression of lymphocyte activation surface phenotype; when vascular permeability leak is induced, it also leads to fluid retention and interstitial oedema (261). After administration, the earliest effect is a drop in the systematic vascular resistance, associated with tachycardia, decreased mean of arterial blood pressure, and an increase in cardiac index. Serum creatinine rises, probably from a prerenal azotemia (273). Weight gain (fluid retention), pulmonary vascular resistances, renal dysfunction, hepatic dysfunction, reversible cholestasi (274), obstructive jaundice (275), oral mucositis, confusion, lethargy, and hallucinations can occur. Rosenberg, Longo and Lotze (276) observed thirty side-effects of *in vivo* rIL-2 administration, from minor symptoms like chills and pruritus to the life-threatening syndromes of anaphylaxis, pulmonary edema, tissue-necrosis, hepatocellular injury, myocardial infarction, coma and two cases of death, which characterize foreign protein reaction.

Changes also occurred in some laboratory parameters, especially in the middle or after clinical trials requiring rIL-2 infusions (277). To assess toxicity, the authors observed enzymatic changes in alanine aminotransferase, gamma-glutamyl transferase, lactate dehydrogenase, alkaline phosphatase, creatine kinase, creatinine, urea nitrogen, and C-reactive protein. There were significant, progressive increases in the results, except for CK until the rIL-2 infusion stopped. Seven tests were related to the liver function and within five days after stopping the IL-2 therapy, results improved and moved toward the baseline value. Creatinine and urea nitrogen concentrations in serum were normal three days after the rIL-2 treatment.

6.6 Combination of rIL-2 treatment with chemotherapy

Preclinical animal models demonstrated the existence of an anti-tumor interaction between cytotoxic chemotherapy and IL-2 mediated immunotherapy in the treatment of cancer and later also in the treatment of the minimal residual disease (278). It is also possible that a population of "suppressor" cells may be interfering with IL-2 directed immunoactivation; the use of immunosuppressive chemotherapy may prevent such development except for suppressor cell activation. Furthermore, chemotherapy directed damage to the tumor mass may make a tumor more susceptible to immunotherapeutic destruction and *vice versa*. Anti-tumor effects are seen in mice receiving tumor associated antigen specific MoABs combined with *in vivo* administration of IL-2. It was also reported that chemoresistant tumor sublines were more susceptible to LAK cell lysis than the parent tumor cell lines. Combining chemotherapy that is able to greatly reduce the tumor burden with IL-2 and LAK may produce a better synergistic effect. Although most anticancer cytotoxic drugs are thought to exert immunosuppressive actions, some can enhance immune responses (279). The authors cited doxorubicin, which can augment the functional activity of NK cells, LAK cells and of antigen specific CTLs, but monocyte differentiation and cytokine production were also enhanced. Mitchell (280) described chemotherapeutic agents like bleomycin, doxorubicin, myeleran and imidazole carboxamide with little immunosuppressive activity. Bryostatin is another chemically defined immunomodulator (279). Bryostatin is a macrocyclic lactone consisting of a 26-carbon ring, which inhibits proliferation of adult myeloid leukemia cells, hemopoietic stem cells, renal cells and melanoma cells, and enhances the effectiveness of cytosine arabinoside. As a biological response modifier, bryostatin induces granulocyte activation and increases expression of interleukin-2 receptors (IL-2R). As we continue to improve our knowledge about different cell structure directed actions of biological response modifiers, it seems likely that this aspect of their biochemical effect will one day be used for high level molecular immunotherapy, achieving therapeutic goals, like alteration of cell surface antigenic or receptor structure, modulation of cytotoxic effect of chemotherapeutic agents or stimulation (restoration) of immune effector cell function after massive irradiation or chemotherapy.

REFERENCES

1. American Cancer Society: Cancer Facts and Figures. New York; American Psychiatric Association, 1980: Diagnostic and Statistical Manual of Mental Disorders, 3rd edition, Washington DC, 1980.
2. Silverberg BS, Lubera JA: Cancer statistics. Cancer *38:* 5-22, 1988.
3. Ries LA, Hankey BF, Edwards BK (eds.): Cancer Statistics Review 1973-1987. DHHS Publ. No. (NIH)90-2789, Bethesda, Md: NCI, 1990.
4. Levin VA: Chemotherapy for brain tumors of astrocytes and oligodendroglial lineage: the past decade and where we are heading. Neuro-oncol *1:* 69-80, 1999.
5. Gurney JG, and Kadan-Lottick N: Brain and other central nervous system tumors: rates, trends, and epidemiology. Curr Opin Oncol *13:* 160-166, 2001.
6. Lokker NA, Sullivan CM, Hollenbach SJ, Israel MA, Giese NA: Platelet-derived growth factor (PDGF) autocrine signaling regulates survival and mitogenic pathways in glioblastoma cells: evidence that the novel PDGF-C and PDGF-D ligands may play a role in the development of brain tumors. Cancer Res *62:* 3729-3735, 2002.
7. Chochinov H.M., Bereavement: a review for oncology health professionals. Cancer Invest *7:* 593-600, 1989.
8. Jako GJ: Presidential keynote address: The road toward 21st century surgery: new strategies and initiatives in cancer treatment. Lasers Surg Med *7:* 217-218, 1987.
9. Onik G, Rubinsky B, Zemel R, Diamond D: Cryosurgical management of hepatic malignancy. Contemp Oncol *1:* 20-24, 1991.
10. Pastan I, Fitzgerald D: Recombinant toxins for cancer treatment. Science *254:* 1173-1177, 1991.
11. Coley WB: Contributions to the knowledge of sarcoma. Ann Surg *14:* 199-220, 1891.
12. Coley WB: The treatment of malignant tumors by repeated inoculations of erysipelas: With a report of ten original cases. Amer. J. Med. Sci *105:* 487-511, 1893.
13. Sinkovics JG: Clinical immunotherapy for tumors. Postgraduate Med *59:* 110-116, 1976.
14. Siegall CB, Chaudhary VK, FitzGerald DJ, Pastan I: Cytotoxic activity of an interleukin 6-Pseudomonas exotoxin fusion protein on human myeloma cells. Proc Natl Acad Sci USA *85:* 9738-9742, 1988.
15. Ogata M, Chaudhary VK, FitzGerald DJ, Pastan I: Cytotoxic activity of a recombinant fusion protein between interleukin 4 and Pseudomonas exotoxin. Proc Natl Acad Sci USA *86:* 4215-4219, 1989.
16. Rozengurt E: Early signals in the mitogenic response. Science *234:* 161-166, 1986.

17. Kelly K, Kane MA, Bunn PA: Growth factors in lung cancer: possible etiologic role and clinical target. Med Pediat Oncol *19:* 449-458, 1991.
18. Todaro GJ, Sporn MB: Autocrine secretion and malignant transformation of cells. New Engl J Med *303:* 878-880, 1980.
19. Riedel H, Schlessinger J, Ullrich A: Chimeric, ligand-binding v-erbB/EGF receptor retains transforming potential. Science *236:* 197-200, 1987.
20. Hunts J, Gamou S, Hirai M, Shimizu N: Molecular mechanisms involved in increasing epidermal growth factor receptor levels on the cell surface. Jpn J Cancer Res *77:* 423-427, 1986.
21. Imanishi K, Yamaguchi K, Suzuki M, Honda S, Yanaihara N, Abe K: Production of transforming growth factor-α in human tumor cell lines. Brit J Cancer *59:* 761-765, 1989.
22. Brinkmann U, Pai LH, FitzGerald DJ, Willingham M, Pastan I: B3(Fv)-PE38KDEL, a single-chain immunotoxin that causes complete regression of a human carcinoma in mice. Proc Natl Acad Sci USA *88:* 8616-8620, 1991.
23. Lloyd KO, Old LJ: Human monoclonal antibodies to glycolipids and other carbohydrate antigens: dissection of the humoral immune response in cancer patients. Cancer Res *49:* 3445-3451, 1989.
24. Golumbek PT, Lazenby AJ, Levitsky HI, Jaffee LM, Karasuyama H, Baker M, Pardoll DM: Treatment of established renal cancer by tumor cells engineered to secrete interleukin-4. Science *254:* 713-716, 1991.
25. Mitchell MS: Combinations of anticancer drugs and immunotherapy. Cancer Immunol Immunother *52:* 686-692, 2003.
26. Taguchi T: Effects of lentinan in advanced or recurrent cases of gastric, colorectal, and breast cancer. Gan To Kagaku Ryoho *10:* 387-393, 1983.
27. Wakui A, Kasai M, Konno K, Abe R, Kanamaru R, Takahashi K, Nakai Y, Yoshida Y, Koie H, Masuda H: Randomized study of lentinan on patients with advanced gastric and colorectal cancer. Tohoku Lentinan Study Group, Gan To Kagaku Ryoho *13:* 1050-1059, 1986.
28. Carswell EA, Old LJ, Kassel RL, Green S, Fiore N, Williamson B: An endotoxin-induced serum factor that causes necrosis of tumors. Proc Natl Cancer Inst USA *72:* 3666-3670, 1975.
29. Old LJ, Benacerraf B, Clarke DA, Carswell EA, Stockert E: The role of the Reticulo-endothelial system in the host reaction to neoplasia. Cancer Res *21:* 1281-1300, 1961.
30. Vitale B, Allegreth N: Influence of BCG infection on the intensity of homograft reaction in rats. Nature (London) *199:* 507-508, 1963.
31. Mitchell MS, Murahata RI: Modulation of immunity by bacillus *Calmette-Guerin.* Pharmacol Ther *4:* 329, 1979.
32. Piessens WF, Heimann R, Legros N, Heuson J-C: Effect of bacillus *Calmette-Guerin* on mammary tumor formation and cellular immunity in dimethylbenz(a)anthracene-treated rats. Cancer Res *31:* 1061-1065, 1971.
33. Mathe G, Amiel JL, Schwarzenberg L, Schneider M, Cattan A, Schlumberger JR, Hayat M, de Vassal F: Active immunotherapy for acute lymphoblastic leukemia. Lancet *1:* 697-699, 1969.
34. Morton DL, Eiber FR, Malmgren RA: Immunological factors which influence response to immunotherapy in malignant melanoma. Surgery *68:* 158-164, 1970.
35. Morales A, Eidinger D, Bruce AW: Intracavitary bacillus *Calmette-Guerin* in the treatment of superficial bladder tumors. J Urology *116:* 180-183, 1976.
36. Brosman SA: Experience with bacillus *Calmette-Guerin* in patients with superficial bladder carcinoma. J Urol *128:* 27-30, 1982.

37. Herr HW, Pinsky CM, Sogani PG, Whitmore WF, Oettgen HF, Melamed HR: Experience with intravesical bacillus *Calmette-Guerin* therapy of superficial bladder tumors. Urology *25:* 119-123, 1985.
38. Mori K, Lamm DL, Crawford ED: A trial of bacillus *Calmette-Guerin* versus Adriamycin in superficial bladder cancer. Urol Int *41:* 254-259, 1986.
39. Stober V, Peter H: BCG immunotherapy for prevention or relapse in patients with bladder cancer. Ther Woche *30:* 60-64, 1980.
40. Biozzo G, Benacerraf B, Grumback F, Halpern B, Levaditi J, Rist N: Etude, de activite: granulopexique systeme reticuloendiothelial au cours de l'infection tuberculease experimentale de la souris. Ann Inst Pasteur *87:* 291-300, 1954.
41. Freund J: The mode of action of immunologic adjuvants. Adv Tuberc Res *7:* 130-148, 1956.
42. Thorbecke GJ, Benacerraf B: The reticuloendothelial system and immunological phenomena. Prog Allergy *6:* 559-598, 1962.
43. Fey F, Arnold W, Fraffi A: Demonstration of the stimulation of the reticulo-histiocytic system (RHS) of mice by treatment with BCG by means of biometric and histochemical techniques. Eur J Cancer *12:* 595-598, 1976.
44. Kavoussi LR, Brown EJ, Ritchey JK, Ratliff TL: Fibronectin-mediated *Calmette-Guerin* bacillus attachment to murine bladder mucosa. J Clin Invest *85:* 62-67, 1990.
45. Bast RC Jr, Bast BS, Rapp HJ: Critical review of previously reported animal studies of tumor immunotherapy with nonspecific immunostimulants. Ann NY Acad Sci *277:* 60-93, 1976.
46. Jahrsdorfer B, Weiner GJ: Immunostimulatory CpG oligodeoxynucleotides and antibody therapy of cancer. Semin Oncol *30:* 476-482, 2003.
47. Terry WD, Rosenberg SA (eds.): Immunotherapy of Human Cancer. Elsevier-North Holland, New York, 1982.
48. Talmadge JE: Thymosin: immunomodulatory and therapeutic characteristics. Prog Clin Biol Res *161:* 457-465, 1984.
49. Garaci E, Mastino A, Favalli C: Enhanced immune response and antitumor immunity with combinations of biological response modifiers. Bull NY Acad Med *65:* 111-119, 1989.
50. Rebar RW, Miyake A, Low TL, Goldstein AL: Thymosin stimulates secretion of lutenizing hormone-releasing factor. Science *214:* 669-671, 1981.
51. Healy DL, Hodgen GD, Schulte HM, Chrousos GP, Loriaux DL: The thymus-adrenal connection: thymosin has corticotropin-releasing activity in primates. Science *222:* 1353-1355, 1983.
52. Bistoni F, Baccarrini M, Puccetti P, Marconi P, Garaci E: Enhancement of natural killer cell activity in mice by treatment with a thymic factor. Cancer Immunol Immunother *17:* 51-55, 1984.
53. Flexman JP, Holt PG, Mayrohfer G, Latham BI, Shellam GR; The role of thymus in the maintenance of natural killer cells *in vivo*. Cell Immunol *90:* 366-377, 1985.
54. Serrate SA, Schulof RS, Leondaridis L, Goldstein AL, Sztein MB: Modulation of human natural killer cell cytotoxic activity, lymphokine production, and IL-2 receptor expression by thymic hormones. J Immunol *139:* 2338-2343, 1987.
55. Fiorilli M, Sirianni MC, Sorrentino V, Testi R, Aiuti F: *In vitro* enhancement of bone marrow natural killer cells after incubation with thymopoetin 32-36 (TP-5). Thymus *5:* 375-382, 1983.
56. Dokhelar MC, Tursz T, Dardenne M, Bach J-F: Effect of synthetic thymic factor (factuer thymique serique) on natural killer cell activity in humans. Int J Immunopharmacol *5:* 277-282, 1983.

57. Kenady DE, Chretien PB, Potvin C, Simon RM: Thymosin reconstitution of T cell deficits *in vitro* in cancer patients. Cancer *39:* 575-580, 1977.
58. Dillman RO, Beauregard J, Royston I, Zavanelli MI: Phase II trial of thymosin fraction 5 and thymosin α 1. J Biol Response Mod *6:* 263-267, 1987.
59. Gravenstein S, Duthie EH, Miller BA, Roecker E, Drinka P, Prathipati K, Ershler WB: Augmentation of influenza antibody responses in efderly men by the thymosin α one: A double blind placebo controlled clinical study. J Amer Geriat Soc *37:* 1-8, 1989.
60. Shen S: Age dependent enhancement of influenza vaccine response by thymosin in chronic hemodialysis patients. *In:* A.L. Goldstein, editor, Biomedical Advances in Aging. New York, Plenum Press pp. 523-530, 1990.
61. Mihara M, Ohsugi Y: The biological response modifier OK-432 (a Streptococcal preparation) inhibits the development of autoimmune kidney disease in NZB/W F1 hybrid mice: possible involvement of tumor necrosis factor. Int Arch Allergy Appl Immunol 90: 37-42, 1989.
62. Sinkovics JG, Shirato E, Shullenberger CC; Bordetella pertussis vaccine as immunological adjuvant in leukaemia and lymphoma (letter). Brit Med J *1:* 565, 1970.
63. Dinney CPN, Bucana CD, Utsugi T, Fidler IJ, von Eschenbach AC, Killion JJ: Therapy of spontaneous lung metastasis of murine renal adenocarcinoma by systemic administration of liposomes containing the macrophage activator CGP 31362. Cancer Res *51:* 3741-3747, 1991.
64. Sone S, Moriguchi S, Shimizu E, Ogushi F, Tsubura E: *In vitro* generation of tumoricidal properties in human alveolar macrophages following interaction with endotosis. Cancer Res *42:* 2227-2231, 1982.
65. Fidler IJ: Targeting of immunomodulators to mononuclear phagocytes for therapy of cancer. Adv. Drug Delivery Rev *1:* 69-106, 1988.
66. Pace JL, Russell SW, Torres BA, Johnson HM, Gray PW: Recombinant mouse gamma-interferon induces the priming step in macrophage activation for tumor cell killing. J Immunol *130:* 2011-2013, 1983.
67. Csatary LK: Viruses in the treatment of cancer. Lancet *2:* 825, 1971.
68. Cassel WA, Murray DR, Phillips HS: A phase II study on the postsurgical management of stage II malignant melanoma with a Newcastle disease virus oncolysate. Cancer *52:* 856-860, 1983.
69. Sinkovics JG: Clinical immunotherapy for tumors. Postgraduate Med *59:* 110-116, 1976.
70. Csatary L, Gergely P: Rosszindulatu daganatok kezelese virus-vakcinaval. Orvosi Hetilap *131:* 2585-2588, 1990.
71. Koprowski H: Acquired tolerance applied to experimental tumors. Ann NY Acad Sci *69:* 806-817, 1957.
72. Lindenmann J, Klein PA: Viral oncolysis: increased immunogenicity of host cell antigen associated with influenza virus. J Exp Med *126:* 93-108, 1967.
73. Vecil GG, Lang FF: Clinical trials of adenoviruses in brain tumors: a review of Ad-p53 and oncolytic adenoviruses. J Neurooncol *65:* 237-246, 2003.
74. Shimizu Y, Fujiwara H, Ueda S, Wakamiya N, Kato S, Hamaoka T: The augmentation of tumor-specific immunity by virus help. II. Enhanced induction of cytotoxic T-lymphocytes and antibody responses to tumor antigens by vaccinia virus-reactive helper T cells. Eur J Immunol *14:* 839-843, 1984.
75. Fujiwara H, Aoki H, Yoshioka T, Tomita S, Ikegami R, Hamaoka T: Establishment of a tumor-specific immunotherapy model utilizing TNP-reactive helper cell activity and its application to the autochthonous tumor system. J Immunol *133:* 509-514, 1984.
76. Csatary LK: Viruses in the treatment of cancer. Lancet *2:* 825, 1971.

77. Csatary LK, Eckhardt S, Bukosza I, Czegledi F, Fenyvesi C, Gergely P, Bodey B, Csatary CM: Attenuated veterinary virus vaccine for the treatment of cancer. Cancer Detect Prev *17:* 619-627, 1993.
78. Csatary KL, Gosztonyi G, Szeberenyi J, Fabian Z, Liszka V, Bodey B, Csatary CM: MTH-68/H Oncolytic Viral Treatment in Human High-Grade Gliomas. J Neuro-Oncology *67:* 83-93, 2004.
79. Chomienne C, Ballerini P, Balitrand N, Daniel MT, Fenaux P, Castaigne S, Degos L: All-trans retinoic acid in acute promyelocytic leukemias. II. *In vitro* studies: structure - function relationship. Blood *76:* 1710-1717, 1990.
80. Marquez VE, Lim MI, Treanor SP, Plowman J, Priest MA, Markovac A, Khan MS, Kaskar B, Driscoll JS: Cyclopentenyl-cytosine. A carbocyclic nucleoside with anti-tumor and anti-viral properties. J Med Chem *31:* 1687-1694, 1988.
81. Ford H Jr, Cooney DA, Ahluwalia GS, Hao Zh, Rommel ME, Hicks L, Dobyns KA, Tomaszewski JE, Johns DG: Cellular pharmacology of cyclopentenyl cytosine in MOLT-4 lymphoblasts. Cancer Res *51:* 3733-3740, 1991.
82. Isaacs A, Lindenmann J: Virus interference: 1. The interferon. Proc Royal Soc Med *147:* 258-267, 1957.
83. Samuel CE: Mechanisms of the antiviral action of interferon. Prog Nucleic Acid Res Mol Biol *35:* 27-72, 1988.
84. Hoskins M: A protective action of neurotropic against viscerotropic yellow fever virus in *macacus rhesus*. Amer J Trop Med *15:* 675-680, 1935.
85. Chebath J, Benech P, Hovanessian A, Galabru J, Revel M: Four different forms of interferon induced 2'5' oligo(A)synthetase identified by immunoblotting in human cells. J Biol Chem *262:* 3852-3857, 1987.
86. Senn CC: Biochemical pathways in interferon action. Pharmacol Ther *24:* 235-257, 1984.
87. Samuel C: Molecular mechanisms of interferon action. *In:* Clinical applications of interferons and their inducers., Second Edition; D. Stringfellow (editor), Marcel Dekker, New York, pp. 1-18, 1986.
88. Byrne GI, Lehmann LK, Kirschbaum JG, Borden EC, Lee CM, Brown RR: Induction of tryptophan degradation *in vitro* and *in vivo*: a gamma interferon stimulated activity. J Interfer Res *6:* 389-396, 1986.
89. Sinkovics JG: Oncogenes and growth factors. CRC Critical Reviews in Immunology *8:* 217-298, 1988.
90. Gresser I: The anti-tumor effects of interferon. Med Oncol Tumor Pharmacother *3:* 223-230, 1986.
91. Friedman RM: Anti-tumor effects of interferons. J Exp Pathol *3:* 203-227, 1987.
92. Doukas J, Shepro D, Hechtman HB: Vasoactive amines directly modify endothelial cells to affect polymorphonuclear leukocyte diapedesis *in vitro*. Blood *69:* 1563-1569, 1987.
93. Glasgow LA: Leukocytes and interferon in host response to virus infections. II. Enhanced interferon response of leukocytes from immune animals. J Bacteriol *91:* 2185-2191, 1966.
94. Wheelock EF, Larke RPB: Efficacy of interferon in the treatment of mice with established Friend virus leukemia. Proc Soc Exp Biol Med *127:* 230-238, 1968.
95. Hilleman MR: Double stranded RNAs (poly I:C) in the prevention of viral infections. Arch Intern Med *126:* 109-124, 1970.
96. Hirsh MS, Black PH, Wood ML, Monaco AP: Immunosuppression, interferon inducers and leukemia in mice. Proc Soc Exp Biol Med *134:* 309-313, 1970.

97. Levy H: Interferon and interferon inducers in the treatment of malignancies. Arch Intern Med *126:* 78-83, 1970.
98. Rhim J, Huebner R: Comparison of the antitumor effects of interferon and interferon inducers. Proc Soc Exp Biol Med *136:* 524-529, 1971.
99. deVilliers EM: Papilloma viruses in cancers and papillomas of the aerodigestive tract. Biomed Pharmacother *43:* 31-36, 1989.
100. Linnavuori K, Hovi T: Herpes simplex virus as an inducer of interferon in human monocyte cultures. Antiviral Res *8:* 201-208, 1987.
101. Goldstein D, Laszlo J: The role of interferon in cancer therapy: a current perspective. CA *38:* 258-277, 1988.
102. Sikora K: Interferon and malignant disease. Brit J Clin Pract *40:* 406-410, 1986.
103. Rashidbaigi A, Langer JA, Jung V, Jones C, Morse HG, Tischfield JA, Trill JJ, Kung HF, Pestka S: The gene for the human immuneinterferon receptor is located on chromosome 6. Proc Natl Acad Sci USA *83:* 384-388, 1986.
104. Spiegel RJ: The alpha interferons: Clinical overview. Semin Oncol *14(2 suppl.):* 1-12, 1987.
105. Revel M: The interferon system in man: nature of the interferon molecules and mode of action. *In:* Antiviral drugs and interferon. Y. Becker (editor), Martinus Nijhoff, New York pp. 358-433, 1984.
106. Lindahl P, Leary P, Gresser I: Enhancement by interferon of the specific cytotoxicity of sensitized lymphocytes. Proc Nat Acad Sci USA *71:* 714-724, 1972.
107. Herberman RB, Ortaldo JR, Bonnard GD: Augmentation by interferon of human natural and antibody dependent cellular cytotoxicity. Nature *277:* 221-223, 1979.
108. Rubin BY, Gupta SL: Differential efficiencies of human type I and type II interferons as anti-viral and anti-proliferative agents. Proc Nat Acad Sci USA *77:* 5928-5932, 1980.
109. Catalona WJ, Ratcliff TL, McCool RE: Gamma interferon induced by *S. aureus* protein A augments natural killing and ADCC. Nature *291:* 77-79, 1981.
110. Borden EC, Holland JF, Dao TL, Gutterman JU, Wiener L, Chang Y-C, Patel J: Leukocyte-derived interferon (α) in human breast carcinoma. Annals Intern Med *97:* 1-6, 1982.
111. Weigent DA, Langford MP, Fleishman WR, Stanton GJ: Potentiation of lymphocyte natural killing by mixtures of alpha- or beta-interferon with recombinant gamma interferon. Infect Immunol *40:* 35-41, 1983.
112. Czarniecki CW, Fennie CW, Powers GB, Estelle DA: Synergistic anti-viral and anti-proliferative activities of E. coli-derived human alpha, beta and gamma interferons. J Virol *49:* 490-496, 1984.
113. Inghirami G, Djeu JY, Balow JE, Tsokos GC: Enhancement of human allogeneic cytotoxic responses by interferons. J Immunopharmacol *7:* 403-415, 1985.
114. Sinkovics JG: Interferons. *In:* Medical Oncology, Vol. 2., 2nd edition, Marcel Dekker, New York, pp. 1453-1459, 1986.
115. Laszlo J, Huang AT, Brenckman WD, Jeffs C, Koren H, Cianciolo G, Metzgar R, Cashdollar W: Phase I study of pharmacological and immunological effects of human lymphoblastoid interferon given to patients with cancer. Cancer Res *43:* 4458-4466, 1983.
116. Damle NK, Doyle LV: Interleukin-2 activated human killer lymphocytes: lack of involvement of interferon in the development of IL-2-activated killer lymphocytes. Int J Cancer *40:* 519-524, 1987.
117. Borden EC, Hawkins MJ: Biologic response modifiers as adjuncts to other therapeutic modalities. Semin Oncol *80:* 148-149, 1986.

118. Borden EC: Augmented tumor-associated antigen expression by interferons. J Natl Cancer Inst *80:* 148-149, 1988.
119. Plaeger-Marshall S, Haas A, Clement LT, Giorgi JV, Chen IS, Quan SG, Gatti RA, Stiehm ER: Interferon-induced expression of Class II major histocompatibility antigens in the major histocompatibility complex (MHC) Class II deficiency syndrome. J Clin Immunol *8:* 285-295, 1988.
120. Vilcek J, Kelke HC, Jumming LE, Yip YK: Structure and function of human interferon gamma. In: Mediators in cell growth and differentiation. Ford R.J. (editor), Raven Press, New York pp. 299-313, 1985.
121. Bonnem ER, Oldham RK: Gamma-interferon: physiology and speculations on its role in medicine. J Biol Res Modif *6:* 275-301, 1987.
122. Sidky YA, Borden EC: Inhibition of angiogenesis by interferons: effects on tumor- and lymphocyte- induced vascular responses. Cancer Res *47:* 5155-5161, 1987.
123. Satoh M, Inagawa H, Shimada Y, Soma G, Oshima H, Mizuno D: Endogenous production of tumor necrosis factor in normal mice and human cancer patients by interferons and other cytokines combined with biological response modifiers of bacterial origin. J Biol Response Mod *6:* 512-524, 1987.
124. Progress in the development and use of antiviral drugs and interferon. Report of WHO Scientific Group. WHO Technic Rep Ser *754:* 1-28, 1987.
125. Friedman RM: Antitumor effects of interferons. J Exp Pathol *3:* 203-227, 1987.
126. Merigan TC, Sikora K, Breeden JH, Levy R, Rosenberg SA: Preliminary observations on the effect of human leukocyte interferon in non-Hodgkin's lymphoma. New Engl J Med *299:* 1449-1453, 1978.
127. Gutterman JU, Blumenschein GR, Alexanian R, Yap H-Y, Buzdar AU, Cabanillas F, Hortobagyi GN, Hersch EM, Rasmussen SL, Harmon M, Kramer M, Pestka S: Leukocyte interferon-induced tumor regression inb human metastatic breast cancer, multiple myeloma, and malignant lymphoma. Annals Intern Med *93:* 399-406, 1980.
128. Mellstedt H, Ahre A, Bjorkholm M, Holm G, Johansson B, Strander H: Interferon therapy in myelomatosis. Lancet *1:* 245-247, 1979.
129. Wiernik PH, Schwartz B, Dutcher JP, Turman N, Adinolfi C: Successful treatment of hairy cell leukemia with β-ser interferon. Amer J Hematol *33:* 244-248, 1990.
130. Watson JV: What does "response" in cancer chemotherapy really mean? Brit Med J *283:* 34-37, 1981.
131. Quesada JR: Interferons in cancer research - an update. Cancer Bull *35:* 30-39, 1983.
132. Real FX, Oettgen HF, Krown SE: Kaposi's sarcoma and the acquired immunodeficiency syndrome: Treatment with high and low doses of recombinant leukocyte α interferon. J Clin Oncol *4:* 544-551, 1986.
133. Jacobs AD, Champlin RT, Golde DW: Recombinant α-2-interferon for hairy cell leukemia. Blood *65:* 1017-1020, 1985.
134. Thompson JA, Brady J, Kidd P, Fefer A: Recombinant α-2 interferon in the treatment of hairy cell leukemia. Cancer Treat Rep *69:* 791-793, 1985.
135. Quesada JR, Hersh EM, Manning J, Reuben J, Keating M, Schnipper E, Itri L, Gutterman JU: Treatment of hairy cell leukemia with recombinant α-interferon. Blood *68:* 493- 497, 1986.
136. Sigal RK, Lieberman MD, Reynolds JV, Williams N, Ziegler MM, Daly JM: Tumor immunization. Improved results after vaccine modified with recombinant interferon gamma. Arch Surg *125:* 308-312, 1990.
137. Samuels BL, Brownstein BH, Golomb HM: *In vitro* induction of proteins by α-interferon in hairy cell leukemia. Cancer Res *46:* 4151-4155, 1986.

138. Branca AA, Baglione C: Evidence that types I and II interferons have different receptors. Nature (London) *294:* 768, 1981.
139. Orchansky P, Novick D, Fischer DG, Rubinstein M: Type I and type II interferon receptors. J Interferon Res *4:* 275-282, 1984.
140. Niederle N, Kloke O, Scheulen ME, Nowrousian MR, and Schmidt CG: Hairy cell leukemia: therapy with recombinant interferon- α and interferon- gamma. Proc Amer Assoc Cancer Res *27:* 1284, 1986.
141. Ludwig H, Cortelezzi A, Scheithauer W, Van Camp BG, Kuzmits R, Fillet G, Peetermans M, Polli E, Flener R: Recombinant interferon α-2c versus polychemotherapy (VMCP) for treatment of multiple myeloma: A prospective randomized trial. Eur J Cancer Clin Oncol *22:* 1111-1116, 1986.
142. Quesada JR, Alexanian R, Hawkins M, Barlogie B, Borden E, Itri L, Gutterman JU: Treatment of multiple myeloma with recombinant α-interferon. Blood *67:* 275-278, 1986.
143. Case DC Jr, Sonneborn HL, Paul SD, Hiebel J, Boyd MA, Shepp MA, Dorsk BM, Bonnem E: Phase II study of rDNA α-interferon in patients with multiple myeloma using an escalation induction phase. Cancer Treat Rep *70:* 1251-1254, 1986.
144. Cooper MR, Welander CE: Interferons in the treatment of multiple myeloma. Cancer *59:* 594-600, 1987.
145. Figlin RA, deKernion JB, Maldazys J, Sarna G: Treatment of renal cell carcinoma with α (human leukocyte) interferon and vinblastine in combination: A phase I-II trial. Cancer Treat Rep *69:* 263-267, 1985.
146. Umeda T, Niijima T: Phase II study of α interferon on renal cell carcinoma. Cancer *58:* 1231-1235, 1986.
147. Vugrin D, Hood L, Laszlo J: A phase II trial of high-dose human lymphoblastoid α interferon in patients with advanced renal carcinoma. J Biol Response Mod *5:* 309-312, 1986.
148. Sarna G, Figlin R, deKernion J: Interferon in renal cell carcinoma. The UCLA experience. Cancer *59:* 610-612, 1987.
149. Vugrin D: Systematic therapy of metastatic renal cell carcinoma. Semin Nephrol *7:* 152-162, 1987.
150. Trump DL, Elson PT, Borden EC, Harris JE, Tuttle RL, Whisnant JK, Oken MM, Carignan JR, Ruckdeschel JC, Davis TE: High dose lymphoblastoid interferon in advanced renal cell carcinoma. Cancer Treat Rep *71:* 165-169, 1987.
151. Quesada JR: Biologic response modifiers in the therapy of metastatic renal cell carcinoma. Semin Oncol *15:* 396-407, 1988.
152. Creagan ET, Ahmann DL, Frytak S, Long HJ, Chang MN, Itri LM: Phase II trials of recombinant leukocyte A interferon in disseminated malignant melanoma: Results in 96 patients. Cancer Treat Rep *70:* 619-624, 1986.
153. Abdi EA, McPherson A, Tan YH: Combination of fibroblast interferon, carboxamide and cimetidine for advanced malignant melanoma. J Biol Response Mod *5:* 423-428, 1986.
154. McLeod GR, Thomson DB, Hersey P: Recombinant interferon alfa-2a in advanced malignant melanoma: A phase I-II study in combination with DTIC. Inter J Cancer *1:* 31-35, 1987.
155. Gelmann EP, Preble OT, Steis R, Lane HC, Rook AH, Wesley M, Jacob J, Fauci A, Masur H, Longo D: Human lymphoblastoid interferon treatment of Kaposi's sarcoma in the acquired immune deficiency syndrome: Clinical response and prognostic parameters. Amer J Med *78:* 737-741, 1985.

156. Rios A, Mansell PW, Newell GR, Reuben JM, Hersch EM, Gutterman JU: Treatment of acquired immunodeficiency syndrome - related Kaposi's sarcoma with lymphoblastoid interferon. J Clin Oncol *3:* 506-512, 1985.
157. Bunn PA, Foon KA, Ihde DC, Longo DL, Eddy J, Winkler CF, Veach SR, Zeffren J, Sherwin S, Oldham R: Recombinant leukocyte A interferon: An active agent in advanced cutaneous T-cell lymphomas. Ann Intern Med *101:* 484-487, 1984.
158. Eriksson B, Oberg K, Alm G, Karlsson A, Lundqvist G, Magnusson A, Wide L, Wilander E: Treatment of malignant endocrine pancreatic tumors with human leukocyte interferon. Cancer Treat Rep *71:* 31-37, 1987.
159. Ziai MR, Imberti L, Tongson A, Ferrone S: Differential modulation by recombinant immune interferon of the expression and shedding of HLA antigens and melanoma associated antigens by a melanoma cell line resistant to the antiproliferative activity of immune interferon. Cancer Res *45:* 5877-5882, 1985.
160. Greiner JW, Fisher PB, Pestka S, Schlom J: Differential effects of recombinant human leukocyte interferons on cell surface antigen expression. Cancer Res *46:* 4984-4990, 1986.
161. Piguet V, Carrel S, Diserens A-C, Mach J-P, deTribolet N: Heterogeneity of the induction of HLA-DR expression by human immune interferon on glioma cell lines and their clones. J Natl Cancer Inst *76:* 223-228, 1986.
162. deKernion JB, Sarna G, Figlin R, Lindner A, Smith RB, The treatment of renal cell carcinoma with human leukocyte α-interferon. J Urology *130:* 1063-1066, 1983.
163. Rinehart JJ, Young D, Laforge J, Colburn D, Neidhart J: Phase I/II trial of recombinant gamma-interferon in patients with renal cell carcinoma: immunologic and biologic effects. J Biol Response Mod *6:* 302-312, 1987.
164. Mittelman A, Huberman M, Puccio C, Fallon B, Tessitore J, Savona S, Eyre R, Gafney E, Wick M, Skelos A, Arnold P, Ahmed T, Groopman J, Arlin Z, Zeffren J, Levitt D: A phase I study of recombinant human interleukin-2 and α-interferon-2a in patients with renal cell cancer, colorectal cancer, and malignant melanoma. Cancer *66:* 664-669, 1990.
165. Oldham RK: Biological response modifiers program and cancer chemotherapy. Int J Tissue React *4:* 173-188, 1982.
166. Quesada JR, Swanson DA, Trindade A, Gutterman JU: Renal ell carcinoma: anti-tumor effects of leukocyte interferon. Cancer Res *43:* 940-947, 1983.
167. Chadha KC, Ambrus JL Jr, Dembinski W, Ambrus JL Sr: Interferons and interferon inhibitory activity in disease and therapy. Exp Biol Med (Maywood) *229:* 285-290, 2004.
168. Beutler B, Greenwald D, Hulmes JD, Chang M, Pany-CE, Mathison J, Ulevitch R, Cerami A: Identity of tumour necrosis factor and the macrophage-secreted factor cachectin. Nature (London) *316:* 552-554, 1985.
169. Beutler B, Cerami A: Cachectin and tumor necrosis factor as two sides of the same biological coin. Nature (London) # 6063, *320:* 584-588, 1986.
170. Shear WJ, Turner FC Chemical treatment of tumors: isolation of the hemorrhage-producing fraction from Serratia marcescens (*Bacillus prodigiosus*) culture filtrate. J Natl Canc Inst *4:* 81-97, 1943.
171. O'Malley WE, Achinstein B, Shear MJ: Action of bacterial polysaccharide on tumors. II. Damage of sarcoma 37 by serum of mice treated with *Serratia marces cens* polysaccharide, and induced tolerance. J Natl Cancer Inst *29:* 1169-1175, 1962.
172. Carswell EA, Old LJ, Kassel RL, Green S, Fiore N, Williamson B: An endotoxin-induced serum factor that causes necrosis of tumors. Proc Natl Cancer Inst USA *72:* 3666-3670, 1975.

173. Sinkovics JG: Oncogenes and growth factors. CRC Critical Reviews in Immunology *8:* 217-298, 1988.
174. Aggarwal BB, Kohr WJ, Hass PE, Moffat B, Spencer SA, Henzel WJ, Bringman TS, Nedwin GE, Goeddel DV, Harkins RN: Human tumor necrosis factor. Production, purification and characterization. J Biol Chem *260:* 2345-2354, 1985.
175. Pennica D, Nedwin GE, Hayflick JS, Seeburg PH, Derynck R, Palladino MA, Kohr WJ, Aggarwal BB, Goeddel DV: Human tumor necrosis factor: precursor structure, expression and homology to lymphotoxin. Nature (London) *312:* 724-729, 1984.
176. Nedospasov SA, Hirt B, Shakhov AN, Dobrynin VN, Kawashima E, Accolla RS, Jongeneel CV: The genes for tumor necrosis factor (TNF-α) and lymphotoxin (TNF-β) are tandemly arranged on chromosome 17 of the mouse. Nucleic Acids Res *14:* 7713-7725, 1986.
177. Goeddel DV, Aggarwal BB, Gray PW, Leung DW, Nedwin GE, Palladino MA, Patton JS, Pennica D, Shepard HM, Sugarman BJ: Tumor necrosis factors: gene structure and biological activities. Cold Spring Harbor Symp Quant Biol *51(Part 1):* 597-609, 1986.
178. Shirai T, Yamaguchi H, Ito H, Todd ChW, Wallace RB: Cloning and expression in *Escherichia coli* of the gene for human tumor necrosis factor. Nature (London) *313:* 803-806, 1985.
179. Nedwin GE, Naylor SL, Sakaguchi AY, Smith D, Jarrett-Nedwin J, Pennica D, Goeddel DV, Gray PW: Human lymphotoxin and tumor necrosis factor genes: structure, homology and chromosomal localization. Nucleic Acid Res *13:* 6361-6373, 1985.
180. Marmenout A, Fransen L, Tavernier J, van der Hayden J, Tizard R, Kawashima E, Shaw A, Johnson MJ, Semon D, Muller R, Ruysschaert M-R, van Vliet A, Fiers W: Molecular cloning and expression of human tumor necrosis factor and comparison with mouse tumor necrosis factor. Eur J Biochem *152:* 515-522, 1985.
181. Kunkel SL, Spengler M, May MA, Spengler R, Larrick J, Remick D: Prostaglandin E2 regulates macrophage-derived tumor necrosis gene expression. J Biol Chem *263:* 5380-5384, 1988.
182. Arakawa T, Yphantis DA: Molecular weight of recombinant human tumor necrosis factor-α. J Biol Chem *262:* 7484-7485, 1987.
183. Haranaka K, Carswell EA, Williamson BD, Prendergast JS, Satomi N, Old L: Purification, characterization, and anti-tumor activity of non-recombinant mouse tumor necrosis factor. Proc Natl Acad Sci USA *83:* 3949-3953, 1986.
184. Smith RA, Baglioni C: The active form of tumor necrosis factor is a trimer. J Biol Chem *262:* 6951-6954, 1987.
185. Wingfield P, Pain RH, Craig S: Tumor necrosis factor is a compact trimer. FEBS Lett *211:* 179-184, 1987.
186. Kinkhabwala M, Sehajpal P, Skolnik E, Smith D, Sharma VK, Vlassara H, Cerami A, Suthanthiran M: A novel addition to the T cell repertory. Cell surface expression of tumor necrosis factor/cachectin by activated normal human T cells. J Exp Med *171:* 941-946, 1990.
187. Aggarwal BB, Henzel WJ, Moffat B, Kohr WJ, Harkins RN: Primary structure of human lymphotoxin derived from 1788 lymphoblastoid cell line. J Biol Chem *260:* 2334-2344, 1985.
188. Schall TJ, Lewis M, Koller KJ, Lee A, Rice GC, Wong GH, Gatanaga T, Granger GA, Lentz R, Raab H, Kohr WJ, Goeddel DV: Molecular cloning and expression of a receptor for human tumor necrosis factor. Cell *61:* 361-370, 1990.
189. Chensue SW, Remick DG, Shmyr-Forsch C, Beals TF, Kunkel SL: Immunohistochemical demonstration of cytoplasmic and membrane-associated tumor necrosis factor in murine macrophages. Amer J Pathol *133:* 564-572, 1988.

190. McCall JL, Yun K, Funamoto S, Parry BR: *In vivo* immunohistochemical identification of tumor necrosis factor/cachentin in human lymphoid tissue. Amer J Pathol *135:* 421-425, 1989.
191. Maury CP, Salo E, Pelkonen P: Elevated circulating tumor necrosis factor-α in patients with Kawasaki disease. J Lab Clin Med *113:* 651-654, 1989.
192. Tsujimoto M, Feinman R, Kohase M, Vilcek J: Characterization and affinity crosslinking of receptors for tumor necrosis factor on human cells. Arch Biochem Biophys *249:* 563-568, 1986.
193. Kull FC Jr, Jacobs S, Cuetrecasas P: Cellular reeptor for 125-I-labeled tumor necrosis factor: specific binding, affinity labeling, and relationship to sensitivity. Proc Natl Sci USA *82:* 5756-5760, 1985.
194. Tsujimoto M, Yip YK, Vilcek J: Tumor necrosis factor: specific binding and internalization in sensitive and resistant cells. Proc Natl Acad Sci USA *82:* 7626-7630, 1985.
195. Israel S, Hahn T, Holtmann H, Wallach D: Binding of human TNF-α to high-affinity cell surface receptors: effect of IFN. Immunol Lett *12:* 217-224, 1986.
196. Imamura K, Ohno T, Spriggs DR: Effects of Botulinum type D toxin on secretion of tumor necrosis factor from human monocytes. Mol Cell Biol *9:* 2239-2243, 1989.
197. Aggarwal BB, Traquina PR, Eessalu TE: Modulation of receptors and cytotoxic response of tumor necrosis factor-α by various lectins. J Biol Chem *261:* 13652-13656, 1986.
198. Holtmann H, Wallach D: Down regulation of the receptors for tumor necrosis factor by interleukin 1 and 4 β-phorbol-12-myristate-13-acetate. J Immunol *139:* 1161-1167, 1987.
199. Darzynkiewicz Z, Williamson B, Carswell EA, Old LJ: Cell cycle specific effects of tumor necrosis factor. Cancer Res *44:* 83-190, 1984.
200. Niitsu Y, Watanabe N, Sone H, Neda H, Yamauchi N, Urushizaki I: Mechanism of the cytotoxic effect of tumor necrosis factor. Jpn J Cancer Res *76:* 1193-1197, 1985.
201. Bonavida B, Granger G (eds.), Tumor Necrosis Factor: structure, mechanism of action, role in disease and therapy. Karger, Basel, pp. 1-252, 1990.
202. Knauer MF, Longmuir KJ, Yamamoto RS, Fitzgerald TP, Granger GA: Mechanism of human lymphotoxin and tumor necrosis factor induced destruction of cells *in vitro*: phospholipa e activation and deacylation of specific-membrane phospholipids. J Cell Physiol *142:* 469-479, 1990.
203. Schirrmacher V, Heicappell R: Prevention of metastatic spread by postoperative immunotherapy with virally modified autologous tumor cells. II. Establishment of specific systemic anti-tumor immunity. Clin Exp Metastasis *5:* 147-156, 1987.
204. Rosenblum MG Donato NJ: Tumor necrosis factor α: a multifaceted peptide hormone. CRC Crit Rev Immunol *9:* 21-44, 1989.
205. Rosenblum MG, Donato NJ, Kessler D, Gutterman JU: Preliminary characterization of human recombinant tumor necrosis factor (rTNF) antiproliferative effects on human cells in culture. Proc Am Assoc Cancer Res *28:* 398, 1987.
206. Creasey AA, Yamamoto R, Vitt CR: A high molecular weight component of the human tumor necrosis factor receptor is associated with cytotoxicity. Proc Natl Acad Sci USA *84:* 3293-3297, 1987.
207. Creasey AA, Reynolds MT, Laird W: Cures and partial regression of murine and human tumors by recombinant human tumor necrosis factor. Cancer Res *46:* 5687-5690, 1986.

208. Vanhaesebroeck B, Cragoe EJ Jr, Pouyssegur J, Beyaert R, van Roy F, Fiers W: Cytotoxic activity of tumor necrosis factor is inhibited by amiloride derivates without involvement of the Na+/H+ antiporter. FEBS Lett *261:* 319-322, 1990.
209. Yoshimura T, Sone S: Different and synergistic actions of human tumor necrosis factor and interferon-gamma in damage of liposome membranes. J Biol Chem *262:* 4597-4601, 1987.
210. Katakami Y, Nakao Y, Koizumi T, Katakami N, Ogawa R, Fujita T: Regulation of tumour necrosis factor production by mouse peritoneal macrophages: The role of cellular cyclic AMP. Immunology *64:* 719-724, 1988.
211. Strieter RM, Remick DG, Ward PA, Spengler RN, Lynch JP III Larrick J, Kunkel SL: Cellular and molecular regulation of tumor necrosis factor-α production by pentoxifylline. Biochem Biophys Res Commun *155:* 1230-1236, 1988.
212. Scuderi P: Suppression of human leukocyte tumor necrosis factor secretion by the serine protease inhibitor p-toluenesulphonyl-L-arginine methyl ester (TAME). J Immunol *143:* 168-173, 1989.
213. Scheurich P, Unglaub R, Maxeiner B, Thoma B, Zugmaier G, Pfizenmaier K: Rapid modulation of tumor necrosis factor membrane receptors by activators of protein kinase C. Biochem Biophys Res Commun *141:* 855-860, 1986.
214. Jetten AM, Ganong BR, Vandenbark GR, Shirley JE, Bell RM: Role of protein kinase C in diacylglycerol-mediated induction of ornithine decarboxylase and reduction of epidermal growth factor binding. Proc Natl Acad Sci USA *82:* 1941-1945, 1985.
215. Nishizuka Y: Studies and perspectives of protein kinase C. Science *233:* 305-312, 1986.
216. Fidler IJ, Kleinerman ES: Lymphokine-activated human blood monocytes destroy tumor cells but not normal cells under cns. J Clin Oncol *2:* 937-943, 1984.
217. Ziegler-Heitbrock HW, Moller A, Linke RP, Haas JG, Rieber EP, Riethmuller G: Tumor necrosis factor as effector molecule in monocyte mediated cytotoxicity. Cancer Res *46:* 5947-5952, 1986.
218. Urban JL, Shepard HM, Rothstein JL, Sugarman BJ, Schreiber H: Tumor necrosis factor: a potent effector molecule for tumor cell killing by activated macrophages. Proc Natl Acad Sci USA *83:* 5233-5237, 1986.
219. Nissen-Meyer J, Austgulen R, Espevik T: Comparison of recombinant tumor necrosis factor and the monocyte-derived cytotoxic factor involved in monocyte-mediated cytotoxicity. Cancer Res *47:* 2251-2258, 1987.
220. Kirstein M, Baglioni C: Tumor necrosis factor induces synthesis of two proteins in human fibroblasts. J Biol Chem *261:* 9565-9567, 1986.
221. Manda T, Shimomura K, Mukumoto S, Kobayashi K, Mizota T, Hirai O, Matsumoto S, Oku T, Nishigaki F, Mori J, Kikuchi H: Recombinant human tumor necrosis factor-α: evidence of an indirect mode of antitumor activity. Cancer Res *47:* 3707-3711, 1987.
222. Patek PQ, Lin Y, Collins JL: Natural cytotoxic cells and tumor necrosis factor activate similar lytic mechanisms. J Immunol *138:* 1641-1646, 1987.
223. Talmadge JE, Herberman RB: The preclinical screening laboratory: evaluation of immunomodulatory and therapeutic properties of biological response modifiers. Cancer Treat Rep *70:* 171-182, 1986.
224. Regenass U, Muller M, Curschellas E, Matter A: Anti-tumor effects of tumor necrosis factor in combination with chemotherapeutic agents. Int J Cancer *39:* 266-273, 1987.
225. Blick M, Sherwin SA, Rosenblum M, Gutterman J: Phase I study of recombinant tumor necrosis factor in cancer patients. Cancer Res *47:* 2986-2989, 1987.
226. Sugarman BJ, Aggarwal BB, Hass PE, Figari IS, Palladino MA, Shepard HM: Recombinant human tumor necrosis factor-α: effect on proliferation of normal and transformed cells *in vitro*. Science *230:* 943-945, 1985.

227. Watanabe N, Niitsu Y, Neda H, Sone H, Yamauchi N, Umetsu T, Urushizaki I: Antitumor effect of tumor necrosis factor against various primarily cultured human cancer cells. Jpn J Cancer Res *76:* 1115-1119, 1985.
228. Lachman LB, Brown DC, Dinarello CA: Growth-promoting effect of recombinant interleukin 1 and tumor necrosis factor for a human astrocytoma cell line. J Immunol *138:* 2913-2916, 1987.
229. Pober JS, Bevilacqua MP, Mendrick DL, Lapierre LA, Fiers W, Gimbrone MA Jr: Two distinct monokines, interleukin 1 and tumor necrosis factor, each independently induce biosynthesis and transient expression of the same antigen on the surface of cultured human vascular endothelial cells. J Immunol *136:* 1680-1687, 1986.
230. Ming WJ, Bersani L, Matovani A: Tumor necrosis factor is chemotactic for monocytes and polymorphonuclear leukocytes. J Immunol *138:* 1469-1474, 1987.
231. Ruggiero V, Tavarnier J, Fiers W, Baglioni C: Induction of the synthesis of tumor necrosis factor receptors by interferon-gamma. J Immunol *136:* 2445-2450, 1986.
232. Hemni H, Nakamura T, Tamura K, Shimizu Y, Kato S, Miki T, Takahashi N, Muramatsu M, Numao N, Sagamura K: Lymphotoxin: induction of terminal differentiation of the human myeloid leukemia cell lines HL-60 and THP-1. J Immunol *138:* 664-666, 1987.
233. Nakamura H, Motoyoshi S, Seto Y, Kadokawa T, Nakata K, Iida M, Taguchi T: Damaging action of human recombinant TNF on tumor vessels as an aspect of its anti-neoplastic action against Meth A sarcoma in mice. Gan To Kagaku Ryoho *14:* 91-99, 1987.
234. Sherwin SA: Early clinical trials of recombinant tumor necrosis factor (RTNF). "Fifth NCl-EORTC Symposium on New Drugs in Cancer Therapy". Amsterdam, The Netherlands (Abstract 6.04), 1986.
235. Feinberg B, Kurzrock M, Talpaz M, Blick M, Saks S, Gutterman JU: A phase I trial of intravenously-administered recombinant tumor necrosis factor-α in cancer patients. J Clin Oncol *6:* 1328-1334, 1988.
236. Steinmetz T, Schaadt M, Gahl R, Schenk V, Diehl V, Pfreundschuh M: Phase I study of 24-hour continuous intravenous infusion of recombinant human tumor necrosis factor. J Biol Response Mod *7:* 417-423, 1988.
237. Spriggs DR, Sherman ML, Michie H, Arthur KA, Imamura K, Wilmore D, Frei E III, Kufe D.W., Recombinant human tumor necrosis factor administered as a 24-hour intravenous infusion. A phase I and pharmacologic study. J Natl Cancer Inst *80:* 1039-1044, 1988.
238. Sherman ML, Spriggs DR, Arthur KA, Imamura K, Frei III E, Kufe DW: Recombinant human tumor necrosis factor administered as a five-day continuous infusion in cancer patients: Phase I toxicity and effects on lipid metabolism. J Clin Oncol *6:* 344-350, 1988.
239. Wiedenmann B, Reichardt P, Rath U, Theilmann L, Schule B, Ho AD, Shlick E, Kempeni J, Hunstein W, Kommerell B: Phase-I trial of intravenous continuous infusion of tumor necrosis factor in advanced metastatic carcinomas. J Cancer Res Clin Oncol *115:* 189-192, 1989.
240. Creaven PJ, Brenner DE, Cowens JW, Huben RP, Wolf RM, Takita H, Arbuck SG, Razack MS, Proefrock AD: A phase I clinical trial of recombinant tumor necrosis factor given daily for five days. Cancer Chemother Pharmacol *23:* 186-191, 1989.
241. Pfreundschuh MG, Steinmetz HT, Tuschen R, Schenk V, Diehl V, Schaadt M: Phase I study of intratumoral application of recombinant human tumor necrosis factor. Eur J Cancer Clin Oncol *25:* 379-388, 1989.

242. Bartsch HH, Pfizenmaier K, Schroeder M, Nagel GA: Intralesional application of recombinant human tumor necrosis factor α induces local tumor regression in patients with advanced malignancies. Eur J Cancer Clin Oncol *25:* 287-291, 1989.
243. Zamkoff K, Newman N, Rudolph A, Poiesz B: A Phase I study of subcutaneously administered recombinant tumor necrosis factor (rTNF) in patients with advanced malignancy. Proc Amer Soc Clin Oncol *7:* 68 (Abstract #259), 1987.
244. Childs B, Kemeny N, Larchian N, Rosodo K, Kelsen D: A phase II trial of recombinant tumor necrosis factor in patients with advanced colorectal carcinoma. Cancer *66:* 659-663, 1990.
245. Schaadt M, Pfreundschuh M, Lorscheidt G, Peters KM, Steinmetz HT, Diehl V: Phase II study of recombinant human tumor necrosis factor in colorectal carcinoma. J Biol Response Mod *9:* 247-250, 1990.
246. Kemeny N, Childs B, Larchian W, Rosado K, Kelsen D: A phase II trial of recombinant tumor necrosis factor in patients with advanced colorectal carcinoma. Cancer *66:* 659-663, 1990.
247. McIntosh JK, Mule JJ, Merino MJ, Rosenberg SA: Synergistic antitumor effects of immunotherapy with recombinant interleukin-2 and recombinant tumor necrosis factor-α. Cancer Res *48:* 4011-4017, 1988.
248. Talmadge JE, Tribble H, Pennington R: Immunotherapeutic properties of recombinant lymphokines (RH RNF, RM IFN-γ, RH IL-2) in the treatment of metastatic disease. Abstract. Proc Amer Assoc Cancer Res *27:* 317, 1986.
249. Rychly J, Knippel E, Krygier-Stojalowska A, Nizze H, Kuchnio M, Kraeft SK: DNA cytometric and histologic findings in mouse tumors (BP and S180) with different response to treatment with tumor necrosis factor. Acta Oncol *29:* 47-51, 1990.
250. Herberman RB: Interleukin-2 therapy of human cancer: potential benefits versus toxicity. J Clin Oncol *7:* 1-4, 1989.
251. Aggarwal BB, Shishodia S, Ashikawa K, Bharti AC: The role of TNF and its family members in inflammation and cancer: lessons from gene deletion. Curr Drug Targets Inflamm Allergy *1:* 327-341, 2002.
252. Morgan DA, Ruscetti FW, Gallo RG: Selective *in vitro* growth of T-lymphocytes from normal bone marrows. Science *193:* 1007-1008, 1976.
253. Ruscetti FW, Morgan DA, Gallo RC: Functional and morphologic characterization of human T cells continuously grown *in vitro*. J Immunol *119:* 131-138, 1977.
254. Strausser JL, Rosenberg SA: *In vitro* growth of cytotoxic human lymphocytes. I. Growth of cells sensitized *in vitro* to alloantigens. J Immunol *121:* 1491-1495, 1978.
255. Yron I, Wood TA Jr, Spiess PJ, Rosenberg SA: *In vitro* growth of murine T cells. V. The isolation and growth of lymphoid cell infiltrating syngeneic solid tumors. J Immunol *125:* 238-245, 1980.
256. Bodey B, Bodey B, Jr, Kaiser HE: Cell culture observations of Postnatal Thymic Epithelium: An *In vitro* Model for Growth and Humoral Influence on Intrathymic T Lymphocyte Maturation. In Vivo *10:* 515-526, 1996.
257. Rosenberg SA: Adoptive immunotherapy of cancer: accomplishments and prospects. Cancer Treat Rep *68:* 233-255, 1984.
258. Doyle MV, Lee MT, Fong S: Comparison of biological activities of human recombinant and native IL-2. J. Biol Response Mod *4:* 96-109, 1985.
259. Sinkovics JG: Oncogenes and growth factors. CRC Critical Reviews in Immunology *8:* 217-298, 1988.
260. Guinan EC, Smith BR, Miller RA, Pober JS: Vascular endothel cells enhance T cell responses by markedly augmenting interleukin-2. 5th Int. Lymphokine Workshop,

Molecular Basis of Lymphokine Action, Jan. 11-15, 1987, Clearwater, FL, Lymphokine Res *6:* 61 (Abstr. 1632), 1987.
261. Rosenstein M, Ettinghausen SE, Rosenberg SA: Extravasation of intravascular fluid mediated by the systemic administration of recombinant interleukin-2. J Immunol *137:* 1735-1742, 1986.
262. Grimm EA, Mazumder A, Zhang HZ, Rosenberg SA: Lymphokine activated killer cell phenomenon: lysis of naturalo killer-resistant fresh solid tumor cells by interleukin 2-activated autologous human peripheral blood lymphocytes. J Exp Med *155:* 1823-1841, 1982.
263. Tsoukas CD, Landgraf B, Bentin J, Valentine M, Lotz M, Vaugham JH, Carson DA: Activation of resting T lymphocytes by anti-CD3 (T3) antibodies in the absence of monocytes. J Immunol *135:* 1719-1723, 1985.
264. Stern JB, Smith KA: Interleukin-2 induction to T cell G1 progression and c-myb expression. Science *233:* 203-206, 1986.
265. Rouse BT, Miller LS, Turtunen L, Moore RN: Augmentation of immunity to herpes simplex virus by *in vivo* administration of interleukin-2. J Immunol *134:* 926-930, 1985.
266. Ettinghausen SE, Lipford EH III, Mule JJ, Rosenberg SA; Recombinant interleukin-2 stimulates *in vivo* proliferation of adoptively transferred lymphokine-activated killer (LAK) cells. J Immunol *135:* 3623-3635, 1985.
267. Clark RL Jr: Introduction to viruses and tumor growth. Texas Rep Biol Med *15:* 449-450, 1957.
268. Bubenik J, Voitenok NN, Kieler J, Prassolov VS, Chumakov PM, Bubenikova D, Simova J, Jandlova T: Local administration of cells containing an inserted IL-2 gene and producing IL-2 inhibits growth of human tumours in nu/nu mice. Immunol Lett *19:* 279-282, 1988.
269. Lin Y, Case PG, Patek PQ: Inhibition of tumour necrosis factor and natural cytotoxic cell lytic activities by a spleen cell-elaborated factor. Immunology *63:* 663-668, 1988.
270. Rosenberg SA, Lotze MT, Yang JC, Linehan WM, Seipp C, Calabro S, Karp SE, Sherry RM, Steinberg S, White DE: Combination therapy with interleukin-2 and α-interferon for the treatment of patients with advanced cancer. J Clin Oncol *7:* 1863-1874, 1989.
271. Rosenberg SA, Mule JJ: Immunotherapy of cancer with lymphokine activated killer cells and recombinant interleukin-2. Surgery *98:* 437-443, 1985.
272. Rosenberg SA: Adoptive immunotherapy of cancer using lymphokine activated killer cells and recombinant IL-2. *In:* Important Advances in Oncol, 1986.
273. Belldegrun A, Muul LM, Rosenberg SA: Interleukin-2 expanded tumor-infiltrating lymphocytes in human renal cell cancer: Isolation, characterization, and antitumor activity. Cancer Res *48:* 206-214, 1988.
274. Fisher B, Keenan AM, Garra BS, Steinberg SM, White DE, DiBisceglie AM, Hoofnagle JH, Yolles P, Rosenberg SA, Lotze MT: Interleukin-2 induces profound reversible cholestasis: a detailed analysis in treated cancer patients. J Clin Oncol *7:* 1852-1862, 1989.
275. Paciucci PA: Antitumor activity of interleukin-2 without toxicity and other desiderata. Cancer Inves *7:* 297-298, 1989.
276. Rosenberg SA, Longo DL, Lotze MT: Principles and applications of biologic therapy. In: Cancer, Principles and practice of oncology, 3rd edition, Eds. V.T. DeVita Jr., Hellman S. and Rosenberg S.A., J.B. Lippincott Company, Philadelphia pp. 301-347, 1989.

277. Huang CM, Elin RJ, Ruddel M, Sliva C, Lotze MT, Rosenberg SA: Changes in laboratory results for cancer patients treated with interleukin -2. Clin Chem *36:* 431-434, 1990.
278. North RJ: Cyclophosphamide-facilitated adoptive immunotherapy of established tumor depends on elimination of tumor induced suppressor cells. J Exp Med *155:* 1063-1074, 1982.
279. Borden EC, Creekmore SP: Cancer therapy with combinations of biological response modifiers and cytotoxics: an update. Cancer Cells *2:* 217-220, 1990.
280. Mitchell MS: Biomodulation in the treatment of cancer. USC J Med *4:* 23-25, 1990.

Chapter 5

THE LYMPHOKINE ACTIVATED KILLER (LAK) CELL PHENOMENON

Lymphokine activated killer (LAK) cells are autologous, not restricted by MHC, and are cytotoxic lymphocytes that develop the ability to kill fresh tumor cells following an incubation in medium that contains T lymphocyte growth factor (currently called IL-2).

The discovery, generation and therapeutic use of immunoactive host effector cells which kill cancer cells is being development, but it bears positive promise based on the immunologic advances that have been made in the dissection of the cell clones of host cellular immunological effector cells. We learned to generate cytotoxic effector cells and we did not hesitate to use them for cellular immunotherapy of human cancer. At first, long term cultures of peripheral blood lymphocytes were used, with a maximum of 5 x 108 being capable of safely being infused in humans.

In the early 1960s, normal spleen cells from allogeneic donors were infused into cancer-bearing patients, without any therapeutic effect. Yamaue and co-investigators (1) reported a unique way of *in vivo* and *in vitro* activation of NK cells with the streptococcal preparation OK-432. The cell surface phenotype of induced killer cells was: Thy 1^{+}, asialo $GM1^{+}$, suggesting activated NK cells (OK-NK). The adoptive cellular therapy with OK-NK cells was 92% successful in mouse preclinical SP2-tumor (myeloma injected intraperitoneally), and the tumor growth of C26 solid tumor (colon adenocarcinoma) was inhibited, with survival rates being increased significantly. The intratumoral remnants of 125I-labelled OK-NK cells were 61.27% and 8% after intratumoral transfer, respectively. By multiple transfer of OK-NK cells, the anti-tumor action was more augmented than that of a single transfer. We suggest that these OK-NK cells could be useful in further clinical trials in the therapy of human cancer.

Cheema and Hersh (2) used a mitogen, phytohemagglutinin (PHA)-activated and killer (PAK) autologous lymphocytes for the treatment of 15 patients with advanced cancer. Rosenberg and co-workers used the same PAK cells in large numbers [up to 1.7 x 10(11)] obtained from up to 15 leukaphereses, as single agents and also in combination with activated macrophages or cyclophosphamide in 21 patients (3, 4).

Grimm and co-workers (5) first reported and later Phillips and Lanier (6) dissected the LAK cell phenomenon: IL-2 induced (with 3-5 x 10^6/mg specific activity, for rIL-2, and rapid (3-4 days) *in vitro* generation and activation of MHC unrestricted cytotoxic effector cells against numerous autologous and allogeneic tumor cells. The used tissue culture medium remains the RPMI-1640 (enriched with 500 IU/ml rIL-2, 10% human serum, antibiotics) medium. The LAK lymphocyte clone represents a non-T and non-B cell clone, and was also quite distinct from peripheral NK cells or CTLs mentioned above. Originated from the bone marrow, located LAK precursors (stem cells) have a special cell surface IP characteristics: ($CD2^+$,$CD11^+$, $CD16^+$, $CD3^-$).

Human LAK effector cells demonstrated the following three different cell surface IPs:

1) major population - $CD3^-$,$CD8^-$,$CD2^+$,$CD16^+$,$CD19^+$;
2) $CD2^+$,$CD3^+$,$CD8^+$,$CD16^-$,$CD19^-$; and
3) $CD3^-$,$CD8^+$,$CD2^+$,$CD16^+$,$CD19^+$.

LAK cells were called "Pinocchio cells" by Paciucci and co-workers (7) because of their single, long uropode that is so morphologically characteristic of them. LAK cells were either CD3 negative or positive, were nonadherent and E-rosette negative and expressed NK-like surface markers such as CD11 and CD19 (NKH-1).

IL-2 is the sole signal required for the generation of LAK, as was demonstrated in experiments using purified homogeneous recombinant IL-2, produced with recombinant DNA technology in *E. coli* (8). IL-2 also activates Fc receptor expression on LAK cells, suggesting that such cells may be useful in antibody dependent cellular cytotoxicity (ADCC) in which antibody molecules bind directly to target cells and enable bridging to effectors bearing Fc receptors.

The exact nature of antigenic determinants recognized on fresh tumor cell targets by LAK cells is not known yet, although they appear to be broadly expressed on fresh and cultured tumor cells and also on cultured normal cells. Is there a TAA structure that LAK cells are capable of recognizing on the membrane of diverse tumor cells? Fresh and normal cells, probably only with the exception of monocytes, do not bear the cell surface markers recognized by LAK cells.

Zychlinsky and co-authors (9) obtained a highly purified population of murine LAK cells by selecting plastic adherent splenocytes after incubation in high doses of rIL-2. The cells were >95% positive for the cell marker asialo-GM1, and negative for both Lyt-1 (CD5) and Lyt-2 (CD8). Morphologically these cells were typical "large granular lymphocytes" and killed NK-sensitive target cells in an exclusively calcium-dependent *Fas*hion. The presence of Hanukkah Factor/ granzyme A/ serine esterase 1, CTLA-1/ granzyme B/ serine esterase 2, and pore forming protein (PFP/ perforin) in these LAK cells was detected with Northern blot analysis, suggesting that these markers are not exclusively associated with CTL.

Immunoblot assays identified that PFR/perforin reacted with a 70 kD protein of LAK cells and was localized in the cell granules. A 50 kD protein antigenically related to macrophage TNF was isolated with immunoblots, and localized in the cell granules and the cytosol. These results suggest that LAK cells may contain a cytotoxic factor which is related to, but distinct from macrophage TNF.

Adherent lymphokine activated killer cells (A-LAK) could be obtained from human peripheral blood, and represent a potent population of anti-tumor effector cells, enriched with rIL-2 activated natural killer (NK) cells (10). Liver cancer patients were not treated with adjuvant chemotherapy or irradiation. The peripheral mononuclear cells were isolated during the time of liver resection. A-LAK were separated by adherence to plastic following prior activation of peripheral blood mononuclear cells in 1000 U/ml rIL-2. A-LAK cells (enriched up to 92% in $CD3^-$ $CD56^+$ cells) demonstrated better growth *in vitro* and 2-6 times higher anti-tumor cytotoxicity than unseparated LAK cells, cultured under the same conditions.

Local, *in situ* administration of poly-effective LAK cells (demonstrating surface IP: $Leu\text{-}1^{++}$, $Leu\text{-}2a^+$, $Leu\text{-}3a^{++}$, $Leu\text{-}7^+$, and $Leu\text{-}11^{++}$) and rIL-2 was reported in 23 patients with recurrent malignant glioma (11). 1.2 to 324 x 10^8/ml LAK cells and 0.8 to 5.4 x 10^3 units of purified rIL-2 (Shionogi Chemical Industries, LTD, Japan) were directly injected into the cavities of these brain tumors by using an Ommaya reservoir. Definite tumor regression occurred in 6/23 patients, improvement of some clinical symptoms was reported by 9/23, and continuous tumor remission over six months was seen in 3/23 patients.

1. ADMINISTRATION OF RIL-2 AND LAK CELLS: PRECLINICAL TRIALS

Laboratory experiments and preclinical animal tumor models demonstrated that combined administration of LAK cells and IL-2 (adoptive immunotherapy) provided more successful anti-tumor immunotherapy than separate use of either component alone (12-14).

Clinical trials using LAK cells together with rIL-2 for the treatment of cancer in advanced stages in humans were developed after 1984 when recombinant IL-2 became available. In the Surgery Branch of National Cancer Institute, 177 patients were treated with IL-2 and LAK cells, and 119 cancer patients received only rIL-2. During the study, patients received rIL-2 by i.v. bolus infusion every eight hours, usually at the dose of 720,000 IU/kg. The immunotherapy of these 296 patients involved mainly renal cell carcinoma and malignant melanoma. In these groups, approximately 10% of patients obtained a complete regression of metastatic cancer and about 20% had objective, partial regressions. In cases of metastatic colorectal cancer, about 15% of patients experienced an objective regression of the tumor. When tumor regression is seen at one site, it tends to occur at all sites, mixed responses are unusual. Of 18 patients who achieved complete regression, 10 have remained in complete remissions for as long as 42 months follow-up (15). Reports have also shown positive results in patients with Hodgkin's and non-Hodgkin's lymphomas, adenocarcinoma of the breast and of the colon, non-small-cell lung adenocarcinoma, soft-tissue sarcoma, and some osteosarcomas.

Human tumor antigens recognized by T cells have been recently identified in various cancers, including pancreatic cancer. With these identified antigens, new immunotherapies can be developed using more efficient immunologic intervention as well as more quantitative and qualitative immunomonitoring. Various immunotherapies for patients with various cancers, including pancreatic cancer, are currently under evaluation in clinical trials. One such trial involves the adoptive transfer of tumor reactive T cells and LAK cells (16). Current research has also found that umbilical cord blood (UCB) is another good source of LAK cells in adults and children suffering from anaplastic astrocytoma or medulloblastoma (17).

2. CONCLUSIONS

1) The positive results from the clinical trials performed in a large number of patients with advanced cancer at the NCI Surgery Branch using the combined treatment with LAK cells and rIL-2 demonstrated that activated, autologous cells could mediate an anti-neoplastic effect.
2) The mechanisms of anti-neoplastic action of LAK cells and rIL-2 are not completely understood. Dense lymphocyte infiltrates are detected

within tumor deposits. The direct cytotoxicity of LAK cells against tumor cell targets is probably the main factor of the anti-tumor action.

3) The LAK cell clone demonstrated in *in vitro* assays cytotoxicity against numerous autologous and allogeneic *ex vivo* cancers and established cancer cell lines. The LAK cells are **not** TAA antigen directed and specific cytotoxic killer cells.
4) *In vivo* or *in vitro* treatment of lymphocytes with rIL-2 in excess generated numerous lymphocytic killer clones, after time lacking a LAK-type specificity. The mechanism and direction of cytotoxic killing could not be controlled in long time cultures.
5) The *in situ* treatment with LAK cells was more successful.
6) Significant LAK cell related toxicity was never observed during the immunotherapeutical clinical trials.
7) To date, LAK cells have been employed as a form of experimental therapy, at stage IV of advanced cancer, when other therapies have failed. Their immediate use after surgical tumor removal or even prior to surgery have strong indications.
8) Early approaches such as lymphokine activator killer (LAK) cells and tumor infiltrating lymphocytes (TILs) have yielded occasional clinical responses. More recently, attempts to stimulate and/or select antigen-specific T-cells *in vitro* have demonstrated that tumor-specific adoptive immunotherapy is possible.

REFERENCES

1. Yamaue H, Tanimura H, Iwahashi M, Tani M, Tsunoda T, Tabuse K: Successful adoptive immunotherapy with OK432-inducible activated natural killer cells on tumor-bearing mice. *24:* 2546-2555, 1989.
2. Cheema AR, Hersh EM: Local tumor immunotherapy with *in vitro* activated autochthonous lymphocytes. Cancer *29:* 982-986, 1972.
3. Rosenberg SA, Lotze MT, Muul LM, Leitman S, Chang AE, Ettinghausen SE, Matory YL, Skibber JM, Shiloni E, Vetto JT, Seipp CA, Simpson C, Reichert CM: Observations on the systemic administration of autologous lymphokine-activated killer cells and recombinant interleukin-2 to patients with metastatic cancer. New Engl J Med *313:* 1485-1492, 1985.
4. Rosenberg SA, Mule JJ: Immunotherapy of cancer with lymphokine activated killer cells and recombinant interleukin-2. Surgery *98:* 437-443, 1985.
5. Grimm EA, Mazumder A, Zhang HZ, Rosenberg SA: Lymphokine activated killer cell phenomenon: lysis of naturalo killer-resistant fresh solid tumor cells by interleukin 2-activated autologous human peripheral blood lymphocytes. J Exp Med *155:* 1823-1841, 1982.
6. Phillips JH, Lanier LL: Dissection of the lymphokine-activated killer phenomenon. Relative contribution of peripheral blood natural killer cells and T lymphocytes to cytolysis. J Exp Med *164:* 814-825, 1986.

7. Paciucci PA, Chesa PG, Fierro MT, Gordon R, Konefal RG, Glidewell O, Holland JF: Pinocchio cells: morphologically atypical immunologically heterogeneous lymphocytes induced by treatment with interleukin 2. Immunol Lett *19:* 313-320, 1988.
8. Rosenberg SA, Grimm EA, McGrogan M, Doyle M, Kawasaki E, Koths K, Mark DF: Biological activity of recombinant human interleukin-2 produced in *E. coli*. Science *223:* 1412-1414, 1984.
9. Zychlinsky A, Joag S, Liu CC, Young JD: Cytotoxic mechanisms of murine lymphokine-activated killer cells: functional and biochemical characterization of homogeneous populations of spleen LAK cells. Cell Immunol *126:* 377-390, 1990.
10. Schwarz RE, Iwatsuki S, Herberman RB, Whiteside TL: Unimpaired ability to generate adherent lymphokine-activated (A-LAK) cells in patients with primary or metastatic liver tumors. Cancer Immunol Immunother *30:* 312-316, 1989.
11. Yoshida S, Tanaka R, Takai N, Ono K: Local administration of autologous lymphokine-activated killer cells and recombinant interleukin 2 to patients with malignant brain tumors. Cancer Res *48:* 5011-5016, 1988.
12. Mule JJ, Shu S, Schwarz SL, Rosenberg SA: Adoptive immunotherapy of established pulmonary metastases with LAK cells and recombinant interleukin-2. Science *225:* 1487-1489, 1984.
13. Lafreniere R, Rosenberg SA: Adoptive immunotherapy of murine hepatic metastases with lymphokine activated killer (LAK) cells and recombinant interleukin-2 (RIL-2) can mediate the regression of both immunogenic and non-immunogenic sarcomas and an adenocarcinoma. J Immunol *135:* 4273-4280, 1985.
14. Rosenberg SA: The development of new immunotherapies for the treatment of cancer using Interleukin-2. Amer Surg *208:* 121-135, 1988.
15. Rosenberg SA, Longo DL, Lotze MT: Principles and applications of biologic therapy. In: Cancer, Principles and practice of oncology, 3rd edition, Eds. V.T. DeVita Jr., Hellman S. and Rosenberg S.A., J.B. Lippincott Company, Philadelphia pp. 301-347, 1989.
16. Kawakami Y, Okada T, Akada M: Development of immunotherapy for pancreatic cancer. Pancreas *28:* 320-325, 2004.
17. Kang SG, Ryu CH, Jeun SS, Park CK, Shin HJ, Kim JH, Kim MC, Kang JK: Lymphokine activated killer cells from umbilical cord blood show higher antitumor effect against anaplastic astrocytoma cell line (U87) and medulloblastoma cell line (TE671) than lymphokine activated killer cells from peripheral blood. Childs Nerv Syst *20:* 154-162, 2004.

Chapter 6

ANGIOGENESIS INHIBITION IN ANTI-NEOPLASTIC THERAPY

Inhibition of human neoplasm related angiogenesis as an attractive therapeutic possibility was first suggested by Folkman in 1971. Recently, it has been established that permanent neoplasm related angiogenesis is a key element in the continuous heterogeneous progression of an increasingly malignant neoplastic CIP from *in situ* transformation of a group of cells to the invasion and organization of distant metastases (1-4). A number of clinical follow-up studies have demonstrated a significant correlation between the degree of NRA and clinical outcome, strongly suggesting that angiogenic properties influence the aggressiveness and metastatic potential of neoplastic cells (5). Occlusion of the neoplasm related and induced microvasculature has been shown to result in tumor necrosis and collapse.

It is important to distinguish between antiangiogenic treatment and vascular targeting. The treatment of human cancer employing angiogenesis-inhibiting agents represents a complex therapy in the sense that multiple, although closely related, processes are affected by a single molecule. However, targeting of angiogenesis using a single inhibitor still represents a type of mono-therapy in the clinical sense and is thus bound to fail due to the resiliency of the malignancy. Vascular targeting, on the other hand, provides rapid destruction of capillaries, which also results in the death of vascular elements. The clinical application of vascular targeting requires the exact identification of cell surface bound target molecules with optimal density on the endothelial cells in solid tumors, but absent from endothelial cells in normal tissues. MoABs against such target molecules could be used to deliver cytotoxic agents or cellular mediators of the immune response (including cytotoxic T lymphocytes) to the endothelial cells within mammalian neoplasms. Endoglin has been listed among the promising candidates for such purposes (as well as endosialin, a fibronectin isoform; VEGF/PF and its receptors KDF and flt-1) (6).

Inhibition of angiogenesis as a form of anti-neoplastic therapy has been extensively studied (7, 8). A number of anti-angiogenic agents have been discovered and observed *in vitro* and *in vivo*, including IL-12, combretastatin A-4, AGM-1470 (TNP-470), anti-endoglin antibody TEC-11, endostatin, angiostatin, suramin, edelfosine, 2-methoxyestradiol, taxol, and thalidomide (9-20). The driving influence of activating and inactivating mutations in oncogenes (ras) and tumor suppressor genes (p53), respectively, on angiogenesis has also been determined and presents further possibilities for gene therapy-based therapeutic protocols.

In February of 2004, the Food and Drug Administration (FDA) approved Avastin, the first drug to battle human neoplasms by blocking their blood supply, developed by Genentech, Inc. Clinical oncologists consider Avastin a drug that is capable of turning cancer into a chronic, manageable illness. The new drug, the first to win approval based on the theory developed by Folkman, blocks the action of a neoplasm produced protein that orders the host body to sprout new blood vessels to supply the neoplastically transformed, malignant cells with nutrients. Without a solid blood supply, solid tumors do not grow. Avastin is one of a broad class of new anti-neoplastic drugs which perform targeted therapies. These targeted therapies are more benign than chemotherapy regiments are and some have proven to be remarkably effective. We are very satisfied that currently, by one count, 74 drugs that block neoplasm related and induced neoangiogenesis have entered human clinical tests.

Folkman and co-workers are also exploring the possibility of preventing cancer before it occurs with the use of anti-angiogenesis inhibitors. The idea is that using such inhibitors, pharmaceuticals could prevent the angiogenetic switch that is stereotypical of cancer and therefore, bypass the development of cancer. People with Down Syndrome are the most protected population from cancer. It has been shown that these individuals possess about twice as much Endostatin than a "normal" individual does. As such, employing Endostatin has been shown to have some success in patients. Celebrex, a popular medication for treating arthritis, increases the level of Endostatin in the body to that of almost the level in individuals with Down Syndrome. Celebrex is now being investigated for its beneficial use in treating cancer and even possibly, preventing it. As the table demonstrates, there are several other inhibitors and by the same token, stimulants, of angiogenesis that need to be further explored for their possible use in anti-angiogenesis therapy.

Encouraging data on two new, so-called "targeted drugs" that basically jam up the cancer's internal signaling circuits without producing major side effects has been reported (21). OSI Pharmaceuticals' Tarceva was found to be robust both as monotherapy and in combination with chemotherapies because it resulted in significantly less tumor growth and in some cases even partial regressions (22). ImClone system's Erbitux had similar results (21).

Table 6-1. Endogenous Regulators of Angiogenesis

I. Stimulators
Vascular endothelial growth factor/ vascular permeability factor (VEGF/PF)
Fibroblast growth factors [acidic (aFGF) and basic (bFGF)]; FGF-4
Platelet-derived endothelial growth factor (PDGF)
Epidermal growth factor (EGF)
Transforming growth factor-α (TGF-α)
Transforming growth factor-β (TGF-β - only In Vivo)
Vascular smooth muscle cell growth promoting substances [bFGF]
Insulin-like growth factor 1 (IGF-1); Interleukin-1 (IL-1)]
Placenta growth factor (PGF)
Hepatocyte growth factor (HGF, also designated scatter factor and hepatopoietin)
Angiogenin
Proliferin
Prostaglandins
H-ras or K-ras oncogene; p53
Interleukin-8 (IL-8)
Matrix metalloproteinase (MMP)
Neurokines (e.g. midkine, pleiotrophin)
II. Inhibitors
Neoplasm generated inhibitory substances (e.g. Glioma derived inhibitory factor)
TGF-β (only *in vitro*)
bFGF soluble receptor
Thrombospondin-1 and -2
Placental proliferin related peptide
Angiostatin
Endostatin (fragment of collagen XVIII)
Platelet factor 4
Prolactin
MMP tissue inhibitors (TIMPs)
CXC-Chemokine gro-α
Interleukin-12 [IL-12/Natural killer cell stimulatory factor (NKSF)/Cytotoxic lymphocyte maturation factor (CLMF)]
Interferons
Microtubule inhibitors (*e.g.* 2-Methoxyestradiol and Taxol)

REFERENCES

1. Holmgren L, O'Reilly MS, Folkman J: Dormancy of micrometastases: Balanced proliferation and apoptosis in the presence of angiogenesis suppression. Nature Med *1:* 149-153, 1995.

2. Folkman J: Tumor angiogenesis: Therapeutic implications. N Engl J Med *285:* 1182-1186, 1971.
3. Folkman J: Clinical applications of research on angiogenesis. N Engl J Med *333:* 1757-1763, 1995.
4. Fidler IJ, Ellis LM: The implications of angiogenesis for the biology and therapy of cancer metastasis. Cell *79:* 185-188, 1994.
5. Gasparini G, Harris AL: Clinical importance of the determination of tumor angiogenesis in breast carcinoma: Much more than a new prognostic tool. J Clin Oncol *13:* 765-782, 1995.
6. Gastl G, Hermann T, Steurer M, Zmija J, Gunsilius E, Unger C, Kraft A: Angiogenesis as a target for tumor treatment. Oncology *54:* 177-184, 1997.
7. Folkman J: Fighting cancer by attacking its blood supply. Sci Am *275:* 150-154, 1996.
8. Folkman J: New perspectives in clinical oncology from angiogenesis research. Eur J Cancer *32A:* 2534-2539, 1996.
9. Dark GG, Hill SA, Prise VE, Tozer GM, Pettit GR, Chaplin DJ: Combretastatin A-4, an agent that displays potent and selective toxicity toward tumor vasculature. Cancer Res *57:* 1829-1834, 1997.
10. D'Amato RJ, Loughman MS, Flynn E, Folkman J: Thalidomide is an inhibitor of angiogenesis. Proc Natl Acad Sci USA *91:* 4082-4085, 1994.
11. Takano S, Gately S, Neville ME, Herblin WF, Gross JL, Engelhard H, Perricone M, Eidsvoog K, Brem S: Suramin, an anticancer and angiosuppressive agent, inhibits endothelial cell binding of basic fibroblast growth factor, proliferation, and induction of urokinase-type plasminogen activator. Cancer Res *54:* 2654-2660, 1994.
12. Takamiya Y, Brem H, Ojeifo J, Mineta T, Martuza RL: AGM-1470 inhibits the growth of human glioblastoma cells *in vitro* and *in vivo*. Neurosurgery *34:* 869-875, 1994.
13. Bosse DC, Parker JT, Vogler WR, Ades EW: Selective inhibition of adhesion molecule expression by edelfosine (ET-18-OCH3) on human umbilical vein or microvascular endothelium. Pathobiology *63:* 109-114, 1995.
14. Voest EE, Kenyon BM, O'Reilly MS, Truitt G, D'Amato RJ, Folkman J: Inhibition of angiogenesis In Vivo by interleukin-12. J Natl Cancer Inst *87:* 581-586, 1995.
15. Konno H, Tanaka T, Kanai T, Maruyama K, Nakamura S, Baba S: Efficacy of an angiogenesis inhibitor, TNP-470, in xenotransplanted human colorectal cancer with high metastatic potential. Cancer *77:* 1736-1740, 1996.
16. Maier JA, Delia D, Thorpe PE, Gasparini G: In vitro inhibition of endothelial cell growth by the antiangiogenic drug AGM-1470 (TNP-470) and the anti-endoglin antibody TEC-11. Anticancer Drugs *8:* 238-244, 1997.
17. Lannutti BJ, Gately ST, Quevedo ME, Soff GA, Paller AS: Human angiostatin inhibits murine hemangioendothelioma tumor growth In Vivo. Cancer Res *57:* 5277-5280, 1997.
18. O'Reilly MS, Boehm T, Shing Y, Fukai N, Vasios G, Lane WS, Flynn E, Birkhead JR, Olsen BR, Folkman J: Endostatin: an endogenous inhibitor of angiogenesis and tumor growth. Cell *88:* 277-285, 1997.
19. Klauber N, Parangi S, Flynn E, Hamel, E, D'Amato RJ: Inhibition of angiogenesis and breast cancer in mice by the microtubule inhibitors 2-methoxyestradiol and taxol. Cancer Res *57:* 81-86, 1997.
20. Sim BKL, O'Reilly MS, Liang H, Fortier AH, He W, Madsen JW, Lapcevich R, Nacy CA: A recombinant human angiostatin protein inhibits experimental primary and metastatic cancer. Cancer Res *57:* 1329-1334, 1997.
21. Haney D: Chemotherapy Improves Lung Cancer Survival. The Associated Press June 6, 2004.

22. Higgins B, Kolinsky K, Smith M, Beck G, Rashed M, Adames V, Linn M, Wheeldon E, Gand L, Birnboeck H, Hoffmann G: Antitumor activity of erlotinib (OSI-774, Tarceva) alone or in combination in human non-small cell lung cancer tumor xenograft models. Anticancer Drugs *15:* 503-512, 2004.

222. Higgins B, Kolinsky K, Smith M, Beck G, Rashed M, Adames V, Linn M, Wheeldon E, Gand L, Birnboeck H, Hoffmann G. Antitumor activity of erlotinib (OSI-774, Tarceva) alone or in combination in human non-small cell lung cancer tumor xenograft models. Anticancer Drugs 15: 503-512 (2004).

Chapter 7

ANTIGEN PRESENTATION BY DENDRITIC CELLS AND THEIR SIGNIFICANCE IN ANTI-NEOPLASTIC IMMUNOTHERAPY

1. ACTIVE ANTIGEN SPECIFIC IMMUNOTHERAPY (TUMOR VACCINES)

Active antigen specific immunotherapy (AASI) is a therapeutic startegy that attempts to boost the immune responses in a tumor bearing host, specifically against its own, individual tumor associated antigens by using the direct transfer of immunologically active products (1). By its very nature, this type of cancer therapy can only be effective against a small tumor burden and is therefore, applicable when the primary tumor has been resected or is below a specified size (2). Tumor cells isolated from the primary cancer are of low immunogenicity when tested both *in vitro* and *in vivo*. Therefore, immunization with tumor cells has alone been demonstrated to have marginal and impractical therapeutic potential. Various types of tumor cell immunogenicity enhancers (vaccines) have been developed for active immunization.

The idea of producing a human cancer vaccine has a history that extends more than 200 years. In 1777, Nooth, surgeon to the Duke of Kent, inoculated himself with cancer tissue obtained from a patient without any effects other than *in situ* inflammation (3). Alibert (1808), physician to Louis XVIII, received an injection of breast cancer tissue extract, which also resulted in local inflammation, but with regional lymph node enlargement. Domagk (4), discoverer of sulfonamides, injected himself repeatedly over a decade with extracts of cancer tissues obtained during autopsies. Treatment of cancer with active immunization using inactivated tumor cells, cell extracts, and filtrates has thus taken its beginning early in the twentieth century. Vaccines prepared from cultured HeLa cells were used in 100 Swedish cancer patients aged 60 to 70 years; and in three years, no cancer occured either in the vaccinated group or in the control group consisting of 20 volunteers (5). In the minds of Leyden and Blumenthal (6) was born the idea of developing a tumor vaccine that employs the patient's own tumor tissue (the first autologous tumor cell vaccine). Later, active immunization of 85 cancer patients with cultured *in vitro* human neoplastic cells was performed (7, 8). The aim of these attempts was to show that cancer cells of the same type share tumor specific common antigens and that the cancer bearing host by rejecting an allogenic tumor may acquire immunity to common antigenic epitopes of both autologous and foreign neoplastic cells. A patient with osteosarcoma with pulmonary metastases rejected the primary tumor and remained free of recurrence for five years. During these five years, 45 injections of cultured homologous osteosarcoma cells were administered (8).

An "ideal vaccine" should elicit strong and long-lasting protective immunity, with the use of as few injections as possible in a short immunization protocol while evoking minimal side effects and being inexpensive, stable, and easily administered. The traditional problem with cancer vaccines is the difficulty of producing sufficiently large and pure preparations of antigenic material and providing specific and appropriate epitopes (immunogenicity) as targets of recognition by the immune system. In the case of small antigens such as synthetic peptides, appropriate carrier molecules are required. The host immunologic reactivity against successful vaccines is within a desired balance between the humoral versus cell-mediated immunity because vaccine epitopes are capable of stimulating an effective T helper lymphocyte response. The antigenic epitopes which induce immune suppression must be eliminated and the relationship between immunodominant and protective epitopes be regulated.

Immunization against autologous tissue occurs upon freezing. Four days following freezing, the rabbit's prostate, organ- and species-specific antibodies appeared in the serum (9). This cryoimmunological response was used in the cryotherapy of one patient with melanoma of the leg and another with basal cell carcinoma of the scalp. Both experimental treatments resulted in temporary regression of the tumors and a significant humoral immune response of the host (10).

The high sensitivity of the host immune system and minimal toxicity make the ASI therapy a desirable approach. The rationale for this therapy was and remains the strong belief that every malignant tumor cell surface contains tumor specific antigens (TSAs) or immunologically recognizable molecules and that these antigens are absent or qualitatively and quantitavely less in normal cells. The great majority of the known human tumor antigens are tumor associated antigens (TAAs), also referred to as oncofetal antigens. They have relative rather than absolute specificity for cancer cells (11, 12). TAAs provide markers for diagnosis through the use of immunocytological techniques as well as allowing for the monitoring of how cancer patients respond to immunotherapy. They could also be used for *in vivo* detection of tumors *via* nuclear imaging.

These antigenic epitopes on the cancer cell surface serve as effective target sites for antibody mediated or cell mediated cytolysis. A more successful approach to the identification and categorization of potential antigenic epitopes on the cell surface of human tumors would be with human-human MoABs (13) or even better with MoABs that originate from cancer patients against their own lymphocytes.

Thus, the "concept of tumor progression" implies that neoplasms are composed of populations (clones) of cells demonstrating heterogeneity of cell surface IP (14, 15). The immunological heterogeneity is in direct correlation with the problems of neoplasm recurrence and metastasis, or practically the grade of the malignancy. Secondary neoplastic masses recurring at the site of primary tumor excision usually are immunologically distinct from the primary tumor. These metastatic cells have different cell surface IP. Thus, it may no longer be possible to excuse clinical failures of active specific immunotherapy based on tumor heterogeneity. Suboptimal methods of tumor vaccine preparation, or the art of immunization, are responsible for these recent failures. Weak tumor antigens are able to produce potent immune reaction when administered with adjuvants. In the early 1970s, Rapp, Hanna and co-workers, discovered that spontan guinea pig tumors demonstrate regression after BCG injection (16). Hanna made the important discovery of noting that under certain conditions, BCG cells are able to interact with living tumor cells by eliciting a massive immune attack, which causes regression of the autologous tumor. Hanna produced his colon cancer vaccine employing irradiated (with 20,000 rad) autologous colon cancer cells mixed with 10(7) BCG organisms in the ration 1:1. The first vaccine was injected one month postoperatively intradermally into the left anterior thigh, the second was similarly prepared and given one week later and the third contained only 10(7) irradiated tumor cells and was used fourteen days after the initial vaccination. Two clinical trials completed in 1985 and 1989, were headed by H.C. Hoover of the Massachusetts General Hospital. The vaccine prevented the development of metastases in colon and rectal cancer in patients, who had undergone surgery.

The types of tumor vaccines employed for active, TAA specific immunotherapy include:

1. Adjuvant - a) BCG lyophilized or frozen forms, dose dependence > 10^6=10^7-10^8; b) Corynebacterium parvum, dose dependence > 7ug=70ug < 700 ug;
2. Autologous tumor cells conjugated to a strongly immunogenic hapten (dinitrophenyl - DNP or trinitrophenyl - TNP) (17, 18). Cyclophosphamide was used prior to immunization to enhance the immune response and to break possible tolerance to numerous exogenous antigens and self-antigens (18-22). The vaccine induced an inflammatory response in superficial melanoma metastases in 14 of the 24 patients. Immunocytochemically marked infiltration with $CD8^+$ and HLA^-DR^+ T lymphocytes was identified. T lymphocyte mediated response against the melanoma-associated antigens was facilitated by the helper effect of the anti-hapten response.
3. Preparation of neoplastically transformed cells for vaccination: a) Enzymatic dissociation; b) Cryopreservation (viability remain >80%); c) X-irradiation treated;
4. Components: ratio of adjuvant to tumor cells (10:1 to 1:1); number of tumor cells (optimum) 10^7;
5. Administration: intradermal (23) or subcutaneous route of vaccination two-three times weekly (the third vaccine contained only pretreated neoplastic cells).

A number of strategies have been taken to develop better quality of tumor vaccines. However, all have employed one or more antigen - adjuvant combination. Usually, the chosen antigen is on the surface of whole cells or in cell extracts.

When whole cell surfaces are employed for immunization, they were produced in several aspects:

1) autologous or allogeneic whole neoplastically transformed cells (18);
2) living or inactivated tumor cells by x-irradiation or by *in vitro* conditions;
3) freeze-thaw immunological alteration;
4) heat alteration (cryopreservation or hyperthermia);
5) drug inactivation; and
6) xenogenization or modification of the neoplastic cell surface to increase its immunogenicity (24-26), a procedure using virus superinfection that changes the surface expression of membrane proteins by secreting new proteins.

Kobayashi (25) clarified the definition of tumor xenogenization by saying it was the “a) production of a new foreign antigen on the surface of intact tumor cells, which brings about the autonomous regression of the tumor; b) the procedure by which it would also be possible to produce very strong resistance thereafter against any further challenge by the parental tumor";

7) addition of carrier proteins, *e.g.* rabbit gamma globulin;
8) viral oncolysates;
9) chemical modification, *i.e.* iodoacetate;
10) neuraminidase;
11) vaccines composed of intracellular organelles: crude extracts, isolated cell membranes, purified solubilized cell surface neoplastic cell antigens; and immune adjuvants: 1) Freund's adjuvant, 2) bacterial: BCG, methanol extraction residue of BCG, *Corynebacterium parvum*, cyclophosphamide.

Active immunization protocols with irradiated whole neoplastically transformed cells or tumor associated antigens (TAA) have been developed and employed in experimental animal preclinical models (27-31). However the use of TAAs is limited because they are not adequately characterized or are not available in sufficient quantities. In the case of B cell tumors, a soluble and cell membrane located form of TAA can be used for immunization (32). In mice, immunization using TAA located on the cell surface of either irradiated 2C3 cells or the Id coupled covalently to syngeneic spleen cells provided greater and longer lasting protection against a subsequent tumor challenge of up to 1 x 10(6) tumor cells than hyperimmunization with the soluble form of TAA.

The combined use of non-specific immune stimulants as adjuvants helps boost immune responses. Mitchell (33,34) used a bacterial derivate called "Detox" for his vaccine against melanoma while McCune (35) employs another microorganism, previously killed, *Corynebacterium parvum* as his non-specific stimulator.

BCG has been used a as non-specific immunostimulator in various therapeutical trials in combination with a single dose of cyclophosphamide three days before immunization in order to repress the activity of the suppressor cells (18,19,22,36), which normally serves to keep immune responses in check. Schirrmacher (26) tried to make tumor cells more immunogenic by infecting them with viruses. In a preliminary trial, he found that nearly 80% of his vaccinated patients developed an immune response against their tumor cells.

Wiseman and co-authors evaluated the method of active, specific "intralymphatic immunization" (37). Thirty-two patients with various tumor types received three sequential series of immunization during this phase I/II study. In the first phase, 13 patients received two or more injections of autologous, cryopreserved, irradiated tumor cells directly into the lymphatic system through a cannulation of a dorsal pedal lymphatic channel. In series 2, seven patients received low-dose, 300 mg/m^2 cyclophosphamide three days before the autologous tumor cell vaccine was given. Series 3 with 12 patients was similar to the second step except that the tumor cells were treated with cholesterol hemisuccinate immediately prior to irradiation. Patients received 2-6 injections of cells, depending on availability, at two week intervals. Overall, 91 treatments were evaluated in the study. Clinical responses occurred in 7/32 patients and were with the same frequency in all 3 series. Responses occurred in melanoma, lung cancer, colon cancer, and sarcoma. In conclusion, active, specific "intralymphatic immunization" is safe and does produce anti-tumor effects, implying that it should be explored further.

Although tumor vaccines have been studied for decades, there is no vaccine approved as a clinical product. Nevertheless, recent advances in immunology and tumor biology justify a renewed interest (38). First, cancer cells express many antigens that can be recognized by the immune system, some with high tumor selectivity. Second, knowledge about immune regulation, including the importance of costimulatory signals, has been successfully applied to the studies of tumors. Third, mechanisms of how tumors can escape from immunological control have been identified, setting the stage to discover agents to decrease their impact.

A renewed interest is actually more than evident as currently; for example, generally encouraging results are being obtained with autologous, genetically engineered to secrete IL-4 tumor cell vaccines, demonstrating that the concept of non-specific immunostimulation is changing the direction of antigen-specific immunoinduction.

In addition, dendritic cells are being used to elicit systemic cytotoxicity and intracranial T-cell infiltration (39). Phase I trial patients' peripheral blood dendritic cells were pulsed with peptides eluted from the surface of autologous glioma cells. Three biweekly intradermal vaccinations of peptide-pulsed dendritic cells were administered to seven patients with glioblastoma multiforme and two patients with anaplastic astrocytoma. Dendritic cell vaccination elicited systemic cytotoxicity in four of seven tested patients. Robust intratumoral cytotoxic and memory T-cell infiltration was detected in two of four patients who underwent reoperation after vaccination. This Phase I study demonstrated the feasibility, safety, and bioactivity of an autologous peptide-pulsed dendritic cell vaccine for patients with malignant glioma. The immune system is thus uniquely qualified to be an instrument for cancer therapy (40). The challenge we face in coming years to take basic science into the clinical realm where we can help patients win the battle with cancer.

2. ANTIGEN PRESENTATION WITHIN CHILDHOOD BRAIN TUMORS

Microglial cells are of hematopoietic origin (from CD34+ hematopoietic stem cells) and represent 5 to 15% of the human brain tissue's cellular microenvironment. These cells display the characteristic features of a dendritic cell, such as morphological appearance and typical cytoplasmic Birbeck granules (41,42). Immunocytochemically, expression of CD1a, MHC class I and II molecules and the so-called accessory molecules, such as B7 (CD80), LFA3 (CD58), and ICAM-1 (CD54) make them professional antigen presenting cells (APCs) within the normal brain tissue.

3. IMMUNOSUPPRESSION WITHIN THE CELLULAR MICROENVIRONMENT OF CHILDHOOD BRAIN TUMORS

Significant immunological changes have been determined in children suffering from brain tumors:

1) low numbers of circulating T lymphocytes; 2) impaired cytotoxicity of T lymphocyte ($CD8^+$, tumor infiltrating and TAA directed lymphocytes express a defective high affinity interleukin-2 receptor (IL-2R); 3) poor mitogenic responsiveness of tumor infiltrating T lymphocytes *in vitro*; 4) decreased antibody responsiveness (result of defective $CD4^+$ T helper lymphocyte functions); 5) abnormal delayed hypersensitivity responses; and 6) $CD4^+$ and $CD8^+$ lymphocytes derived from TIL are killed by autologous glioma cells *in vitro* (43, 44).

Soluble factors, secreted by brain tumor cells, are responsible for the impaired anti-neoplastic cellular immune responses. *In vitro* experiments have revealed that T lymphocytes from normal individuals cultured in the presence of brain tumor cell supernatants exhibit the very same immunological abnormalities mentioned above.

A dominant suppressor growth factor involved in the intratumoral defense mechanisms of neoplastically transformed cells is transforming growth factor-β2 (TGF-β2). TGF-β2 is derived from primary brain tumors (autocrine secretion), and has been shown to suppress the *in vitro* generation of cytotoxic T lymphocytes (CTL) from TIL derived from peripheral blood lymphocytes (45), and thus its *in vivo* secretion at the tumor site may well be responsible for the intense suppression of $CD8^+$ CTL (46, 47). Brain tumors have been shown to express predominantly low levels of MHC class I molecules, as well as ICAM-1 and to produce TGF-β2, all of which serve as explanations of the inability to effectively isolate and expand infiltrating immunological effectors (CTL) from these neoplasias (77). This combination of factors probably represents a common tumor biological phenomenon and apparently renders the infiltrating cells incapable of proliferation and considerably lowers their immunological (cytotoxic) efficacy, and may be crucial in the inability of the infiltrating immunological effector cells to overcome tumor progression.

Other tumor derived soluble factors, such as prostaglandin E_2 (PGE_2), interleukin-1 receptor (IL-1R) antagonist, and interleukin-10 (IL-10) may also contribute to the observed immunosuppression in childhood brain tumors.

4. THE DENDRITIC CELL NETWORK

Immunophenotypically and functionally, heterogeneous DCs are distributed throughout the human lymphatic, and the majority of non-lymphatic, tissues and represent the most potent, "professional" antigen presenting cell meshwork (APCs), functioning as an integral part of the immune system (48-56). Folliculo-stellate cells (FSC) in the anterior pituitary (AP), Langerhans cells (LCs) in stratified epithelia such as, the skin, oral cavity, upper airways, urethra, female reproductive tissues and lymphatic system (57), "veiled" cells, lympho-dendritic and interdigitating cells (IDCs) in a number of tissues comprising the lymphatic system are the cell types of the DC meshwork. Most of these cells express the immunocytochemical markers S-100, CD1, CD45, CD54, F418, MHC class I and II antigens, Fc and complement receptors.

IP differences among DCs have been detected employing immunocytochemistry. Bednar (58) studied the immunoreactivity of resident DCs in twenty tissues, including lymphatic, skin, bone, soft, nervous, myocardial, lung, esophagus, stomach, and intestinal, among others, employing antibodies against CD34, F XIIIa, F VIII, actin, CD68, S-100 protein, HLA-DR, CD3, and OPD4. Three characteristic immunoreactivities of resident DCs were established: a $CD34^+$ subset of cells, a $CD68^+$ subset of cells which were phagocytic, and an $S\text{-}100^+$ subset of cells. The ability of dendritic cells (DCs and LCs) in epithelial tissue structures to interact on a cell to cell level with epithelial cells was also noted (in the thymic microenvironment), which may provide these cells with increased regulatory functions. DCs can also provide all of the known co-stimulatory signals required for activation of unprimed T lymphocytes and are the most effective, specialized APCs in the induction of primary T lymphocyte-mediated immune responses (59, 60).

4.1 Interdigitating cells (IDCs)

Interdigitating cells (IDCs) are specialized, bone marrow derived DCs, located in T cell domains of various human tissues (61-64). Von Gaudecker and Müller-Hermelink (65) morphologically identified the IDC precursors, originally discovered in the human thymus by Kaiserling and co-workers (66) (first appearance during the 10th week of intrauterine life). The same authors described the TEM characteristics of IDCs in an 85 mm long (3rd lunar month) human fetus:

> "Large electronlucent cells with irregularly shaped nuclei are found in both the mesenchymal septa and in the presumptive medullary regions of the thymus primordium. These cells appear to be precursors of the IDCs, which have been described in the medulla of the prenatal thymus."

When the already committed, lympho-hematopoietic progenitors come into contact with these IDC precursors, the IDCs develop small finger-like protrusions to assure more efficient contact between these cell types.

A histochemical osmium-zinc iodide impregnation procedure has been reported for the identification of IDCs on semithin sections (67). IDCs in such a preparation appear as large dendritic elements containing numerous cytoplasmic protrusions with established structural contacts with lymphatic cells. Differentiating thymocytes have also been observed to be in intimate contact with IDCs. Kaiserling (66) was the first to describe this significant intercellular contact: "The surfaces of the IDC were in close contact with those of small lymphocytes, sometimes polysomal lymphatic cells, epithelial cells, and occasionally with those of lymphatic cells containing ergastoplasm." As stated above, IDCs are large cells with several cytoplasmic processes that form an extensive three-dimensional network, which further envelopes maturating thymocytes (68). Aggregates between IDCs, differentiating thymocytes and functionally complete T lymphocytes represent the specialized microanatomical and functional units within the thymic microenvironment that are crucial for T lymphocyte maturation and education (68).

The most significant TEM features of IDCs are an irregularly shaped, euchromatic nucleus, with very loosely arranged chromatin except for a small rim of heterochromatin along the inner surface of the nuclear envelope, and a nucleolus (not readily visible in all cases) often in an eccentric position. The cytoplasm contains single or groups of flat cisternae of the rough endoplasmic reticulum. Secretory vesicles are always present near the tubules of the Golgi complex. A number of mitochondria are distributed throughout the cytoplasm. Their characteristic cytoplasmic projections regularly *"interdigitate"* with other IDCs and several other cell types in the developing thymic medulla. One of the most important characteristics of IDCs remains that they are never connected to each other or to other cells by desmosomes (69-71). Immunocytochemical observations established the expression of MHC class I and II molecules, as well as several adhesion molecules, including intercellular adhesion molecule-1 (ICAM-1), and lymphocyte function-associated molecules (LFA-3) on the surface of IDCs (72, 73).

The well determined localization of IDCs within the thymus provides significant insight into their function. IDCs have been described in all of the many observations (including our material in dogs) to be located predominantly at the level of the thymic cortico-medullary junction, as well as in the deep cortex, but never in the subcapsular region. In guinea pig thymuses, Klug and Mager (74) found IDCs located mostly in the inner cortex, with kidney-shaped nuclei (with finely dispersed chromatin) and a thin layer of marginally located heterochromatin. The nucleolus in these cells was usually eccentric and small. These IDCs also contained all of the organelles in their cytoplasm, including lysosome-like bodies, and few ergastoplasmic and tubulovesicular structures. The most interesting cytoplasmic features were large electron-dense bodies (phagosomes) containing fragments of picnotic lymphocytes. Many finger-like processes for contact with other thymic cell types were also observed. Glycoprotein synthesis by IDCs has been discussed, but contact with thymocytes has always been described as their main physiological function.

The phagocytic activity of IDCs is not of great importance when compared to macrophages, but still numerous ingested cells can be present in their cytoplasm. Klug (75) described such findings following irradiation of the rat thymus. Duijvestijn and co-workers (76-78) observed the phagocytic activity and the population development of medullary IDC and cortical macrophages following irradiation-induced acute tissue necrosis in the thymus. IDCs clearly demonstrated phagocytic activity sixteen hours after treatment, but this activity could be attributed to the fact that at this stage, the number of necrotic thymocytes was maximal and the total number of phagocytic cells was few and thus, IDCs were recruited to phagocytize the necrotic material. In discussing their findings, the authors speculated on the possible similarities between IDCs and epidermal Langerhans cells. Miyazawa and co-workers (79) reported a twelve-fold increase in the thymic macrophage population in mice eight hours after 3 Gy irradiation. The mammalian cell surface is rich in acidic sugars, such as sialic, hyaluronic, and chondroitic acids. The surface charges caused by these and similar molecules play an important role in cell to cell interactions, such as phagocytosis. The negative surface charge of irradiated thymocytes was found to be enhanced within a few hours, as illustrated by the attachment of these cells to esterase positive thymic phagocytic cells through the reduction of the electrostatic repulsive forces. The irradiated thymocytes may thus be recognized as foreign proteins (antigens) and phagocytized by thymic phagocytic cells. Higley and O'Morchoe (80) published a morphometric analysis of the thymic non-lymphoid cells within the medulla of rats, between the ages of one and 65 days. They found that the largest depot of thymic non-lymphoid cells is within the medulla or the so-called central part. They described the presence of macrophages, IDCs, and LCs at this location.

Von Gaudecker and Müller-Hermelink (65) mentioned diapedesis from the mesenchymal septa and perivascular spaces as a possible mechanism of IDC entrance into the thymus. Once inside the thymus, these IDCs construct a cellular microenvironment at the cortico-medullary junction that is necessary for the migration pattern, differentiation, and maturation of thymocytes into T lymphocytes (65, 81, 82). Monoclonal antibodies (MoABs) reactive with thymic macrophages and IDCs have been developed using these particular cell types isolated from peripheral lymphatic organs (Mac-1, Mac-2, and ER-BMDMl for both macrophages and IDCs; F4/80, BM8, MOMA-2 exclusively for macrophages; and NLDC-145, MIDC-8, M1-8 and ER-TR6 only for IDCs). IDCs have, however, never been identified in B lymphocyte regions.

4.2 Langerhans cells (LCs)

Another type of DC within the developing cellular microenvironment of various tissues is the so-called Langerhans cells (LCs). Thoughts concerning the origin and function of LCs have historically been based on two differing hypotheses. One group of investigators maintained that LCs perform neural functions (83-90), while others followed Masson's (91) theory that regarded LCs as the last stage in the life cycle of an active melanocyte (the theory of the *"worn out effect"* melanocyte) (92-94). LCs have also been described as lymphatic cells that are capable of forming antibodies (95-96). The limited degree to which LCs phagocytize foreign particles suggests that under normal conditions this does not represent their primary function. Tissue culture experiments established that murine epidermal LCs can mature into potent immunostimulatory DCs (97). The necessity of GM-CSF for ensuring the *in vitro* viability and function of epidermal LCs has also been described (98). Novel experimental evidence identified that the LCs originate from the $CD34^+$ hematopoietic stem cells located in the yolk sac, liver and later in bone marrow (99). *In vitro* stimulation of $CD34^+$ cells with GM-CSF and TNF-α led to their rapid proliferation and differentiation into a cell clone with the following IP: $CD45^+/CD68^+/CD3^-/CD19^-/CD56^-$, and also expressing CD1a, CD4 and MHC class II. The $CD1a^+$ cells included three cell populations: *1)* LCs identified by the presence of Birbeck granules; *2)* Birbeck granule negative DCs; and *3)* $CD14^+$ monocytes. The addition of IL-4 interfered with the generation of monocytes, but also resulted in an increased percentage of $CD1a^+$ LCs (<24%) that are potent stimulators of the primary mixed leukocyte reaction and as such, are promising candidates for the generation or augmentation of host responses against different pathogens following vaccination. Studies in the last decade have identified the lymphoid origin of LCs (57, 100-103). Lymphoid committed, $CD4^{low}$ stem cells, as well as $CD25^+$ and $CD44^+$ pro-T lymphocyte progenitors are capable of differentiating into T lymphocyte, B lymphocyte, natural killer (NK), $CD8^+$ DC and LC lineages in many mammalian tissues, including the epidermis.

The presence of these specialized DCs in the epidermis was first described in 1868 by Paul Langerhans following gold chloride staining of skin sections (104). It was not until 1973 that the first experimental evidence for antigen presentation by LCs was reported (105). Today it is well known that human LCs are $CD1a^+$ DCs that function as very potent APCs for primary and secondary immune responses (99). TEM observations of LCs were first reported by Birbeck and co-workers (106) who described these cells as DCs that can be distinguished by a relatively clear cytoplasm, a lobulated nucleus, cored tubules and unique organelles, and the so-called Birbeck or LC granules. Although the LC granule is the ultimate marker for these cells, the same granules have occasionally been detected inside phagocytic vacuoles within activated macrophages (107-110). In addition, IDCs also occassionally possess Birbeck granules, which serves as the basis for the similarity between LCs and IDCs. Birbeck granules originate from the cell membrane and play a role in receptor mediated endocytosis, intracellular processing and presentation of CD1a, MHC class II, granule-associated marker (Lag) and other antigens (111-114). The notion that LCs may represent epidermal macrophages has also been expressed (68,69).

LCs have also been reported within the rat thymic microenvironment (117-120). Olah and co-workers (117) described them as cells of a special type within the thymic medulla:

> "...new kind of cells, found in the medulla of the rat thymus is described. The special structure of their cytoplasm which points to intensive cellular activity, as well as the characteristic granules contained in these cells, justify their classification as a separate cell type. The cells in question should be distinguished from the cells contained in the

> epithelial reticulum of the thymus, from the macrophages containing phagocyted fragments and lysosomes as also from those cells whose granules include a rod-shaped dark structure. It is, on the other hand, possible that the electron-microscopically observed cells of the present study are identical with Ito's inclusion cells (1959)."

As noted previously, the presence of these same cytoplasmic granules has been a defining characteristic of LCs of the skin (121, 122).

In routine tissue sections stained with hematoxylin-eosin (HE), LCs cannot be identified. However, they can be easily demonstrated by impregnation with gold salts (120, 123, 124). Histochemically, the ATPase method has proved most useful for the detection of LCs, provided that proper fixation and cutting techniques are employed. This reaction seems to be specific for LCs.

Questions arise concerning the specificity and significance of the LC granules. Hashimoto (116) in his histoenzymatic study described that the granules, not only those attached to the plasma membrane, but also those entirely enclosed within the cytoplasm, were permeated passively by lanthaum complex during a post-vital incubation procedure. These results suggest that these intracytoplasmic granules have a direct connection with the extracellular spaces. In his opinion, the granules are endocytic in nature, but he discussed that these typical organelles may also be involved in the secretory functions of LCs. The extrusion of specific molecules from the interior of the LCs into the intercellular space within the thymic microenvironment may be the mechanism by which these DCs are involved in the regulation of thymocyte differentiation and also provides a way for interaction induction between DCs and RE cells located in the medulla.

Just last year, Plzak and collegues (125) defined the molecular structure of the Birbeck granules in LCs:

> "In LCs langerin (CD207), a type II transmembrane protein with a single C type carbohydrate recognition domain attached to a heptad repeat in the neck region, which is likely to establish oligomers with an α-coiled-coil stalk, has been implicated in endocytosis and the formation of Birbeck granules. The langerin structure harbours essential motifs for Ca^{2+}-binding and sugar accommodation."

It is significant that the access to the carbohydrate recognition domain of langerin is impaired in tissues showing high cell proliferative activity (*i.e.* neoplasms).

Rowden (126-128) described the expression of Ia-like antigens on the surface of LCs. The relationship between LCs and IDCs in the T lymphocyte areas of lymph nodes and spleen (129), the dendritic reticulum cells of germinal centers (130), and other areas of the spleen (131) appears to lie in their common function in antigen presentation rather than in antigen processing because when trypsin-treated LC suspensions were observed, C3 receptors were detected on LCs in humans (121,122). Trypsin has been reported to destroy C3 receptors on B lymphocyte surfaces, but Berman and Gigli (132) have found that, as in macrophages, this receptor is not sensitive to enzyme digestion in LCs.

It is known that in mice, the H-2 system is on chromosome 17 and has two parts: an H-2K and an H-2D region, comparable to human HLA-A and HLA-B. These regions code for a set of serologically detectable cell surface alloantigens, known as the Ia antigens. These have α and β polypeptide subunits of approximately 33kD and 28kD, respectively, and they lack a common β_2-microglobulin subunit (on primate B cells, HLA-DR antigens). In mice, the distribution of Ia antigens is as follows: I-A, I-E, and I-C region products are present on B cells, macrophages, sperm, and fetal liver cells, but not on T cells (133). Certain subsets of T cells, however, appear to possess excess products of the I-J region (suppressor cells) and it has also been shown that a high percentage of the cells in the epidermis express Ia antigen (30-90%) (134). Quantitative absorption studies have found that these cells express 2-4 times less Ia antigen than B lymphocytes (134, 135). The investigations of Rowden (126, 127) demonstrated that only LCs express HLA-DR antigens on their surface in human skin. Naturally, the expression of MHC (Ia) antigens on the LCs have a physiological role, but it has yet to be studied in great detail. Perhaps in skin, LCs play a role in the recognition of viral and microbial intrusions into the body. Viruses have been found within LCs (136), but it is not known whether viral antigens are displayed in association with the Ia antigens. In recent experiments was demonstrated that a single ultraviolet solar simulated radiation (UV-SSR) exposure of mammalian skin induced a dose dependent reduction in LC density with only slight morphological alterations of the another skin cells (137). The location of LCs was modified: they were present in the spinous rather than in the usual suprabasal layer. Morphologically a cell body rounding and significant reduction of dendricity was identified. All these alterations on cell and tissue level in the skin may impair the antigen presenting function of LCs and the effectivity of immune responses.

The classical experiments of Katz and co-workers (138) on radiation chimeras and the investigations by Frelinger and co-workers (139) show that bone marrow grafting with appropriate stain differences in the Ir gene region permit the tracing of the arrival of Ia^+ cells in the epidermis. Since bone marrow transplantation is now a clinical procedure in the treatment of patients with various pathologic hematologic conditions, there must be many examples of the repopulation of LCs in the organism. A detailed study on these transplantations should prove illuminating since graft-versus-host (GvH) disease is a common complication in such procedures. Although LCs may arise from bone marrow monoblasts, nothing is known concerning the relative pool sizes of other possible precursor cells. Evidence is accumulating of the existence of cells within the lymph nodes and spleen with structural, antigenic and functional characteristics similar to LCs (81, 140-142). These localizations of LCs also represent a symbiotic union of epithelial (in origin) and mesenchymal cells, but the intimate role played in T cell maturation and in phagocytosis is not yet clear.

4.3 Folliculo-stellate cells (FSCs)

Folliculo-stellate cells (FSCs) are a key set of DCs in the AP that are involved in cell to cell interactions and regulations between the endocrine and immune systems (143-145). The stellate-shaped FSCs are organized in a cellular network in the AP and are positive for S-100, produce interleukin-6 (IL-6), and are in intercellular contact with hormone producing cells, their stimulation generally results in an increase in secretory responses (144, 146-149). The presence of a network of lymphoid DCs, expressing a lymphoid DC specific aminopeptidase and MHC class II determinants, has been reported in mouse, rat and human pituitaries (150). Since S-100 immunoreactivity is also typical for lymphoid DCs (151), the subpopulation of S-100 positive pituitary FSCs may represent members of this class of DCs. This would mean that the pituitary FSCs derive from three distinct anlagen: neuroectodermal, ectodermal (oral anlage) and mesenchymal (lymphoid anlage). The multiple ontogenetic origins of the pituitary FSCs and their intermingling with lymphoid DCs allow us to distinguish a DC-FSC cell population at the level of the AP. Since these cells form a morphologically distinct cellular network and since a close functional inter-relationship between the cell groups has been documented (*i.e.* the synchronized increase of S-100 protein and MHC class II determinants during ontogenesis), it is probably better to consider them as a distinguishable cell clone. The increased levels of S-100 and MHC class II antigen expression in the pituitaries of autoimmune-prone BB/R rats raises further questions concerning a possible involvement of DC-FSCs in the development of autoimmune endocrine disorders (152). Several cytokines are now known to influence the release of AP hormones by acting on the hypothalamus and on the pituitary gland. IL-1, IL-2, IL-6, TNF-α and IFN-τ are the most important cytokines involved in the stimulation of the hypothalamic-pituitary-adrenal axis and the suppression of the hypothalamic-pituitary-thyroid and gonadal axes, as well as the release of growth hormone (153). The effects of acute and chronic (systemic) diseases on growth regulation, thyroid, adrenal and reproductive functions may, at least partially, be explained by the numerous important interactions between the neuroendocrine and immune systems.

5. ANTIGEN PRESENTATION BY DCS

Denritic cells patrol throughout the peripheral blood, lymph and peripheral tissues, including secondary lymphatic organs (154). Processing exogenous and endogenous proteins for presentation by MHC molecules to T lymphocyte's receptor repertoire is the defining function of "professional" antigen-presenting cells (APC) as major regulatory cells in antigen specific immune responses (155, 156). DCs respond to two types of "signals": 1) direct recognition of foreign antigen, pathogens (called "danger signal") through specific receptors (named pattern recognition receptors) and 2) indirectly sensed inflammatory mediators such as TNF-α, IL-1β, and PGE-2 of infection. Both of these two pathways induce a well integrated program called "maturation process" which transforms peripheral DCs into the most efficient APCs and T lymphocyte activators (157-159). In a detailed review, the research group of Guermonprez (154) functionally characterized five types of surface receptors triggering the maturation process in DCs: 1) toll-like receptors - TLR2 to TLR4 (160, 161); 2) cytokine receptors; 3) TNF receptor family molecules; 4) receptors for immunoglobulins (most of FcR) by immune complexes or specific antibodies (162-164); and 5) sensors for cell death. T cells through CD40 dependent and independent signaling and endothelial cells *via* cell to cell contacts and secretion of cytokines are helping in the maturation of DCs (165, 166).

The complex process of antigen uptake, internalization by receptor mediated endocytosis, degradation and specific loading on MHC class I and II molecules was named "antigen presentation". DCs present complexes between peptides derived from exogenous antigens and MHC antigens expressed on their surfaces to resting T cells, thereby initiating several immune responses such as the sensitization of MHC-restricted T lymphocytes, the rejection of organ transplants, and the formation of T-dependent antibodies (167-172).

The T lymphocytes are capable of recognizing a number of self and non-self antigens, but they also need an immunostimulatory microenvironment to become activated and initiate an immune response (173-176). $CD4^+$ and $CD8^+$ T lymphocytes can initiate a systemic immune response when the

TAAs are presented to them in the form of short peptides connected to the surface of MHC class I and II molecules together with costimulatory B7 or other molecules (173, 177, 178). MHC class II-restricted antigen presentation to $CD4^+$ T lymphocytes is achieved by an essentially common, multimolecular pathway, the so-called "immunological synapse" (179-182). This multimolecular interaction is subject to variation with regard to the location and extent of degradation of protein antigens and the site of peptide binding to MHC class II molecules (156). To provoke a $CD4^+$ lymphocyte activation, antigens connected to MHC class II molecules (cytoskeletally accumulated) should be presented together with costimulatory molecules (B7.1, B7.2) in case of APC interaction with CD28 in lymphocytes and adhesion molecules (usually ICAM-1) if the APC lymphocyte interaction is fulfilled *via* lymphocyte function associated antigens 1 and 3 (LFA-1 and LFA-3) (183). At the same time, activated $CD4^+$ lymphocytes are able "to help" or activate maturing (licensing) APCs, a functionally ready stage, directly to initiate $CD8^+$ lymphocytes. This activation step requires the presence of CD40 receptor on the APCs and the CD40-ligand (CD40-L or CD154), expressed on the surface of $CD4^+$ lymphocytes (179, 180). The same results can be achieved employing a similar interaction between TRANCE, present on the surface of activated $CD4^+$ T lymphocytes and the TNF superfamily receptor RANK, expressed on APCs (184, 185). There are a number of "molecular couples" that influence DC and T lymphocyte interaction during antigen presentation: CD11/CD18 integrins, CD80/B7-1, CD86/B7-2, and heat-stable antigen (185). Antigens presented by MHC class I molecules on previously activated APCs to $CD8^+$ lymphocytes is recognized by the T lymphocyte receptor (TCR). After the TCR-antigen interaction, cytotoxic lymphocytes (CTLs) recirculate through lymphatic organs and peripheral tissues to initiate the death of cells expressing the same antigen (185-187).

DEC-205 (gp200MR6) is a type I protein, the multilectin receptor for adsorptive endocytosis mostly expressed on DCs (including IDCs and LCs) in T lymphocyte areas of lymphatic tissues, but also, at low levels, on macrophages and T lymphocytes and on the cortical reticulo-epithelial (RE) cells of the thymic cellular microenvironment (188-191). The results indicate that thymic RE cells participate in clearance of apoptotic thymocytes through the DEC-205 protein. DEC-205 belongs to a family of C-type multilectins that also include the macrophage mannose receptor (MMR) and the phospholipase A_2 receptor (PLA_2R) (192). The antigen-presentation function is associated with the high-level expression of DEC-205, an integral membrane protein homologous to the MMR and related receptors that are able to bind carbohydrates and mediate endocytosis. DEC-205 and MMR can serve as potential antigen-uptake receptors. After binding to DEC-205, antigens (proteins) are internalized, processed, and presented in a complex with MHC class II molecules to $CD4^+$ lymphocytes (193). A number of receptors for endocytosis recycle enter into and out of cells through early endosomes. New research defined that the DEC-205 receptor targets late endosomes or lysosomes rich in MHC II products, whereas the homologous macrophage mannose receptor (MMR), as expected, is found in more peripheral endosomes.

It is well known that not all C-type (type II) lectins on DCs serve as antigen receptors recognizing pathogens through carbohydrate structures. The ICAM-3-grabbing, non-integrin, type II lectin DC-SIGN is unique in that it regulates adhesion processes of interstitial DCs, such as DC trafficking and T-cell synapse formation, as well as antigen capture (194-196). DC-SIGN does not only capture HIV-1 but instead protects it in early endosomes, allowing HIV-1 transport by DCs to lymphoid tissues, where it enhances transinfection of T lymphocytes. This recent article discusses the carbohydrate/protein recognition profile and other features of DC-SIGN that contribute to the potency of DCs to control immunity. Another type II lectin, the langerin induces the formation of a unique endocytic compartment of LCs, the so-called Birbeck granules (194).

Proteasomes are multisubunit enzyme complexes that reside in the cytoplasm (mostly in the endoplasmic reticulum) and nucleus of eukaryotic cells. Employing selective protein degradation, proteasomes regulate a number of cytoplasmic biochemical functions including MHC class I antigen processing (197). Three constitutively expressed catalytic subunits are responsible for proteasome regulated proteolysis. These subunits are exchanged for three homologous subunits, the immunosubunits, in IFN-γ-exposed cells and in cells with "professional" and specialized antigen presenting function. Both constitutive and immunoproteasomes degrade the endogenous proteins into small peptide fragments that are capable of binding *via* specialized transporters TAP1 and TAP2 to MHC class I molecules for presentation on the cell surface to cytotoxic T lymphocytes. However, immunoproteasomes seem to fulfill this function more efficiently. IFN-γ further induces the expression of a proteasome activator, PA28, which can also enhance antigenic peptide production by proteasomes. The ubiquitin-proteasome system also plays an important role in antigen presentation. However, unfortunately, neoplasms employ different ways to target the proteasome system to avoid MHC class I presentation of their CAAs.

5.1 Neuroendocrine-immune interactions in antigen presentation

During mammalian ontogenesis, the thymic "pure" endodermal epithelial anlage develops and differentiates into a complex cellular and humoral microenvironment. Beginning the 7-8th week of intrauterine development, thymic reticulo-epithelial (RE) cells chemotactically regulate (induce) numerous waves of migration of stem cells into the thymus, including the $CD34^+$, yolk sac-derived, pluripotent hematopoietic stem cells (198). *In vitro* experiments have established that $CD34^+CD38^{dim}$ human thymocytes differentiate into T lymphocytes when co-cultured with mouse fetal thymic organs. Hematopoietic stem cells for myeloid and thymic DCs, IDCs and LCs are present within the minute population of $CD34^+$ progenitors in the mammalian thymus. The RE cells express the DEC-205 receptor that is so important for antigen presentation. In addition, DCs and specialized epithelial tissue structures (such as "the nursing" thymic epithelial cells - TNCs) may also be involved in direct, cryptocrine-type cell to cell interactions with the epithelial cells of the thymus. TNCs regulate the development of immature thymocytes into immunocompetent T lymphocytes by *emperipolesis*, a highly specialized form of cell-cell interaction in which immature thymocytes are engulfed by large thymic reticulo-epithelial (RE) cells. TNCs *in vitro* are capable of rescuing an early subset of $CD4^+CD8^+$ thymocytes from programmed cell death at 32°C, the temperature at which binding and internalization were identified. This thymocyte subpopulation later matured into a characteristic IP at the double positive stage of T lymphocyte differentiation that is indicative of positive selection. Antigen presentation by thymic RE cells to thymocytes that undergo differentiation is one of the most important events in the selection of the T lymphocyte repertoire (199).

Interactions between the neuroendocrine and immune systems have been reported in various regions of the mammalian body including the anterior pituitary (AP), the skin, and the central (thymus) and peripheral lymphatic tissue (200). The network of bone marrow derived DCs is part of the reticuloendothelial system (RES), with DCs representing the cellular mediators of these regulatory endocrine-immune interactions. Folliculo-stellate cells (FSCs) are non-hormone secreting cells that communicate directly with hormone producing cells, which is a form of neuro-endocrine-immune regulation. As a result, an attenuation of secretory responses follows stimulation of these cells. FSCs are also the cells in the AP producing interleukin-6 (IL-6) and have been identified as the interferon-γ responsive elements. FSCs express lymphatic DC markers, such as DC specific aminopeptidase, leucyl-β-naphthylaminidase, non-specific esterase, MHC class I and II molecules and various other lymphatic immunological determinants [platelet derived growth factor-α chain (PDGF-α chain), CD13, CD14 and L25 antigen]. There is strong evidence that such DCs in the AP, and similar ones in the developing thymus and peripheral lymphatic tissue, are the components of a powerful "professional" antigen presenting DC network. These APCs contain a specialized late endocytic compartment, MIIC (MHC class II-enriched compartment), that harbors newly synthesized MHC class II antigens *en route* to the cell membrane. The limiting membrane of MIIC can fuse directly with the cell membrane, resulting in the release of newly secreted intracellular MHC class II antigen containing vesicles (exosomes). DCs possess the ability to present foreign peptides complexed with the MHC molecules expressed on their surfaces to naive and resting T cells. There are a number of "molecular couples" that influence DC and T lymphocyte interaction during antigen presentation: CD11/CD18 integrins, intercellular adhesion molecules (ICAMs), lymphocyte function associated antigen 3 (LFA-3), CD40, CD80/B7-1, CD86/B7-2, and heat-stable antigen. The "molecular couples" are involved in adhesive or costimulatory regulations, mediating an effective binding of DCs to T lymphocytes and the stimulation of specific intercellular communications. DCs also provide all of the known co-stimulatory signals required for activation of unprimed T lymphocytes. It has been defined that DCs initiate several immune responses, such as the sensitization of MHC-restricted T lymphocytes, resistance to infections and neoplasms, rejection of organ transplants, and the formation of T-dependent antibodies.

Neuroimmunologic aspects of skin inflammation are also regulated by several interacting systems (201). The modulating influence of autonomic and sensory nerves has been known for a long time. Neurokinines derived from these nerves have recently been shown to interact with antigen presentation in dermal LCs and other key functions of allergic skin disease. While some connections between the afferent function and local reflex are known, the nature of efferent regulatory effects (from brain to periphery) remains to be discovered. New research topics include the involvement of the brain-derived neurotrophic factor, in addition to, the autonomic nervous system in mental stress response and insight in the immunomodulation by proopiomelanocortins.

6. THE SIGNIFICANCE OF DCS IN ANTI-NEOPLASTIC IMMUNOTHERAPY

The growth and metastatic spread of neoplasms, to a large extent, depends on their capacity to evade host immune surveillance and overcome host defenses. All tumors express antigens that are recognized, to a variable extent, by the immune system, but in many cases, an inadequate immune response is elicited because of partial antigen masking or ineffective activation of effector cells (202, 203). Tumor Associated Antigens (TAAs) presented in the context of MHC class I complexes on either the neoplastic cell itself or on antigen-presenting cells are only inducing immunological tolerance but not the production of CAA specific cytotoxic T lymphocytes (204). The presence of co-stimulatory molecules, such as B7-1 and B7-2, on antigen-presenting cells and the secretion of IL-2 promote the differentiation of recruited $CD8^+$ lymphocytes into cytotoxic T lymphocytes (205).

Dynamic and permanent changes in the IP of cells following their neoplastic transformation have been explored in numerous immunocytochemical studies (206-212). Expression of novel antigens not usually expressed in the surrounding normal cells and re-expression of developmental antigens has also been well established. Many further alterations in the physiology of cells following neoplastic transformation are contained within the cytoplasm and nucleus of the cell itself, thus making molecules associated with such changes less than ideal targets for immunotherapy. Furthermore, these regulatory molecules are of quite low immunogenicity due to the fact that their presence usually represents more of a quantitative dysregulation than a qualitative one as such antigens are components of normal cells. The downregulation of major histocompatibility complex (MHC) molecule expression and antigen processing in such cells has also been described. The establishment and maintenance of a humoral milieu unfavorable for *in situ* immune activation and neoplastic cell lysis poses yet another difficulty. Neoplastic cell escape from the host's immune effectors is most often caused by weak immunogenicity of TAAs, antigen masking, or overall immunosuppression, a characteristic of advanced stage neoplastic disease. Failure of antigen processing or binding to MHC molecules, inadequate or low-affinity binding of MHC complexes to T lymphocyte receptors, or inadequate expression of co-stimulatory adhesion molecules in conjunction with the antigen-presenting MHC complex may all lead to poor immunogenicity of neoplasm associated peptides and impaired anti-tumor response (213-215). Neoplasm induced defects are known to occur in all major branches of the immune system (216). Antigen presentation to T lymphocytes appears to be one of the steps that is highly deficient in a number of human neoplasms, thereby making it that much more difficult to generate specific cellular mediators of the host anti-neoplastic immune response (217).

DCs are crucial orchestrators of the adaptive immune response (218). Antigen presentation is a very important regulatory element for the induction of cellular immune responses, which is why one of the main goals of tumor immunotherapy is to control and enhance tumor antigen presentation (219). Thus, the basic immunobiology of DCs (220) is still being widely investigated to allow the development of effective DC based immunotherapy protocols for the treatment of human malignancies (221, 222).

Efficient MHC class I and II presentation is developed when antigens are synthesized endogenously by the DCs. This is fully achieved in patients with acute or chronic myelogenous leukemia, which upon cytokine stimulation, undergo cellular differentiation into DCs that contain the full repertoire of neoplastic antigens (223-225). In other neoplasms, desired immunotherapy antigens must be delivered to the DCs *ex vivo*. The identification of a large number of CAAs has suggested new possibilities for more effective, individualized anti-neoplastic immunotherapy. However, multiple mechanisms may contribute to the ability of tumors to escape anti-tumor immune responses. Tumor antigen heterogeneity, modulation of HLA expression, and immune suppressive mechanisms may occur at any time during tumor cell progression, and can affect the outcome of therapeutic immune intervention. In particular, the appearance of altered HLA class I phenotypes during the progression of neoplasms may have important biological and medical implications due to the role of these molecules in T and NK cell functions. Exhaustive tumor tissue studies are necessary before deciding whether a particular patient is suitable for inclusion in T cell-based immunotherapy protocols. However, immune suppressive mechanisms may occur at any time during neoplastic cell progression, and can affect the outcome of immunotherapeutic intervention (226). To overcome this deficit, it is possible to employ "professional" antigen presenting DCs. DCs can be purified from the spleen, bone marrow and peripheral or cord blood. This *in vitro* propagation of tumor specific CTLs is hampered by the necessity for large amounts of professional APCs used for periodical cycles of restimulation (227). The principle of the treatment is to prime the purified DCs with CAAs, and to reinject them into the neoplasm-bearing patient. The sensitization to the CAAs could be obtained employing the crude extract of neoplastic cells or purified antigen, which will lead to MHC class II restricted antigen presentation to $CD4^+$ T lymphocytes (1). Unfortunately, these peptides have rapid turnover on the surface of DCs and their efficacy is subsequently limited (228, 229). Recently, several computer assisted approaches have been developed to predict CTL epitopes within larger protein sequences based on proteasome cleavage specificity. The availability of such programs, as well as, a general insight into the proteasome mediated steps in MHC class I antigen processing, provides us with a rational basis for the design of new anti-neoplastic T lymphocyte vaccines (230).

Survivin is present during embryonic and fetal ontogenesis, but it is downregulated in normal adult cell and tissues, and has dual effects:

antiapoptotic and as a regulator of the cell cycle. However, it becomes re-expressed during oncogenesis. Almost all types of neoplastic cells contain survivin, a TAA and member of the inhibitor of the apoptosis protein (IAPs) family, and considered to play a pivotal role in early oncogenesis (overexpressed during G2/M phase in wide variety of neoplastically transformed cells). The expression (detection) of survivin (almost exclusively in proliferating neoplastic cells) is considered an important prognostic factor of many human neoplasms. Therefore, survivin is an attractive target for the development of broadly applicable tumor vaccines (231). Interestingly, the authors revealed that the upregulation of survivin by DCs was carried out upon stimulation with TNF-α.

Employment of neopastic cell derived exosomes (cell secretory compartment), a defined source of tumor rejection antigens released in membrane vesicles for vaccination, could be the most efficient approach (232). Exosomes transfer a number of neoplastic cell specific antigens to DCs and induce peptide specific, MHC class I restricted presentation to T lymphocytes clones and induce neoplastic antigens specific CTL responses in cancer patients.

Dendritic cells with a MHC class $II^+/CD1a^+/CD3^-/CD14^-/CD20^-$ membrane IP (typical for DCs) can be isolated from peripheral blood mononuclear cells (PBMC) *in vitro* by the addition of IL-4 and GM-CSF (233-235). Interleukin-13 (IL-13) is as effective as IL-4, and combined with GM-CSF, influences the differentiation pathway of DCs in a comparable manner. Human mature DCs have also been generated in large numbers *in vitro* by culturing $CD34^+$ hematopoietic progenitors in GM-CSF and tumor necrosis factor-α (TNF-α) enriched medium for 12 days (215 ,236). On days 5 to 7, the DC progenitors differentiated into two subsets ($CD1a^+$ and $CD14^+$) in tissue culture, both of which matured between days 12 and 14 into cells with a typical DC morphology and IP: CD80, CD83, CD86, CD58, and high HLA class II expression. $CD1a^+$ progenitors give rise to LCs characterized by the presence of typical Birbeck granules and the expression of Lag antigen and E-cadherin. In contrast, $CD14^+$ progenitors mature into DCs lacking Birbeck granules and the other two LC markers, but expressing CD2, CD9, CD68 and the coagulation factor XIIIa, described in dermal DCs (237). Both mature DCs have been shown to be equally efficient in stimulating allogeneic $CD45RA^+$ naive T lymphocytes. $CD14^+$ progenitors are bipotent in nature, as demonstrated by their ability to differentiate into macrophage-like cells, lacking accessory function to T lymphocytes in response to M-CSF. Strobl and co-investigators (238) have recently reported that the factors necessary for DC growth *in vitro* are poorly characterized and that the cytokine combination of GM-CSF and TNF-α, and stem cell factor (SCF) in the absence of serum supplementation, is inefficient in inducing DC maturation (differentiation). The authors further demonstrated that transforming growth factor-β1 (TGF-β1) supplementation is required for substantial DC development in the absence of serum. Thus, $CD34^+$ [best known as an endothelial cell marker (239)] stem cells in serum free conditions required the following cytokine combination for differentiation into DCs: GM-CSF + TNF-α + SCF + TGF-β. A significant number (21% ± 7%) of DCs grown in TGF-β1 supplemented, but not plasma supplemented, showed medium presence of Birbeck granules and their marker, the Lag molecule. The presence of TGF-β1 in the culture medium also diminished the number of cells with monocytic features. Human DCs, with 80-85% purity, were also isolated from peripheral blood mononuclear cells (PBMCs) by negative selection for T lymphocytes, B lymphocytes, NK cells, monocytes and granulocytes (240, 241).

Passive immunotherapeutic interventions involve efforts to augment the neoplastic cells' antigenicity through vaccination with immunogenic peptides, administration of tumor infiltrating immune effector cells *in vitro* expanded and activated, *in vivo* effector cell expansion with cytokine therapies, or genetic modification of either immune effectors or neoplastic cells with cytokine genes or genes encoding costimulatory molecules to effectively activate the immune response (242-246). A variety of adoptive cellular strategies, aimed at boosting the patient's immune system, have been tested in the management of human neoplastic diseases. Despite the drawbacks associated with *ex vivo* cell manipulation and upscaling, several such approaches have been assessed in the clinic (247, 248).

The disclosure of the human genome sequence and rapid advances in genomic expression profiling has revolutionized our knowledge about molecular changes in neoplastic diseases (249). Rapidly growing gene expression databases and improvements in bioinformatics tools set the stage for new approaches using large-scale molecular information to develop specific therapeutics in cancer (250). On the one hand, the ability to detect clusters of genes differentially expressed in normal and malignant tissue may lead to widely applicable targeting of defined molecular structures. On the other hand, analyzing the "molecular fingerprint" of an individual neoplasm raises the possibility of developing customized immunotherapeutical regiments. One approach to using the emerging new datasets for the development of novel therapeutics is to identify genes that are specifically expressed in neoplasms as targets for immune intervention employing the method of "reverse immunology" for the screening of potential candidate genes by bioinformatics in order to identify the immunogenicity of candidate CAAs (249). Gene transfer can also be employed in the case of a cloned antigen (like the co-stimulatory molecule B7.1) and would lead to the MHC class I restricted priming of $CD8^+$ CTLs (227). Unfortunately, following gene transfer, the efficacy of transduction is still very low, but with an increase in our understanding, this difficulty has the possibility of being overcome. Reinjection of DCs primed with neoplastic cell lysate leads to the protection of mice against a tumor challenge. The mode and place of administration, the nature of employed DCs, and the technology of sensitization may all depend on the malignancy and metastatic potential.

What kind of whole tumor cells, apoptotic or necrotic, are more efficient for tumor vaccines? A group of scientists in Hannover, Germany compared the effect of apoptosis versus necrosis on the effective antigen presentation by DCs, employing various tumor models (251). Their experimental data determined that only apoptotic whole tumor cell vaccines were capable of inducing a DC mediated, potent anti-neoplastic immune response. In contrast, necrotic cell vaccines produced a strong, localized macrophage response.

Gene-engineered DCs are currently being tested in anti-neoplastic immunotherapy throughout the world. Genetic immunotherapy with DCs can also be engineered to express tumor antigens, having the potential advantages of endogenous epitope presentation by both major histocompatibility complex (MHC) class I and II molecules. Another advantage of using DCs is that DCs can be gene-modified to express immunostimulatory molecules that further enhance their antigen-presenting function (252).

Numerous routes to employing DCs for anti-neoplastic immunotherapy are being tested, ranging from direct *in situ* expansion and activation of DCs to adoptive transfer of *ex vivo* generated DCs. Numerous techniques have also been designed to optimize DC maturation and their migratory abilities, for effective tumor antigen delivery to DCs, and induction of tumor-specific, as well as, helper immune responses, *in vivo*. However, the results of recent preclinical studies and the diversity of the clinical phase I trials that are currently underway indicate that little is still known about the exact mechanisms by which DCs modulate tumor immunity. This lack of knowledge brings with it the concern that premature clinical trials might not yield the desired results and might even be harmful to, rather than promote, the concept of DC based tumor immunotherapy.

DCs have also been used to help improve the efficacy of tumor vaccines (253). Certainly we are still far from the "ideal vaccination" which should be cell free, employing immunorelevant, neoplasm specific antigens, regardless of patient's HLA haplotype and the production allow large volume manufacturing. Recently clinical investigators have developed genetically modified tumor cell vaccines (254, 255). Irradiated tumor cells transduced with and expressing various cytokines, such as IL-2, IL-6, IL-12, lymphotactin, IFN-γ, or GM-CSF, or co-stimulatory molecules, such as B7-

1, were capable of eliciting their trafficking into lymph nodes and their interaction with T lymphocytes (256-258). These gene-modified DC vaccines also produced a direct regression of pre-existing neoplasms, thereby curing the experimental animals. Induction of strong immune responses in neoplasm-bearing animals against non-immunogenic or weakly immunogenic malignancies supports the research view that active immunization of cancer patients deserves further consideration.

Another possibility to ensure endogenous antigen synthesis has been to fuse DCs to neoplastically transformed cells. Gong and co-workers (259) demonstrated that murine DCs fused to colon cancer cells enhanced the formation of neoplasm specific CTL both in *in vitro* and *in vivo*. In a recent experimental study, a transgenic murine model expressing polyomavirus middle T oncogene and mucin 1 TAA was used to determine the preventive effect of a vaccine containing DCs fused to breast carcinoma cells (260). The animals developed mammary carcinoma between the ages of 65 to 108 days with 100 percent penetrance. No spontaneous CTL were identified. The prophylactic vaccination of these genetically modified mice induced polyclonal CTL activity and rendered 60 percent of the animals free of neoplastic disease by the end of the 180 days experiment. These results indicate that prophylactic vaccination with dendritic/tumor fusion cells is capable of inducing sufficient anti-neoplastic immunity to counter the process of oncogenesis of powerful oncogenic products. A similar type of tumor vaccine was successful in the prevention of human cervical cancer (261). The fusion cell vaccine experimental approach has been applied in several other anti-neoplastic immunotherapy protocols, including the immunoprevention of colorectal cancer, confirming the immunogenicity of such tumor vaccines (262, 263).

In vivo DCs are found in a number of tissues and reside in direct proximity to extracellular matrix proteins (264-266). Since extracellular matrix proteins affect differentiation and location of cells in tissues, a number of research observations investigate potential effects of extracellular matrix proteins on differentiation and maturation of dendritic cells. DCs were isolated and enriched from $CD34^+$ human cord blood stem cells in the presence of GM-CSF and tumor necrosis factor (TNF)-α for 6-d and subsequently cultured for an additional 6-d period on tissue culture plates coated with various extracellular matrix proteins. Among the extracellular matrix proteins tested, exposure to fibronectin stimulated DC/LC cell differentiation as indicated by the 50% increase of the number of cells expressing the Birbeck granule-associated marker Lag and displaying numerous Birbeck granules. Adhesion on fibronectin was shown to be specifically mediated by the integrin α5β. Because laminin and collagen were unable to cause similar changes in LC development, these results suggest that fibronectin may cause changes affecting cellular differentiation of progenitors. Hematopoietic progenitors may exhibit maturational regulated differences in response to both matrix molecules and cytokines.

Human cord blood, $CD34^+$ progenitors cultured in the presence of GM-CSF and TNF-α generate a heterogeneous population of DCs including Langerhans-like DCs (LLDCs) and monocytes (267). The authors noticed that IL-4 exerts different effecs in cultures according to the cells considered. Thus, IL-4 favors DC components at the expense of monocytic development, and permits long-time persistence of DCs that can be maintained up to one month in culture. These results show an IL-4-dependent inhibition of proliferation and emergence of $CD14^+$ cells. Notably, however, IL-4 also acts on the DC precursors. Thus, IL-4 enhances survival and delays maturation of LLDCs from $CD1a^+$ $CD14^-$ precursors. In addition, IL-4 also favors orientation of $CD14^+CD1a^-$ DC/monocyte precursors towards dermal-type $CD1a^+$ DC. DCs recovered from IL-4 treated cultures displayed reduced allostimulatory capacity, but this function is restored upon IL-4 weaning. A significant discovery suggested that a short (48h) IL-4 pulse is sufficient to favor DC development. These *in vitro* experiments demonstrate that IL-4 positively regulates DC development at several levels on distinct precursor cells.

6.1 DC-based tumor vaccines in clinical trials

DC-based immunotherapy was employed in clinical trials in melanoma, lymphoma, myeloma, renal and prostate cancer patients (268-272). Numerous clinical trials testing DC-based vaccines against neoplasms are in progress and partial clinical efficacy has been already proved. In some cases, bone marrow-derived DCs have been employed to treat established experimental neoplasms by unleashing a cellular immune response against TAAs (273). An *ex vivo* gene transfer technique with viral and non-viral vectors provided such antigens into DCs' antigen-presenting molecules. This gene transfer technique is often used to obtain expression of TAAs and hence, used thereby to formulate the anti-neoplastic immunotherapeutic vaccines. Efficacy of the approaches is greatly enhanced if DCs are transfected with more than one gene encoding immunostimulating factors. In some cases, such as with IL-12, IL-7 and CD40L genes, injection inside experimental malignancies of thus transfected DCs induces complete tumor regression in experimental animal models. In this case, TAAs are captured by DCs by still unclear mechanisms and transported to lymphatic organs where productive antigen presentation to T lymphocytes takes place. Transfection of genes will further strengthen the immunogenicity of CAAs and the antigen presentation efficacy of DCs. New immunotherapeutical strategies will soon join the clinical research being conducted.

Because of cases mostly resistant to chemotherapy, renal cell carcinoma (RCC) has been a testing ground for immunotherapy in decades (274, 275). The approval of IL-2 for immuno- treatment of RCC was a landmark "proof of principle" showing that agents working solely *via* the immune system can cause long lasting neoplasm remission. *In vitro* strategies to expand and load DCs with antigens have now led to human vaccine trials in RCC (at the University of California at Los Angeles, a phase 1 clinical trial) and a number of other malignancies.

Malignant lymphomas (MLs) are clonal neoplasms of lymphatic origin. By definition, all cells of the malignant clone have undergone the same rearrangement of antigen receptor genes and express identical antigen receptor molecules (immunoglobulin for B lymphocytes, T cell receptor for T lymphocyte MLs). The hypervariable that stretches within the variable regions of these receptors are considered true tumor-specific antigens ('idiotypes') (276). In several animal models, protective humoral or cellular immunity was induced against the ML by vaccination with the neoplasm-derived idiotype. Successful experimental immunization strategies in animals include idiotype protein vaccines combined with various adjuvants, genetically or immunologically modified ML cells, idiotype-presenting DCs, idiotype-encoding viral vectors, and DNA immunization. Firm evidence for the induction of lymphoma-specific immunity has also been obtained from human idiotype vaccination trials. Furthermore, some trials have provided strong but hitherto formally unproven evidence for clinical benefit of idiotype-vaccinated patients. Alternative vaccination approaches are based on immunologically modified neoplastic cells. Future research efforts should involve efforts for identifying the most efficacious vaccination route, on definitive proof of clinical efficacy, and on the development of new protocols to produce individually designed idiotype vaccines.

Over the last decade, the incidence of malignant melanomas (MMs) has been continuously increasing worldwide (277). Surgery is a treatment of choice in the early stages of primary lesions. Advanced MM, however, is resistant to chemotherapy and radiotherapy. Therefore, there is an essential need for new, possibly more effective treatments. In the last few years, biotherapy such as immunotherapy, has been receiving quite a bit of attention. Unfortunately, systemic administration of immunostimulatory factors is very often associated with severe side effects. Thus, concepts of specific immunotherapies, such as immunogene therapy, have been developed. Currently, various gene therapy strategies of MM are being evaluated in multiple clinical trials carried out all over the world (268, 277). As was mentioned before the new tendencies include gene modified tumor vaccines (GMTV) modified with genes encoding cytokines or co-stimulatory molecules and DCs modified with genes encoding CAA or immunostimulatory factors. Since January 1996, in the Department of Cancer Immunology USOMS, (at Great Poland Cancer Center in Poznan, Poland), a GMTV has been tested in MM patients. More than 220 patients were enrolled in this study of GMTV consisting of melanoma cells modified with genes encoding IL-6 and its agonistic soluble receptor (sIL-6R). More than 25% of the patients were observed to have objective clinical responses and significant life extensions. The encouraging results formed a basis for design of a phase III prospective, randomized clinical study.

7. CONCLUSIONS

- Dendritic cells (DCs) are the most effective, "professional" antigen presenting cells that capture antigens in the periphery, migrate centrally, and present the processed antigens in the context of MHC and appropriate co-stimulatory molecules to T lymphocytes for the initiation of an immune response.
- Dendritic cells (DCs) and probably the thymic reticulo-epithelial (RE) cells are capable to perform important immunoregulatory functions by presenting antigens in the form of peptides bound to cell-surface MHC molecules to T lymphocytes. It is a fact that the intimate events of intrathymic T lymphocyte maturation regulated by DCs and other professional APCs support the action of the various putative thymic hormones, driving the immature thymocytes (prethymocytes) to a stage of maturity.
- Antigen presentation is a critical regulatory element for the induction of cellular immune responses. Therefore, one of the principal current goals of anti-neoplastic immunotherapy is to control and enhance tumor antigen presentation. In this respect, dendritic cells (DC) are now being widely investigated as immunotherapeutic agents for the treatment of disseminated malignancies.
- During the progression of neoplastic disease and the constant IP changes of tumor cells, the appearance of altered HLA class I phenotypes may have important immunobiological and immunotherapeautical implications due to the role of these molecules in T and NK cell functions.
- The fundamental immunobiology of DCs is still being widely investigated to allow the development of effective DC based immunotherapy protocols for the treatment of human malignancies. In the last decade, experimental protocols for *in vitro* growth and

maturation of large quantities of DCs from their $CD34^+$ bone marrow derived stem cells have been carried out employing several cytokine cocktails. This strategy enables the generation of functionally mature DCs even from advanced neoplasm patients whose antigen presentation is suppressed by neoplasm derived molecules. The ability to culture autologous DCs *ex vivo* has influenced the development of protocols of genetically engineering them. The era of active, specific immunotherapy is born.

- There is scientific evidence showing that DCs and thymic RE cells share surface receptors that play a crucial role in antigen presentation. There are also tight regulation connections between the neuroendocrine and the immune system that should be investigated in the process of antigen presentation. Without neuroendocrine regulation, there is no proper immune function. Neuroendocrine regulation is currently not encompassed in the tumor vaccine design, which is a suggestion that I have that future research into the neuroendocrine question could improve the quality of DC based tumor vaccines.
- *In vitro* strategies to expand and load DCs with CAA antigens have now led to human vaccine trials in renal cell carcinoma and a number of other malignancies such as malignant melanomas, breast carcinomas and lymphomas. I am sure that new immunotherapeutical strategies and protocols will soon join the clinical research being conducted worldwide.

REFERENCES

1. Old LJ: Cancer immunology: The search for specificity - G.H.A. Clowes Memorial Lecture, Cancer Res *41:* 361-375, 1981.
2. Shinitzky M, Skornick Y: Cancer immunotherapy with autologous and allogeneic vaccines: A practical overview. EORTC Genitourinary Group Monograph 9: Basic Research and Treatment of Renal Cell Carcinoma Metastasis, Vol.348, Wiley-Liss Inc., pp. 95-125, 1990.
3. Altman LK: Who Goes First? New York, Random House, 1986, p.287.
4. Domagk G: Die Bedeutung korpereigener Abwehrkrafte fur die Ansiedlung von Geschwulstzellen. Zschr Krebsforsch *56:* 247-252, 1949.
5. Bjorklund B: Early data promising on cancer vaccines (interview). Med World News p. 48, 1965.
6. von Leyden E, Blumenthal F: Vorlaufige Mittheilungen uber einige Ergebnisse der Krebsforschung auf der I. medizinischen Klinik. Deutsche Med Wschr *28:* 637-638, 1902.
7. Moore GE: Cancer immunity: fact or fiction? Texas Med *64:* 54-59, 1968.
8. Mendoza CB Jr, Moore GE, Watna AL, Hiramoto R, Jurand J: Immunologic response following homologous transplantation of cultured human tumor cells in patients with malignancy. Surgery *64:* 897-900, 1968.

9. Shulman S, Yantorno C, Bronson P: Cryo-immunology: a method of immunization to autologous tissue. Proc Soc Exp Biol Med *124:* 658-661, 1967.
10. Moore FT, Blackwood J, Sanzenbacher L, Pace WG: Cryotherapy for malignant tumors. Immunologic response. Arch Surgery *96:* 527-529, 1968.
11. Hellstrom I, Hellstrom KE: Tumor vaccines--a reality at last? J Immunother *21:* 119-126, 1998.
12. Hellstrom KE, Hellstrom I: Oncogene-associated tumor antigens as targets for immunotherapy. *FAS*EB J *3:*1715-1722, 1989.
13. Lloyd KO, Old LJ: Human monoclonal antibodies to glycolipids and other carbohydrate antigens: dissection of the humoral immune response in cancer patients. Cancer Res *49:* 3445-3451, 1989.
14. Bodey B, Zeltzer PM, Saldivar V, Kemshead J: Immunophenotyping of childhood astrocytomas with a library of monoclonal antibodies. Int J Cancer *45:* 1079-1087, 1990.
15. Miller FR: Intratumor immunologic heterogeneity. Cancer Metastasis Rev *1:* 319-334, 1982.
16. Hanna MG Jr, Zbar B, Rapp HJ: Histopathology of tumor regression after intralesional injection of Mycobacterium bovis. II. Comparative effects of vaccinia virus, oxazolone, and turpentine. J Natl Cancer Inst *48:* 1697-1703, 1972.
17. Mitchison NA: Immunologic approach to cancer. Transplant Proc *11:* 92-103, 1970.
18. Berd D, Murphy G, Maguire HC Jr, Mastrangelo MJ: Immunization with haptenized, autologous tumor cells induces inflammation of human melanoma metastases. Cancer Res *51:* 2731-2734, 1991.
19. Maguire HC Jr, Ettore VL: Enhancement of dinitrochlorobenzene (DNCB) contact sensitization by cyclophosphamide in the guinea pig. J Invest Dermatol *48:* 39-42, 1967.
20. Schwartz A, Askenase PW, Gershon RK: Regulation of delayed-type hypersensitivity reactions by cyclophosphamide-sensitive T cells. J Immunol *121:* 1573-1577, 1978.
21. Yoshida S, Nomoto K, Himeno K, Takeya K: Immune response to syngeneic or autologous testicular cells in mice. I. Augmented delayed footpad reaction in cyclophosphamide-treated mice. Cancer Res *41:* 2163-2167, 1981.
22. Berd D, Mastrangelo MJ, Engstrom PF, Paul A, Maguire H: Augmentation of the human immune response by cyclophosphamide. Cancer Res *42:* 4862-4866, 1982.
23. Gross L: Experimental immunization against implantation of cancer. Quart Bull Polish Inst Arts and Sc America *1:* 418-430, 1943.
24. Sinkovics JG, Howe CD: Superinfection of tumors with viruses. Experientia *25:* 733-734, 1969.
25. Kobayashi H: Viral xenogenization of intact tumor cells. Adv Cancer Res *30:* 279-299, 1979.
26. Schirrmacher V, Heicappell R: Prevention of metastatic spread by postoperative immunotherapy with virally modified autologous tumor cells. II. Establishment of specific systemic anti-tumor immunity. Clin Exp Metastasis *5:* 147-156, 1987.
27. Wagner H, Rollinghoff M: *In vitro* induction of tumor-specific immunity. I. Parameters of activation and cytotoxic reactivity of mouse lymphoid cells, immunized *in vitro* against syngeneic and allogeneic plasma cell tumors. J Exp Med *138:* 1-15, 1973.
28. Rollinghoff M: Secondary cytotoxic tumor immune response induced *in vitro.* J Immunol *112:* 1718-1725, 1974.
29. Flood PM, Kripke ML, Rowley DA, Schreiber H: Suppression of tumor rejection by autologous anti-idiotypic immunity. Proc Natl Acad Sci USA *77:* 2209-2213, 1980.

30. George KC, van Beuningen D, Streffer C: Growth, cell proliferation and morphological alterations of a mouse mammary carcinoma after exposure to x rays and hyperthermia. Rec Res Cancer Res *107:* 113-117, 1988.
31. George RE, Loudon WG, Moser RP, Bruner JM, Stock PA, Grimm EA: *In vitro* cytolysis of primitive neuroectodermal tumors of the posterior fossa (medulloblastoma) by lymphokine-activated killer cells. J Neurosurgery *69:* 403-409, 1988.
32. Ghosh AK, Cerny T, Wagstaff J, Thatcher N, Moore M: Effect of *in vivo* administration of interferon gamma on expression of MHC products and tumor associated antigens in patients with metastatic melanoma. Eur J Cancer Clin *11:* 1637-1643, 1989.
33. Mitchell MS, Kan-Mitchell J, Kempf RA, Harel W, Shau H, Lind S: Active specific immunotherapy for melanoma: Phase I trial of allogeneic lysates and a novel adjuvant. Cancer Res *48:* 5883-5893, 1988.
34. Mitchell MS, Harel W, Kempf RA, Hu E, Kan-Mitchell J, Boswell WD, Dean G, Stevenson L: Active-specific immunotherapy for melanoma. J Clin Oncol *8:* 856-869, 1990.
35. McCune CS, Marquis D.M: Interleukin 1 as an adjuvant for active specific immunotherapy in a murine tumor model. Cancer Res *50:* 1212-1215, 1990.
36. Sinkovics JG: Suppressor cells in human malignant disease. Brit Med J *1:* 1072-1073, 1976.
37. Wiseman CL, Rao VS, Kennedy PS, Presant CA, Smith JD, McKenna RJ: Clinical responses with active specific intralymphatic immunotherapy for cancer - a phase I-II trial. West J Med *151:* 283-288, 1989.
38. Hellstrom KE, Hellstrom I: Novel approaches to therapeutic cancer vaccines. Expert Rev Vaccines *2:* 517-532, 2003.
39. Yu JS, Wheeler CJ, Zeltzer PM, Ying H, Finger DN, Lee PK, Yong WH, Incardona F, Thompson RC, Riedinger MS, Zhang W, Prins RM, Black KL: Vaccination of malignant glioma patients with peptide-pulsed dendritic cells elicits systemic cytotoxicity and intracranial T-cell infiltration. Cancer Res *61:* 842-847, 2001.
40. Zeltzer PM, Moilanen B, Yu JS, Black KL: Immunotherapy of malignant brain tumors in children and adults: from theoretical principles to clinical application. Childs Nerv Syst *15:* 514-528, 1999.
41. Davis EJ, Foster TD, Thomas WE: Cellular forms and functions of brain microglia. Brain Res Bull *34:* 73-78, 1994.
42. Krivit W, Sung JH, Shapiro EG, Lockman L: Microglia: the effector cell for reconstitution of the central nervous system following bone marrow transplantation for lysosomal and peroxisomal storage diseases. Cell Transplant *4:* 385-392, 1995.
43. Walker PR, Sikorska M: New aspects of the mechanism of DNA fragmentation in apoptosis. Biochem Cell Biol *75:* 287-299, 1997.
44. Walker PR, Saas P, Dietrich P-Y: Role of *Fas* ligand (CD95L) in immune escape: the tumor cells strikes back. J Immunol *158:* 4521-4524, 1997.
45. Mule JJ, Schwarz SL, Roberts AB, Sporn MB, Rosenberg SA: Transforming growth factor-beta inhibits the in vitro generation of lymphokine-activated killer cells and cytotoxic T cells. Cancer Immunol Immunother *26:* 95-100, 1988.
46. Sporn MB, Roberts AB, Wakefield LM, Assoian RK: Transforming growth factor-beta: biological function and chemical structure. Science *233:* 532-534, 1986.
47. Rivoltini L, Arienti F, Orazi A, Cefalo G, Gasparini M, Gambacorti-Passerini C, Fossati-Bellani F, Parmiano G: Phenotypic and functional analysis of lymphocytes infiltrating paediatric tumours, with a characterization of the tumour phenotype. Cancer Immunol Immunother *34:* 241-251, 1992.

48. Wu L, Scollay R, Egerton M, Pearse M, Spangrude GJ, Shortman K: CD4 expressed on earliest T-lineage precursor cells in the adult murine thymus. Nature *349:* 71-74, 1991.
49. Wu L, Li CL, Shortman K: Thymic dendritic cell precursors: relationship to the T lymphocyte lineage and phenotype of the dendritic cell progeny. J Exp Med *184:* 903-911, 1996.
50. Shortman K, Caux C: Dendritic cells development: multiple pathways to nature's adjuvants. Stem Cells *15:* 409-419, 1997.
51. Shortman K, Wu L: Parentage and heritage of dendritic cells. Blood *97:* 3325, 2001.
52. Wu L, D'Amico A, Hochrein H, O'Keeffe M, Shortman K, Lucas K: Development of thymic and splenic dendritic cell populations from different hemopoietic precursors. Blood *98:* 3376-3382, 2001.
53. Shortman K, Liu YJ: Mouse and human dendritic cell subtypes. Nat Rev Immunol *2:* 151-161, 2002.
54. O'Keeffe M, Hochrein H, Vremec D, Caminschi I, Miller JL, Anders EM, Wu L, Lahoud MH, Henri S, Scott B, Hertzog P, Tatarczuch L, Shortman K: Dendritic cell precursor populations of mouse blood: identification of the murine homologues of human blood plasmocytoid pre-DC2 and CD11+ DC1 precursors. Blood *101:* 1453-1459, 2003.
55. Tan PS, Gavin AL, Barnes N, Sears DW, Vremec D, Shortman K, Amigorena S, Mottram PL, Hogarth PM: Unique monoclonal antibodies define expression of FcgammaRI on macrophages and mast cell lines and demonstrate heterogeneity among subcutaneous and other dendritic cells. J Immunol *170:* 2549-2556, 2003.
56. Kim YJ, Broxmeyer HE: 4-IBB ligand stimulation enhances myeloid dendritic cell maturation from human umbilical cord blood $CD34^+$ progenitor cells. J Hematother Stem Cell Res *11:* 895-903, 2002.
57. Anjuere F, Martinez Del Hoyo G, Martin P, Ardavin C: Langerhans cells develop from a lymphoid-committed precursor. Blood *96:* 1633-1637, 2000.
58. Bednar B: Dendritic resident cells and their immunohistologic determination. Ceskoslov Patologie *31:* 9-16, 1995.
59. Paglia P, Girolomoni G, Robbiati F, Granucci F, Ricciardi-Castagnoli P: Immortalized dendritic cell line fully competent in antigen presentation initiates primary T cell responses In Vivo. J Exp Med *178:* 1893-1901, 1993.
60. Lenz A, Heine M, Schuler G, Romani N: Human and murine dermis contain dendritic cells, Isolation by means of a novel method and phenotypical and functional characterization. J Clin Invest *92:* 2587-2596, 1993.
61. Steinman RM, Cohn ZA: Identification of a novel cell type in peripheral lymphoid organs of mice. I. Morphology, quantitation, tissue distribution. J Exp Med *137:* 1142-1162, 1973.
62. Markgraf R, von Gaudecker B, Müller-Hermelink HK: The development of the human lymph node. Cell Tissue Res *225:* 387-413, 1982.
63. Witmer MD, Steinman RM: The anatomy of peripheral lymphoid organs with emphasis on accessory cells: light-microscopic immunocytochemical studies of mouse spleen, lymph node and Peyer's patch. Am J Anat *170:* 4655-4681, 1984.
64. Bodey B, Bodey B Jr, Kaiser HE: Dendritic type, accessory cells within the mammalian thymic microenvironment. Antigen presentation in the dendritic neuro-endocrine-immune cellular network. In Vivo *11:* 351-370, 1997.
65. Gaudecker B von, Müller-Hermelink HK: Ontogeny and organization of the stationary non-lymphoid cells in the human thymus. Cell Tissue Res *207:* 287-306, 1980.
66. Kaiserling E, Stein H, Müller-Hermelink HK: Interdigitating reticulum cells in the human thymus. Cell Tiss Res *155:* 47-55, 1974.

67. Crivellato E, Mallardi F, Basa M, Zweyer M: Osmium-zinc iodide reacts with interdigitating cells in the mouse lymph nodes and spleen. Z mikroskop-anat Forsch *104:* 476-484, 1990.
68. Crivellato E, Baldini G, Basa M, Fusaroli P: The three-dimensional structure of interdigitating cells. Italian J Anat Embryol *98:* 243-258, 1993.
69. Kelly RH, Balfour BM, Armstrong JA, Griffiths S: Functional anatomy of lymph nodes. II. Peripheral lymph-borne mononuclear cells. Anat Rec *190:* 5-22, 1978.
70. Kamperdijk EWA, de Leeuw JHS, Hoefsmit ECM: Lymph node macrophages and reticulum cells in the immune response; the secondary response to paratyphoid vaccine. Cell Tissue Res *227:* 277-290, 1982.
71. Fossum S, Vaalard JL: The architecture of rat lymph nodes. I. Combined light and electronmicroscopy of lymph node cell types. Anat Embryol *167:* 229-246, 1983.
72. Hart DNJ, Fabre JW: Demonstration and characterization of Ia-positive dendritic cells in the interstitial connective tissue of rat heart and other tissues, but not brain. J Exp Med *154:* 347-361, 1981.
73. Hart DNJ, McKenzie JL: Interstitial dendritic cells. Int Rev Immunol *6:* 128-149, 1990.
74. Klug H, Mager B: Ultrastructure and function of interdigitating cells in the guinea pig thymus. Acta Morph Acad Sci Hung *27:* 11-9, 1979.
75. Klug H: Elektronenmikroskopische Untersuchungen zur Phagocytose strahlengeschädigter Lymphozyten im Thymus von Ratten. Z Zellforsch *68:* 43-56, 1965.
76. Duijvestijn AM, Kamperdijk EW: Birbeck granules in interdigitating cells of thymus and lymph node. Cell Biol Int Rep *6:* 655, 1982.
77. Duijvestijn AM, Sminia T, Kohler YG, Janse EM, Hoefsmit EC: Rat thymus microenvironment: an ultrastructural and functional characterization. Adv Exp Med Biol *149:* 441-446, 1982.
78. Duijvestijn AM, Kohler YG, Hofsmit EC: Interdigitating cells and macrophages in the acute involuting rat thymus. An electron-microscopic study on phagocytic activity and population development. Cell Tiss Res *224:* 291-301, 1982.
79. Miyazawa T, Sato C, Kojima K: Thymic phagocytosis and reduction in the negative surface charge of thymocytes after X irradiation. Radiat Res *79:* 622-629, 1979.
80. Higley HR, O'Morchoe CC: Morphometric analysis of thymic medullary non-lymphoid cell changes during postnatal development. Dev Comp Immunol *8:* 711-719, 1984.
81. Ewijk W van, Verzijden JH, Kwast TH van der, Luijcx-Meijer SW: Reconstitution of the thymus dependent area in the spleen of lethally irradiated mice. A light and electron microscopical study of the T-cell microenvironment. Cell Tiss Res *149:* 43-60, 1974.
82. Heusermann U, Stutte HJ, Müller-Hermelink HK: Interdigitating cells in the white pulp of the human spleen. Cell Tiss Res *153:* 415-417, 1974.
83. Ferreira-Marques J: Systema sensitivum intra-epidermicum. Die Langerhansschen Zellen als Rezeptoren des hellen Schmerzes: Doloriceptores. Arch Dermatol Syph *193:* 191-250, 1951.
84. Niebauer G: Über die interstitiellen Zellen der Haut. Hautarzt *7:* 123-126, 1956.
85. Richter R: Studien zur Neurohistologie der nervösen vegetativen Peripherie der Haut bei verschiedenen chronischen infektiösen Granulomen mit besonderer Berücksichtigung der Langerhansschen Zellen; Tuberkolosen der Haut. Arch Klin Exp Dermatol *202:* 466-495, 1956.
86. Richter R: Studien zur Neurohistologie der nervösen vegetativen Peripherie der Haut bei verschiedenen chronischen infektiösen Granulomen mit besonderer Berücksichtigung der Langerhansschen Zellen; tertiäre Syphilide der Haut. Arch Klin Exp Dermatol *202:* 496-508, 1956.

87. Richter R: Studien zur Neurohistologie der nervösen vegetativen Peripherie der Haut bei verschiedenen chronischen infektiösen Granulomen mit besonderer Berücksichtigung der Langerhansschen Zellen; Leishmaniosis cutis. Arch Klin Exp Dermatol *202:* 509-517, 1956.
88. Richter R: Studien zur Neurohistologie der nervösen vegetativen Peripherie der Haut bei verschiedenen chronischen infektiösen Granulomen mit besonderer Berücksichtigung der Langerhansschen Zellen; Lepra. Arch Klin Exp Dermatol *202:* 518-555, 1956.
89. Niebauer G: Über die Dendritenzellen bei Vitiligo. Dermatologica *130:* 317-324, 1965.
90. Niebauer G, Sekido N: Über die Dendritenzellen der Epidermis. Eine Studie über die Langerhans-Zellen in der normalen und ekzematösen Haut des Meerschweinchens. Arch Klin Exp Dermatol *222:* 23-42, 1965.
91. Masson P: My conception of cellular nevi. Cancer *4:* 9-38, 1951.
92. Billingham RE, Medawar PB: "Desensitization" to skin homografts by injections of donor skin extracts. Ann Surg *137:* 444-449, 1953.
93. Fan J, Hunter R: Langerhans cells and the modified technic of gold impregnation by Ferreira-Marques. J Invest Dermatol *31:* 115-121, 1958.
94. Fan J, Schoenfeld RJ, Hunter R: A study of the epidermal clear cells with special references to their relationship to the cells of Langerhans. J Invest Dermatol *32:* 445-450, 1959.
95. Billingham RE, Silvers WK: Re-investigation of the possible occurrence of maternally induced tolerance in guinea pigs. J Exp Zool *160:* 221-224, 1965.
96. Billingham RE, Silvers WK: Some biological differences between thymocytes and lymphoid cells. Wistar Inst Sympos Monogr *2:* 41-51, 1964.
97. Schuler G, Steinman RM: Murine epidermal Langerhans cells mature into potent immunostimulatory dendritic cells in vitro. J Exp Med *161:* 526-546, 1985.
98. Witmer-Pack MD, Olivier W, Valinsky J, Schuler G, Steinman RM: Granulocyte/macrophage colony-stimulating factor is essential for the viability and function of cultured murine epidermal Langerhans cells. J Exp Med *166:* 1484-1498, 1987.
99. Strunk D, Rappersberger K, Egger C, Strobl H, Kromer E, Elbe A, Maurer D, Stingl G: Generation of human dendritic cells/Langerhans cells from circulating CD34+ hematopoietic progenitor cells. Blood *87:* 1292-1302, 1996.
100. Ardavin C, Wu L, Li CL, Shortman K: Thymic dendritic cells and T cells develop simultaneously in the thymus from a common precursor population. Nature *362:* 761-763, 1993.
101. Galy A, Travis M, Cen D, Chen B: T, B, natural killer, and dendritic cells arise from a common bone marrow progenitor cell subset. Immunity *3:* 459-473, 1995.
102. Marquez C, Trigueros C, Fernandez E, Toribio ML: The development of T and non-T cell lineages from CD34+ human thymic precursors can be traced by the differential expression of CD44. J Exp Med *181:* 475-483, 1995.
103. Martinez-Caceres E, Jaleco AC, Res P, Noteboom E, Weijer K, Spits H: CD34+CD38dim cells in the human thymus can differentiate into T, natural killer, and dendritic cells but are distinct from pluripotent stem cells. Blood *87:* 5196-5206, 1996.
104. Langerhans P: Über die Nerven der menschlichen Haut. Virchow's Arch A (Pathol Anat) *44:* 325-338, 1868.
105. Silberberg I: Apposition of mononuclear cells to Langerhans cells in contact allergic reactions: an ultrastructural study. Acta Dermatol Venereol *53:* 1-12, 1973.

106. Birbeck MS, Breathnach AS, Everall JD: An electron microscope study of basal melanocytes and high-level clear cells (Langerhans cells) in vitiligo. J Invest Dermatol *37:* 51-64, 1961.
107. Silberberg I, Baer RL, Rosenthal SA: Circulating Langerhans cells in a dermal vessel. Acta Dermato-Venereol *54:* 81-85, 1974.
108. Silberberg-Sinakin I, Fedorko ME, Baer RL, Rosenthal SA, Berezowsky V, Thorbecke GJ: Langerhans cells: target cells in immune complex reactions. Cell Immunol *32:* 400-416, 1977.
109. Silberberg-Sinakin I, Gigli I, Baer RL, Thorbecke GJ: Langerhans cells: role in contact hypersensitivity and relationship to lymphoid dendritic cells and to macrophages. Immunol Rev *53:* 203-232, 1980.
110. Bucana CD, Munn CG, Song MJ, Dunner K Jr, Kripke ML: Internalization of Ia molecules into Birbeck granule-like structures in murine dendritic cells. J Invest Dermatol *99:* 365-373, 1992.
111. Henkes W, Syha J, Reske K: Nucleotide sequence of rat invariant gamma chain cDNA clone pLRgamma34.3. Nucleic Acids Res *16:* 11822, 1988.
112. Bakke O, Dobberstein B: MHC class II associated invariant chain contains a sorting signal for endosomal compartments. Cell *63:* 707-716, 1990.
113. Lotteau V, Teyton L, Peleraux A, Nilsson T, Karlsson L, Schmid SL, Quaranta V, Peterson PA: Intracellular transport of class II MHC molecules directed by invariant chain. Nature *348:* 600-605, 1990.
114. Naujokas MF, Morin M, Anderson MS, Peterson M, Miller J: The chondroitin sulfate form of invariant chain can enhance stimulation of T cell responses through interaction with CD44. Cell *74:* 257-268, 1993.
115. Hashimoto K, Tarnowski WM: Some new aspects of the Langerhans cell. Arch Dermatol *97:* 450-464, 1968.
116. Hashimoto K: Langerhans' cell granule. An endocytotic organelle. Arch Dermatol 104: 148-160, 1971.
117. Olah I, Dunay C, Rohlich P, Toro I: A special type of cells in the medulla of the rat thymus. Acta Biol Acad Sci Hung *19:* 97-113, 1968.
118. Haelst U van: Light and electron microscopic study of the normal and pathological thymus of the rat. I. The normal thymus. Zeitschr Zellforsch Mikroskop Anat *77:* 534-553, 1967.
119. Haelst U van: Light and electron microscopic study of the normal and pathological thymus of the rat. II. The acute thymic involution. Zeitschr Zellforsch Mikroskop Anat *80:* 153-182, 1967.
120. Warchol JB, Brelinska R, Jaroszewski J: Granules of Langerhans cells in the thymus contain gold. Experientia *40:* 75-76, 1984.
121. Zelickson AS: The Langerhans cell. J Invest Dermatol *44:* 201-212, 1965.
122. Breathnach AS, Wyllie LM: Electron microscopy of melanocytes and Langerhans cells in human fetal epidermis at fourteen weeks. J Invest Dermatol *44:* 51-60, 1965.
123. Wolff K: The fine structure of the Langerhans cell granule. J Cell Biol 35: 468-473, 1967.
124. Wolff K: The Langerhans cell. Curr Prob Dermatol *4:* 79-145, 1971.
125. Plzak J, Holikova Z, Dvorankova B, Smetana K Jr, Betka J, Hercogova J, Saeland S, Bovin NV, Gabius HJ: Analysis of binding of mannosides in relation to langerin (CD207) in Langerhans celld of normal and transformed epithelia. Histochem J *34:* 247-253, 2002.
126. Rowden G: Immuno-electron microscopic studies of surface receptors and antigens of human Langerhans cells. Br J Dermatol *97:* 593-608, 1977.

127. Rowden G, Lewis MG, Sullivan AK: Ia antigen expression on human epidermal Langerhans cells. Nature *268:* 247-248, 1977.
128. Rowden G: Expression of Ia antigens on Langerhans cells in mice, guinea pigs, and man. J Invest Dermatol *75:* 22-31, 1980.
129. Veerman AJ: On the interdigitating cells in the thymus-dependent area of the rat spleen: a relation between the mononuclear phagocyte system and T-lymphocytes. Cell Tiss Res *148:* 247-257, 1974.
130. Nossal GJ, Abbot A, Mitchell J, Lummus Z: Antigens in immunity. XV. Ultrastructural features of antigen capture in primary and secondary lymphoid follicles. J Exp Med *127:* 277-290, 1968.
131. Steinman RM, Witmer MD: Lymphoid dendritic cells are potent stimulators of the primary mixed leukocyte reaction in mice. Proc Natl Acad Sci USA *75:* 5132-5136, 1978.
132. Berman B, Gigli I: Complement receptors on guinea pig epidermal Langerhans cells. J Immunol *124:* 685-690, 1980.
133. Hammerling GJ, McDevitt HO: Antigen-binding structures on the surface of T lymphocytes. Israel J Med Sci *11:* 1331-1341, 1975.
134. Klein J, Hauptfeld V: Ia antigens: their serology, molecular relationships, and their role in allograft reactions. Transplant Rev *30:* 83-100, 1976.
135. Shreffler DC, David CS: The H-2 major histocompatibility complex and the I immune response region: genetic variation, function, and organization. Adv Immunol *20:* 125-195, 1975.
136. Nagao S, Inaba S, Ijima S: Langerhans cells at the sites of vaccinia virus inoculation. Arch Dermatol Res/Archiv fur Dermatol Forsch *256:* 23-31, 1976.
137. Seite S, Zucchi H, Moyal D, Tison S, Compan D, Christiaens F, Gueniche A, Fourtanier A: Alterations in human epidermal Langerhans cells by ultraviolet radiation: quantitative and morphological study. Br J Dermatol *148:* 291-299, 2003.
138. Katz SI, Tamaki K, Sachs DH: Epidermal Langerhans cells are derived from cells originating in bone marrow. Nature *282:* 324-326, 1979.
139. Frelinger JG, Hood L, Hill S, Frelinger JA: Mouse epidermal Ia molecules have a bone marrow origin. Nature *282:* 321-323, 1979.
140. Rausch E, Kaiserling E, Goos M: Langerhans cells and interdigitating reticulum cells in the thymus-dependent region in human dermatopathic lymphadenitis. Virchows Archiv - B Cell Pathol *25:* 327-343, 1977.
141. Hoffman-Fezer G, Rodt H, Thierfelder S: Immunohistochemical identification of T- and B-lymphocytes delineated by the unlabeled antibody enzyme method. II. Anatomical distribution of T- and B-cells in lymphoid organs of nude mice. Beitr Pathol *161:* 17-26, 1977.
142. Hoffman-Fezer G, Rodt H, Götze D, Thierfelder S: Anatomical distribution of T and B lymphocytes identified by immunohistochemistry in the chicken spleen. Int Arch Allergy Appl Immunol *55:* 86-95, 1977.
143. Cocchia D, Miani N: Immunocytochemical localization of the brain-specific S-100 protein in the pituitary gland of adult rat. J Neurocytol *9:* 771-782, 1980.
144. Baes M, Allaerts W, Denef C: Evidence for functional communication between folliculo-stellate cells and hormone-secreting cells in perfused anterior pituitary cell aggregates. Endocrinology *120:* 685-691, 1987.
145. Vankelecom H, Carmeliet P, van Damme J, Billiau A, Denef C: Production of interleukin-6 by folliculo-stellate cells of the anterior pituitary gland in a histiotypic cell aggregate culture system. Neuroendocrinology *49:* 102-106, 1989.

146. Nakajima T, Yamaguchi H, Takahashi K: S-100 protein in folliculo-stellate cells of the rat of the pituitary anterior lobe. Brain Res *191:* 523-531, 1980.
147. Allaerts W, Denef C: Regulatory activity and topological distribution of folliculo-stellate cells in rat anterior pituitary cell aggregates. Neuroendocrinology *49:* 409-418, 1989.
148. Allaerts W, Jeucken PHM, Hofland LJ, Drexhage HA: Morphological, immunohistochemical and functional homologies between pituitary folliculo-stellate cells and lymphoid dendritic cells. Acta Endocrinol *125:* 92-97, 1991.
149. Carmeliet P, Vankelecom H, van Damme J, Billiau A, Denef C: Release of interleukin-6 from anterior pituitary cell aggregates: developmental pattern and modulation by glucocorticoids and forskolin. Neuroendocrinology *53:* 29-34, 1991.
150. Allaerts W, Jeucken PHM, Bosman FT, Drexhage HA: Relationship between dendritic cells and folliculo-stellate cells in the pituitary: immunohistochemical comparison between mouse, rat and human pituitaries. In: Dendritic Cells in Fundamental and Clinical Immunology (Kamperdijk et al., eds), Plenum Press, New York, pp 637-642, 1993.
151. Takahashi K, Yamaguchi H, Ishizeki J, Nakajima T, Nakazato Y: Immunohistochemical and immunoelectron microscopic localization of S-100 protein in the interdigitating reticulum cells of the human lymph node. Virchows Arch [Cell Pathol *37:* 125-135, 1981.
152. Allaerts W, Jeucken PHM, Bosman FT, Drexhage HA: Relationship between dendritic cells and folliculo-stellate cells in the pituitary: immunohistochemical comparison between mouse, rat and human pituitaries. Adv Exp Med Biol *329:* 637-642, 1993.
153. Jones TH, Kennedy RL: Cytokines and hypothalamic-pituitary function. Cytokine *5:* 531-538, 1993.
154. Guermonprez P, Valladeau J, Zitvogel L, Thery C, Amigorena S: Antigen presentation and T cell stimulation by dendritic cells. Annu Rev Immunol *20:* 621-667, 2002.
155. Steinman RM: Dendritic cells and immune-based therapies. Exp Hematol *24:* 859-862, 1996.
156. Robinson JH, Delvig AA: Diversity in MHC class II antigen presentation. Immunology *105:* 252-262, 2002.
157. Kalinski P, Hilkens CM, Wierenga EA, Kapsenberg ML: T-cell priming by type-1 and type-2 polarized dendritic cells: the concept of a third signal. Immunol Today *20:* 561-567, 1999.
158. Banchereau J, Briere F, Caux C, Davoust J, Lebecque S, Liu YJ, Pulendran B, Palucka K: Immunobiology of dendritic cells. Annu Rev Immunol *18:* 767-811, 2000.
159. Lutz MB, Assmann CU, Girolomoni G, Ricciardi-Castagnoli P: Different cytokines regulate antigen uptake and presentation of a precursor dendritic cell line. Eur J Immunol *26:* 586-594, 1996.
160. Aderem A, Ulevitch RJ: Toll-like receptors in the induction of the innate immune response. Nature *406:* 782-787, 2000.
161. Visintin A, Mazzoni A, Spitzer JH, Wyllie DH, Dower SK, Segal DM: Regulation of Toll-like receptors in human monocytes and dendritic cells. J Immunol *166:* 249-255, 2001.
162. Jurgens M, Wollenberg A, Hanau D, de la Salle H, Bieber T: Activation of human epidermal Langerhans cells by engagement of the high affinity receptor for IgE, Fc epsilon RI. J Immunol *155:* 5184-5189, 1995.
163. Regnault A, Lankar D, Lacabanne V, Rodriguez A, Thery C, Rescigno M, Saito T, Verbeek S, Bonnerot C, Ricciardi-Castagnoli P, Amigorena S: Fcgamma receptor-mediated induction of dendritic cell maturation and major histocompatibility complex

class I-restricted antigen presentation after immune complex internalization. J Exp Med *189:* 371-380, 1999.
164. Geissmann F, Launay P, Pasquier B, Lepelletier Y, Leborgne M, Lehuen A, Brousse N, Monteiro RC: A subset of human dendritic cells expresses IgA Fc receptor (CD89), which mediates internalization and activation upon cross-linking by IgA complexes. J Immunol *166:* 346-352, 2001.
165. Thery C, Amigorena S: The cell biology of antigen presentation in dendritic cells. Curr Opin Immunol *13:* 45-51, 2001.
166. Galluci S, Matzinger P: Danger signals: SOS to the immune system. Curr Opin Immunol *13:* 114-119, 2001.
167. Steinman RM, Nussenzweig MC: Dendritic cells: features and functions. Immunol Rev *53:* 127-148, 1980.
168. Steinman RM: The dendritic cell system and its role in immunogenicity. Annual Rev Immunol *9:* 271-296, 1991.
169. Monaco JJ: Structure and function of genes in the MHC class II region. Curr Opin Immunol *5:* 17-20, 1993.
170. Neefjes JJ, Momburg F: Cell biology of antigen presentation. Curr Opin Immunol *5:* 27-34, 1993.
171. Germain RN: MHC-dependent antigen processing and peptide presentation: providing ligands for T lymphocyte activation. Cell *76:* 287-299, 1994.
172. Ossevoort MA, Kleijmeer MJ, Nijman HW, Geuze HJ, Kast WM, Melief CJM: Functional and ultrastructural aspects of antigen processing by dendritic cells. Adv Exp Med Biol *378:* 227-231, 1995.
173. Matzinger P: Tolerance, danger, and the extended family. Annu Rev Immunol *12:* 991-1045, 1994.
174. Nanada NK, Sercarz E: A truncated T cell receptor repertoire reveals underlying immunogenicity of an antigenic determinant. J Exp Med *184:* 1037-1043, 1996.
175. Lee PP, Yee C, Savage PA, Fong L, Brockstedt D, Weber JS, Johnson D, Swetter S, Thompson J, Greenberg PD, Roederer M, Davis MM: Characterization of circulating T cells specific for tumor-associated antigens in melanoma patients. Nat Med *5:* 677-685, 1999.
176. Pardoll DM: Inducing autoimmune disease to treat cancer. Proc Natl Acad Sci USA *96:* 5340-5342, 1999.
177. Bodey B, Bodey B Jr, Siegel SE, Kaiser HE: Failure of cancer vaccines: the significant limitations of this approach to immunotherapy. Anticancer Research *20:* 2665-2676, 2000.
178. Ribas A, Butterfield LH, Glaspy JA, Economou JS: Cancer immunotherapy using gene-modified dendritic cells. Curr Gene Ther *2:* 57-78, 2002.
179. Bennett SR, Carbone FR, Karamalis F, Flavell RA, Miller JF, Heath WR: Help for cytotoxic-T-cell responses is mediated by CD40 signalling. Nature *393:* 478-480, 1998.
180. Ridge JP, Di Rosa F, Matzinger P: A conditioned dendritic cell can be a temporal bridge between a CD4+ T-helper and a T-killer cell. Nature *393:* 474-478, 1998.
181. Grakoui A, Bromley SK, Sumen C, Davis MM, Shaw AS, Allen PM, Dustin ML: The immunological synapse: a molecular machine controlling T cell activation. Science *285:* 221-227, 1999.
182. Malissen B: Dancing the immunological two-step. Science *285:* 207-208, 1999.
183. Wulfing C, Davis MM: A receptor/cytoskeletal movement triggered by costimulation during T cell activation. Science *282:* 2266-2269, 1998.

184. Bachmann MF, Wong BR, Josien R, Steinman RM, Oxenius A, Choi Y: TRANCE, a tumor necrosis factor family member critical for CD40 ligand-independent T helper cell activation. J Exp Med *189:* 1025-1031, 1999.
185. Lu Z, Yuan L, Zhou X, Sotomayor E, Levitsky HI, Pardoll DM: CD40-independent pathways of T cell help for priming of $CD8^+$ cytotoxic T lymphocytes. J Exp Med *191:* 541-550, 2000.
186. Chinnaiyan AM, Hanna WL, Orth K, Duan H, Poirier GG, Froelich CJ, Dixit VM.: Cytotoxic T-cell-derived granzyme B activates the apoptotic protease ICE-LAP3. Curr Biol *6:* 897-899, 1996.
187. Froelich CJ, Dixit VM, Yang X: Lymphocyte granule-mediated apoptosis: matters of viral mimicry and deadly proteases. Immunol Today *19:* 30-36, 1998.
188. Inaba K, Swiggard WJ, Inaba M, Meltzer J, Mirza A, Sasagawa T, Nussenzweig MC, Steinman RM: Tissue distribution of the DEC-205 protein that is detected by the monoclonal antibody NLDC-145. I. Expression on dendritic cells and other subsets of mouse leukocytes. Cell Immunol *163:* 148-156, 1995.
189. Guo M, Gong S, Maric S, Misulovin Z, Pack M, Mahnke K, Nussenzweig MC, Steinman RM: A monoclonal antibody to the DEC-205 endocytosis receptor on human dendritic cells. Hum Immunol *61:* 729-738, 2000.
190. Kato M, Neil TK, Fearnley DB, McLellan AD, Vuckovic S, Hart DN: Expression of multilectin receptors and comparative FITC-dextran uptake by human dendritic cells. Int Immunol *12:* 1511-1519, 2000.
191. Small M, Kraal G: In vitro evidence for participation of DEC-205 expressed by thymic cortical epithelial cells in clearance of apoptotic thymocytes. Int Immunol *15:* 197-203, 2003.
192. Jiang W, Swiggard WJ, Heufler C, Peng M, Mirza A, Steinman RM, Nussenzweig MC: The receptor DEC-205 expressed by dendritic cells and thymic epithelial cells is involved in antigen processing. Nature *375(6527):* 151-155, 1995.
193. Mahnke K, Guo M, Lee S, Sepulveda H, Swain SL, Nussenzweig M, Steinman RM: The dendritic cell receptor for endocytosis, DEC-205, can recycle and enhance antigen presentation via major histocompatibility complex class II-positive lysosomal compartments. J Cell Biol *151:* 673-684, 2000.
194. Valladeau J, Ravel O, Dezutter-Dambuyant C, Moore K, Kleijmeer M, Liu Y, Duvert-Frances V, Vincent C, Schmitt D, Davoust J, Caux C, Lebecque S, Saeland S: Langerin, a novel C-type lectin specific to Langerhans cells, is an endocytic receptor that induces the formation of Birbeck granules. Immunity *12:* 71-81, 2000.
195. Geijtenbeek TB, Krooshoop DJ, Bleijs DA, van Vliet SJ, van Duijnhoven GC, Grabovsky V, Alon R, Figdor CG, van Kooyk Y: DC-SIGN-ICAM-2 interaction mediates dendritic cell trafficking. Nat Immunol *1:* 353-357, 2000.
196. Geijtenbeek TB, Engering A, van Kooyk Y: DC-SIGN, a C-type lectin on dendritic cells that unveils many aspects of dendritic cell biology. J Leukoc Biol *71:* 921-931, 2002.
197. Sijts A, Zaiss D, Kloetzel PM: The role of the ubiquitin-proteasome pathway in MHC class I antigen processing: implications for vaccine design. Curr Mol Med *1:* 665-676, 2001.
198. Bodey B: Neuroendocrine influence on thymic haematopoiesis *via* the reticulo-epithelial cellular network. Expert Opinion Therapeutical Targets *6:* 57-72, 2002.
199. Kasai M, Hirokawa K, Kajino K, Ogasawara K, Tatsumi M, Hermel E, Monaco JJ, Mizuochi T: Difference in antigen presentation pathways between cortical and medullary thymic epithelial cells. Eur J Immunol *26:* 2101-2107, 1996.

200. Bodey B, Bodey B Jr, Kaiser HE: Dendritic type, accessory cells within the mammalian thymic microenvironment. Antigen presentation in the dendritic neuro-endocrine-immune cellular network. In Vivo*11:* 351-370, 1997.
201. Darsow U, Ring J: Neuroimmune interactions in the skin. Curr Opinion Allergy Clin Immunol *1:* 435-439, 2001.
202. Timmerman JM, Levy R: Dendritic cell vaccines for cancer immunotherapy. Annu Rev Med *50:* 507-529, 1999.
203. Foss FM: Immunologic mechanisms of anti-tumor activity. Semin Oncol *29:* 5-11, 2002.
204. Matzinger P: Tolerance, danger, and the extended family. Annu Rev Immunol *12:* 991-1045, 1994.
205. Porgador A, Gilboa E: Bone marrow-generated dendritic cells pulsed with a class I-restricted peptide are potent inducers of cytotoxic T lymphocytes. J Exp Med *182:* 255-260, 1995.
206. Bodey B, Bodey B Jr, Kaiser HE: Apoptosis in the mammalian thymus during its normal histogenesis and under various in vitro and In Vivo experimental conditions. In Vivo *12:* 123-134, 1998.
207. Bodey B, Bodey B Jr, Siegel SE, Kaiser HE: Over-expression of endoglin (CD105): A marker of breast carcinoma-induced neo-vascularization. Anticancer Research *18:* 3621-3628, 1998.
208. Bodey B, Bodey B Jr, Siegel SE, Kaiser HE: Immunophenotypical (IP) analysis and immunobiology of childhood primary brain tumors. Anticancer Research *19:* 2973-2992, 1999.
209. Bodey B, Bodey B Jr, Siegel SE, Kaiser HE: Immunocytochemical detection of MMP-3 and -10 expression in hepatocellular carcinomas. Anticancer Research *20:* 4585-4590, 2000.
210. Bodey B, Bodey B Jr, Gröger AM, Siegel SE, Kaiser HE: Immunocytochemical detection of Homeobox B3, B4, and C6 gene product expression in lung carcinomas. Anticancer Research *20:* 2711-2716, 2000.
211. Bodey B, Bodey B Jr, Siegel SE, Kaiser HE: Matrix metalloproteinase expression in malignant melanomas: tumor-extracellular matrix interactions in invasion and metastasis. In Vivo *15:* 57-64, 2001.
212. Bodey B, Siegel SE, Kaiser HE: MAGE-1, a Cancer-Testis Antigen, Expression in Childhood Astrocytomas as an Indicator of Tumor Progression. In Vivo *16:* 583-588, 2002.
213. Nair SK, Snyder D, Rouse BT, Gilboa E: Regression of tumors in mice vaccinated with professional antigen-presenting cells pulsed with tumor extracts. Int J Cancer *70:* 706-715, 1997.
214. Paquette RL, Hsu NC, Kiertscher SM, Park AN, Tran L, Roth MD, Glaspy JA: Interferon-α and granulocyte-macrophage colony-stimulating factor differentiate peripheral blood monocytes into potent antigen-presenting cells. J Leukocyte Biol *64:* 358-367, 1998.
215. Carbone JE, Ohm DP: Immune dysfunction in cancer patients. Oncology (Huntingt) *16:* 11-18, 2002.
216. Pioche C, Salomon B, Klatzmann D: Cellules dendritiques et therapie cellulaire anti-tumorale. Pathologie Biologie *43:* 904-909, 1995.
217. Turnbull E, MacPherson G. Immunobiology of dendritic cells in the rat. Immunol Rev *184:* 58-68, 2001.
218. Gunzer M, Grabbe S: Dendritic cells in cancer immunotherapy. Crit Rev Immunol *21:* 133-145, 2001.

219. Gallucci S, Lolkema M, Matzinger P: Natural adjuvants: endogenous activators of dendritic cells. Nat Med *5:* 1249-1255, 1999.
220. Gabrilovich DI, Ciernik IF, Carbone DP: Dendritic cells in anti-tumor immune responses. I. Defective antigen presentation in tumor-bearing hosts. Cell Immunol *170:* 101-110, 1996.
221. Gabrilovich DI, Corak J, Ciernik IF, Kavanaugh D, Carbone DP: Decreased antigen presentation by dendritic cells in patients with breast cancer. Clin Cancer Res *3:* 483-490, 1997.
222. Eibl B, Ebner S, Duba C, Bock G, Romani N, Erdel M, Gachter A, Niederwieser D, Schuler G: Dendritic cells generated from blood precursors of chronic myelogenous leukemia patients carry the Philadelphia translocation and can induce a CML-specific primary cytotoxic T-cell response. Genes Chromosomes Cancer *20:* 215-223, 1997.
223. Choudhury BA, Liang JC, Thomas EK, Flores-Romo L, Xie QS, Agusala K, Sutaria S, Sinha I, Champlin RE, Claxton DF: Dendritic cells derived in vitro from acute myelogenous leukemia cells stimulate autologous, antileukemic T-cell responses. Blood *93:* 780-786, 1999.
224. Charbonnier A, Gaugler B, Sainty D, Lafage-Pochitaloff M, Olive D: Human acute myeloblastic leukemia cells differentiate in vitro into mature dendritic cells and induce the differentiation of cytotoxic T cells against autologous leukemias. Eur J Immunol *29:* 2567-2578, 1999.
225. Ruiz-Cabello F, Cabrera T, Lopez-Nevot MA, Garrido F: Impaired surface antigen presentation in tumors: implications for T cell-based immunotherapy. Semin Cancer Biol *12:* 15-24, 2002.
226. Iezzi G, Pprotti MP, Rugarli C, Bellone M: B7.1 expression on tumor cells circumvents the need of professional antigen presentation for in vitro propagation of cytotoxic T cell lines. Cancer Res *56:* 11-15, 1996.
227. Amoscato AA, Prenovitz DA, Lotze MT: Rapid extracellular degradation of synthetic class I peptides by human dendritic cells. J Immunol *161:* 4023-4032, 1998.
228. Ludewig B, McCoy K, Pericin M, Ochsenbein AF, Dumrese T, Odermatt B, Toes RE, Melief CJ, Hengartner H, Zinkernagel RM: Rapid peptide turnover and inefficient presentation of exogenous antigen crutically limit the activation of self-reactive CTL by dendritic cells. J Immunol *166:* 3678-3687, 2001.
229. Sijts A, Zaiss D, Kloetzel PM: The role of the ubiquitin-proteasome pathway in MHC class I antigen processing: implications for vaccine design. Curr Mol Med *1:* 665-676, 2001.
230. Altieri DC: Validating survivin as a cancer therapeutic target. Nat Rev Cancer *3:* 46-54, 2003.
231. Wolfers J, Lozier A, Raposo G, Regnault A, Thery C, Masurier C, Flament C, Pouzieux S, Faure F, Tursz T, Angevin E, Amigorena S, Zitvogel L: Tumor-derived exosomes are a source of shared tumor rejection antigens for CTL cross-priming. Nat Med *7:* 297-303, 2001.
232. Piemonti L, Bernasconi S, Luini W, Trobonjaca Z, Minty A, Allavena P, Mantovani A: IL-13 supports differentiation of dendritic cells from circulating precursors in concert with GM-CSF. Eur Cytokine Network *6:* 245-252, 1995.
233. Barratt-Boyes SM, Henderson RA, Finn OJ: Chimpanzee dendritic cells with potent immunostimulatory function can be propagated from peripheral blood. Immunology *87:* 528-534, 1996.
234. Hanada K, Tsunoda R, Hamada H: GM-CSF-induced In Vivo expansion of splenic dendritic cells and their strong costimulation activity. J Leukocyte Biol *60:* 181-190, 1996.

235. Caux C, Vanbervliet B, Massacrier C, Dezutter-Dambuyant C, de Saint-Vis B, Jacquet C, Yoneda K, Imamura S, Schmitt D, Banchereau J: CD34$^+$ hematopoietic progenitors from human cord blood differentiate along two independent dendritic cell pathways in response to GM-CSF and TNF α. J Exp Med *184:* 695-706, 1996.
236. Hochrein H, Jahrling F, Kreyschh HG, Sutter A: Immunophenotypical and functional characterization of bone marrow derived dendritic cells. Adv Exp Med Biol *378:* 61-63, 1995.
237. Strobl H, Riedl E, Scheinecker C, Bello-Fernandez C, Pickl WF, Rappersberger K, Majdic O, Knapp W: TGF-β 1 promotes *in vitro* development of dendritic cells from CD34$^+$ hemopoietic progenitors. J Immunol *157:* 1499-1507, 1996.
238. Yamazaki K, Eyden BP: Ultrastructural and immunohistochemical observations on intralobular fibroblasts of human breast, with observations on the CD34 antigen. J Submicr Cytol Pathhol *27:* 309-323, 1995.
239. Nijman HW, Kleijmeer MJ, Ossevoort MA, Oorschot VM, Vierboom MP, van de Keur M, Kenemans P, Kast WM, Geuze HJ, Melief CJ: Antigen capture and major histocompatibility class II compartments of freshly isolated and cultured human blood dendritic cells. J Exp Med *182:* 163-174, 1995.
240. Ossevoort MA, Kleijmeer MJ, Nijman HW, Geuze HJ, Kast WM, Melief CJM: Functional and ultrastructural aspects of antigen processing by dendritic cells. Adv Exp Med Biol *378:* 227-231, 1995.
241. Kaplan JM, Yu Q, Piraino ST, Pennington SE, Shankara S, Woodworth LA, Roberts BL: Induction of anti-tumor immunity with dendritic cells transduced with adenovirus vectorencoding endogenous tumor-associated antigens. J Immunol *163:* 699-707, 1999.
242. Kirk CJ, Mule JJ: Gene-modified dendritic cells for use in tumor vaccines. Hum Gene Ther *11:* 797-806, 2000.
243. Furumoto K, Arii S, Yamasaki S, Mizumoto M, Mori A, Inoue N, Isobe N, Imamura M: Spleen-derived dendritic cells engineered to enhance interleukin-12 production elicit therapeutic anti-tumor immune responses. Int J Cancer *87:* 665-672, 2000.
244. Hirschowitz EA, Weaver JD, Hidalgo GE, Doherty DE: Murine dendritic cells infected with adenovirus vectors show signs of activation. Gene Ther *7:* 1112-1120, 2000.
245. Jenne L, Schuler G, Steinkasserer A: Viral vectors for dendritic cell-based immunotherapy. Trends Immunol *22:* 102-107, 2001.
246. Paul S, Calmels B, Acres RB: Improvement of adoptive cellular immunotherapy of human cancer using ex-vivo gene transfer. Curr Gene Ther *2:* 91-100, 2002.
247. Bodey B: Spontaneous regression of neoplasms: new possibilities for immunotherapy. Expert Opinion Biological Therapy *2:* 459-476, 2002.
248. Maecker B, von Bergwelt-Baidon, Anderson KS, Vonderheide RH, Schultze JL: Linking genomics to immunotherapy by reverse immunology--'immunomics' in the new millennium. Curr Mol Med *1:* 609-619, 2001.
249. Onaitis M, Kalady MF, Pruitt S, Tyler DS: Dendritic cell gene therapy. Surg Oncol Clin N Am *11:* 645-660, 2002.
250. Scheffer SR, Nave H, Korangy F, Schlote K, Pabst R, Jaffee EM, Manns MP, Greten TF: Apoptotic, but not necrotic, tumor cell vaccines induce a potent immune response In Vivo. Int J Cancer *103:* 205-211, 2003.
251. Ribas A, Butterfield LH, Glaspy JA, Economou JS: Cancer immunotherapy using gene-modified dendritic cells. Curr Gene Ther *2:* 57-78, 2002.
252. Steinman RM, Pope M: Exploiting dendritic cells to improve vaccine efficacy. J Clin Invest *109:* 1519-1526, 2002.
253. Gilboa E: Immunotherapy of cancer with genetically modified tumor vaccines. Semin Oncol *23:* 101-107, 1996.

254. Topf N, Schmiegel WH: Immuntherapie mit genetisch modifizierten Tumorzellen. Internist *37:* 374-381, 1996.
255. Zhang W, He L, Yuan Z, Xie Z, Wang J, Hamada H, Cao X: Enhanced therapeutic efficacy of tumor RNA-pulsed dendritic cells after genetic modification with lymphotactin. Hum Gene Ther *10:* 1151-1161, 1999.
256. Klein C, Bueler H, Mulligan RC: Comparative analysis of genetically modified dendritic cells and tumor cells as therapeutic cancer vaccines. J Exp Med *191:* 1699-1708, 2000.
257. Armstrong TD, Jaffee EM: Cytokine modified tumor vaccines. Surg Oncol Clin N Am *11:* 681-696, 2002.
258. Gong J, Avigan D, Chen D, Wu Z, Koido S, Kashiwaba M, Kufe D: Activation of anti-tumor cytotoxic T lymphocytes by fusions of human dendritic cells and breast carcinoma cells. Proc Natl Acad Sci USA *97:* 2715-2718, 2000.
259. Xia J, Tanaka Y, Koido S, Liu C, Mukherjee P, Gendler SJ, Gong J: Prevention of spontaneous breast carcinoma by prophylactic vaccination with dendritic/tumor fusion cells. J Immunol *170:* 1980-1986, 2003.
260. Schultz J: Success of vaccine offers promise of cervical cancer prevention. J Natl Cancer Inst *95:* 102-104, 2003.
261. Shu S, Cohen P: Tumor-dendritic cell fusion technology and immunotherapy strategies. J Immunother *24:* 99-100, 2001.
262. Lollini PL, De Giovanni C, Nicoletti G, Di Carlo E, Musiani P, Nanni P, Forni G: Immunoprevention of colorectal cancer: a future possibility? Gastroenterol Clin N Am *31:* 1001-1014, 2002.
263. Bell D, Young JW, Banchereau J: Dendritic cells. Adv Immunol *72:* 255-324, 1999.
264. Banchereau J, Briere F, Caux C, Davoust J, Lebecque S, Liu YJ, Pulendran B, Palucka K: Immunobiology of dendritic cells. Annu Rev Immunol *18:* 767-811, 2000.
265. Staquet MJ, Jacquet C, Dezutter-Dambuyant C, Schmitt D: Fibronectin upregulates in vitro generation of dendritic Langerhans cells from human cord blood CD34+ progenitors. J Invest Dermatol *109:* 738-743, 1997.
266. Rougier N, Schmitt D, Vincent C: IL-4 addition during differentiation of CD34 progenitors delays maturation of dendritic cells while promoting their survival. Eur J Cell Biol *75:* 287-293, 1998.
267. Nestle FO, Alijagic S, Gilliet M, Sun Y, Grabbe S, Dummer R, Burg G, Schadendorf D: Vaccination of melanoma patients with peptide- or tumor lysate-pulsed dendritic cells. Nat Med *4:* 328-332, 1998.
268. Zitvogel L, Angevin E, Tursz T: Dendritic cell-based immunotherapy of cancer. Ann Oncol *11(suppl.)3:* 199-205, 2000.
269. Panelli MC, Wunderlich J, Jeffries J, Wang E, Mixon A, Rosenberg SA, Marincola FM: Phase I study in patients with metastatic melanoma of immunization with dendritic cells presenting epitopes derived from the melanoma-associated antigens MART-1 and gp100. J Immunother *23:* 487-498, 2000.
270. Small EJ, Fratesi P, Reese DM, Strang G, Laus R, Peshwa MV, Valone FH: Immunotherapy of hormone-refractory prostate cancer with antigen-loaded dendritic cells. J Clin Oncol *18:* 3894-3903, 2000.
271. Cohen L, De Moor C, Parker PA, Amato RJ: Quality of life in patients with metastatic renal cell carcinoma participating in a phase I trial of an autologous tumor-derived vaccine. Urol Oncol *7:* 119-124, 2002.
272. Tirapu I, Rodriguez-Calvillo M, Qian C, Duarte M, Smerdou C, Palencia B, Mazzolini G, Prieto J, Melero I: Cytokine gene transfer into dendritic cells for cancer treatment. Curr Gene Ther *2:* 79-89, 2002.

273. Kugler A, Stuhler G, Walden P, Zoller G, Zobywalski A, Brossart P, Trefzer U, Ullrich S, Muller CA, Becker V, Gross AJ, Hemmerlein B, Kanz L, Muller GA, Ringert RH: Regression of human metastatic renal cell carcinoma after vaccination with tumor cell-dendritic cell hybrids. Nat Med *6:* 332-336, 2000.
274. Gitlitz BJ, Figlin RA, Pantuck AJ, Belldegrun AS: Dendritic cell-based immunotherapy of renal cell carcinoma. Curr Urol Rep *2:* 46-52, 2001.
275. Veelken H, Osterroth F: Vaccination strategies in the treatment of lymphomas. Oncology *62:* 187-200, 2002.
276. Perales MA, Wolchok JD: Melanoma vaccines. Cancer Invest *20:* 1012-1026, 2002.
277. Wysocki PJ, Karczewska A, Mackiewicz A: Gene modified tumor vaccines in therapy of malignant melanoma. Otolaryngol Pol *56:* 147-153, 2002.

Chapter 8

GENETICALLY ENGINEERED ANTIBODIES FOR DIRECT ANTI-NEOPLASTIC TREATMENT AND SYSTEMATIC DELIVERY OF VARIOUS THERAPEUTIC AGENTS TO CANCER CELLS

1. INTRODUCTION

The clinical use of mouse MoABs in humans is limited due to the development of a foreign anti-globulin immune response by the human host. Genetically engineered chimeric human-mouse MoABs have been developed by replacing the mouse Fc region with the human constant region. Moreover, the framework regions of variable domains of rodent immunoglobulins were also experimentally replaced by their human equivalents. These antibodies can also be designed to have specificities and effector functions determined by researchers, which may not appear in nature. The development of antibodies with two binding ends (bispecific antibodies) provided a great improvement in targeting neoplastic cells. The existing inadequacies of MoABs in immunotherapy may also be improved by increasing their efficiency with chemical coupling to various agents such as bacterial or plant toxins, radionuclides or cytotoxic drugs.

In writing this chapter, we want to encourage further clinical research with the use of genetically engineered rodent MoABs and various immunoconjugates in the treatment of human cancer, as well as the combination of such immunotherapy with the three conventional modalities of therapy. Finally, after three decades of MoAB research, we have the right to suggest that MoAB-based immunotherapy be accepted as a conventional modality of anti-neoplastic therapy and employed not only in terminal cancer patients but also, for instance, during and following surgical resection.

2. HUMAN CANCER CELL RELATED ANTIGENS

At the beginning of the twentieth century, Paul Ehrlich, the father of tumor immunology, postulated that the mammalian immune system is capable of recognizing neoplastically transformed cells as foreign (1, 2). The antigenic differences between normal and neoplastic cells are in great majority only quantitative rather than qualitative. Moreover, a number of antigenic epitopes identified on solid human tumors also appear to be normal cell differentiation antigens. It is well known, that neoplastic cell heterogeneity may well be due to a dynamic mechanism of antigenic modulation, i.e. the turning on and off of the expression of desired (screened) tumor antigens by regulatory processes, such as the cell cycle or intratumoral environmental conditions (3-5). Control mechanisms of normal cell differentiation have been shown to be disrupted in neoplastic cells (6, 7). It is also well documented that malignant tumor cells, during their characteristic dedifferentiation, develop an anaplastic IP, *i.e. de novo* expression of oncodevelopmental (oncofetal) antigens, such as altered or incomplete carbohydrate structures (8, 9). A number of MoABs that recognize carbohydrate determinants on glycolipids and glycoproteins have been developed by immunizing mice with various human neoplastic cells (10). None of the antibodies developed to human tumor cells detects tumor specific (restricted) antigens (TSAs) (11). Expression of these mucin-like glycoproteins may reflect aberrant profiles of glycosyl transferases. During the mid-1990's, a number of observations determined the ability of interferon-γ (IFN-γ) to upregulate the expression of some cancer-associated glycoproteins on human neoplastic cells (12).

3. ONCOGENES AND GROWTH FACTORS IN NEOPLASTIC CELLS

The term oncogene was coined by Huebner and Todaro (13) to denote genes found in endogenous retroviral genomes that might play a role in oncogenesis. Reviving Darlington's hypothesis (14), these authors suggested that carcinogens of various kinds might act by inducing the expression of otherwise latent retroviral genomes, endogenous in the target cells. The viral oncogene hypothesis is no longer regarded as strictly correct. Understanding the genetic mechanisms that regulate the expression of the alternately quiescent and activated phenotypes is a fundamental necessity for both immunology and oncology (15).

In the last decade, scientists have formally established an "oncogene concept" of carcinogenesis on the basis of the sequence homology between viral *onc* genes and cellular *proto-onc* genes (16-20). Viral *onc* genes have been postulated to be transduced cellular cancer genes and *proto-onc* genes have been implicated as latent cancer genes or oncogenes (21). Oncogenes were first isolated as part of the small genomes of rare RNA tumor viruses, named acute transforming retroviruses (15). The resident cellular gene, or proto-oncogene, is a normal cellular gene that is highly conserved in Metazoan evolution (22). Proto-oncogenes have been determined to lie within or near various chromosomal rearrangements that are characteristic of several human tumors, thereby establishing their involvement in carcinogenesis. For instance, the c-*myc* proto-oncogene has been recognized at the sites of three different chromosomal translocations that occur in Burkitt's lymphoma (23, 24). A proto-oncogene becomes a transforming cellular oncogene (c-*onc*) after some genetic alteration in the DNA has occurred. Proto-oncogenes are thereby capable of bypassing the normal regulatory mechanisms that guide cell division and differentiation and may eventually lead to the dedifferentiation of cells that comprise the malignant tumor towards embryonic epithelial cells, and consequently a very anaplastic cell surface IP (25).

A great number of oncogenes have already been identified and their presence in various tumours studied. Oncogenes are divided into several families: *1)* the growth factor family including the *sis* oncogene; *2)* the protein kinase family including the oncogenes *src*, *abl*, and *erb B*; *3)* the p21 RAS family comprised of the oncogenes N-*ras* and H/K-*ras*; and *4)* the nuclear protein family including oncogenes c-*myb*, N-*myc*, c-*myc*, and FOB. Other oncogenes present in human neoplasms include *ets-1* and *rel*.

4. ANTIBODIES AND NEOPLASTIC CELLS

The strict specificity of MoABs was immediately recognized, and has subsequently revolutionized our knowledge concerning all facets of human cancer (26). MoABs are normally produced by the technique of somatic hybridization between non-secreting myeloma cells and immunological spleen lymphocytes, both of mouse or rat origin (27). Monoclonal antibodies of defined specificity are capable of detecting a single antigenic determinant (epitope) on neoplastic cells in a heterogeneous, very complex and dense cell surface antigenic distribution. To be effective, an *ideal* specific, anti-cancer MoAB has to reach tumor sites in the parts of the human body where the primary and metastatic tumors are located, but never target the great majority of growing and differentiating normal cells.

A variety of MoAB libraries have been used to identify and search for a wide range of antigens:

1) it is possible to use MoABs to establish the cellular origin of undifferentiated metastatic neoplastic cells;
2) MoABs can also be employed in the search for tumor-specific antigens (TSAs), which is still not very productive, but some current results encourage future research in this area (28).

The MoAB producing hybridoma technique developed by Kohler and Milstein (26) is one of the most important discoveries in biotechnology in the last 100 years, and has been enhanced by the rapid developments in molecular biology (29). MoABs can be produced in large enough quantities to permit their employment in the biochemical characterization, diagnosis, research and immunotherapy of neoplasms (30, 31) Differentiation antigens represent a special tool in understanding not only ontogenetic development, but also the mechanism of neoplastic transformation which is most usually characterized by dedifferentiation of the affected cells. Cancer can be viewed as a disease resulting from an uncoupling of gene expression that controls cell proliferation and differentiation (32).

Numerous preclinical studies (assays) are necessary to characterize the real effects of a mouse anti-human MoABs on target neoplastic cells (33). The *ideal* preclinical observations will guide the therapeutical strategies that should be employed in human anti-neoplastic clinical trials. Repeated administration of murine foreign protein (immunoglobulins) almost uniformly resulted in the development of human anti-mouse antibody (HAMA) responses and the production of high titers of human anti-mouse antibodies. Prior to planning a clinical trial, one should be able to answer the following questions:

1) Does the investigated MoAB kill only autologous or also allogeneic tumor target cells, either used alone or together with some components of the immune system?
2) How effective is the MoAB when conjugated to a potentially cytotoxic/cytostatic agent?

In vitro assays used in assessing the efficacy of a MoAB could include:

1) observation of the ability of a MoAB to mediate destruction of target neoplastic cells in the presence of human complement (complement-mediated cytotoxicity);

2) observation of the ability of a MoAB to induce killing of the neoplastic target cells in the presence of human immune effector (cytotoxic) cells (antibody dependent cellular cytotoxicity- ADCC);
3) since complete MoAB molecules have limited access to some cells due to their size, the production of biologically active Fab and Fv fragments by genetic engineering has been achieved in myeloma cells and *E. coli.* (34-36). The variable region genes of both heavy and light chains can be genetically linked to peptides, and expressed in *E. coli* systems. Functional, single chain polypeptides (single chain Fv) are composed of only the heavy and light chain variable regions and have a molecular size that is 1/6 of the original, natural antibody molecule;
4) unfortunately, a key factor limiting the effectiveness of antibodies is their inadequate and non-uniform localization in solid neoplasms;
5) examination of the direct biological/biochemical effects of the MoAB on neoplastic target cells, *i.e.* block cell proliferation, lysis after internalization, inhibition of protein synthesis, *etc.*;
6) analysis of the cytotoxic/cytostatic effects of the MoAB conjugated to anticancer agents;
7) toxin A chain-MoAB conjugates have been observed primarily by differential toxicity between target and nontarget cells. The clonogenic assay of log killing is even more sensitive and significant in the *in vitro* screening of intact toxin-MoAB conjugates;
8) conduction of *in vivo* studies in experimental, preclinical animal models, using nude, athymic animals and human neoplastic xenografts; the interpretation of the results of such studies should be made very carefully since: a) the use of mouse anti-human MoAB in such a mouse experimental host model does not predict how toxic a mouse, anti-human MoAB will behave in a human host. *In vivo* cancer treatment with mouse MoABs have been severely restricted by the development of a strong anti-idiotypic antibody response directed against the mouse immunoglobulin (37, 38); and b) the nude animals only accepted the xenografted, human tumour tissue because the reticulo-epithelial cell network of their thymus was not secreting the various thymic hormones required for T cell development and thus the peripheral immune tissues were also undeveloped. Anti-neoplasm effects that rely on immunological cooperation sometimes cannot be achieved in such an experimental system.

5. ANTI-NEOPLASTIC IMMUNOTHERAPEUTICAL REGIMENTS INFLUENCED BY IMMUNOHISTOCHEMISTRY

Treatment of human malignancies with ultraviolet light, radiomimetic drugs, γ-irradiation, and a variety of cancer chemotherapeutic drugs which damage DNA (39, 40) results in the accumulation of wild-type p53, through a post-translational stabilization mechanism (41). This accumulation has been shown to mediate the arrest of the cell cycle in G1, which is in accord with the growth-inhibitory activities of high levels of normal p53 (40). Mutations in the p53 gene result in dramatically decreased sensitivity of normal and transformed cells to DNA damaging agents, which would normally activate p53-mediated apoptosis (42-45). The mutation of the p53 gene is associated with the formation of tetramers of mutant p53 protein of unusual stability with a half-life of several hours rather than the 6-20 minutes associated with normal p53. The binding of host proteins, such as MDM2, or viral oncoproteins, such as SV40 T antigen, can also result in an increase of the half-life of the normal p53 protein by forming hetero-oligomers which inhibit wild-type p53 from carrying out its normal function (46). Consequently, loss of p53 function is associated with accumulation of p53 (both wild-type and mutant, depending on the mechanism of loss) in the cell, which is detectable by sensitive immunocytochemical methods (47).

Neoplasm associated cell surface antigens are the traditional targets for IC detection and antibody guided anti-neoplastic immunotherapy. Data on the targeted treatment with two newly developed “smart bomb” drugs named Iressa and ImClone’s Cetuximab or C225 (produced by AstraZeneca Plc. and ImClone Systems Inc), which are the first in a new class of drugs that block EGFR, a biochemical switch which promotes the growth of neoplastic cells was reported at the American Association for Cancer Research (AACR) conference held in Miami. Both drugs are capable of knocking out neoplastic cells without damaging healthy tissues. Dr. Baselga (Vall d’Hebron University Hospital in Barcelona, Spain) repoted that Iressa was a significant advance in the therapy of non-small-cell lung cancer. The Phase II trial involved 209 patients. Shrinking of lung tumors was present in at least half of the 18.7 percent of patients who failed to respond to conventional chemotherapy, in 52.9 percent the disease stabilized, while in 34 percent the tumors had not grown after four months. ImClone has completed the submission for its antibody drug to the U.S. Food and Drug Administration. After approval, it will be employed in the treatment of colon cancer.

Another targeted immunotherapy employing Herceptin (Genentech Inc., San Francisco, CA), a humanized MoAB specific for p185HER-2/neu, present in BCs and is well tolerated compared to the classical chemotherapy but often does not produce high response rates alone (48). Consequently, there is a great deal of interest in combining targeted immunotherapy with conventional chemotherapy for the treatment of metastatic BCs. In gastric adenocarcinoma, the wide expression range may truly reflect patient selection because HER-2/neu positivity appears linked to advanced rather than early disease with limited invasion. The majority of studies favor a significant prognostic value of HER-2/neu status for this cancer. In colorectal cancer, HER-2/neu overexpression also appears to be a significant adverse outcome indicator as judged by current published literature. Note that HER-2/neu protein overexpression or gene amplification is associated with approximately one-fourth of all gastrointestinal tract malignancies, making strategies designed to employ this marker in targeted therapy selection appear warranted (49). During the next two decades I would not be surprised to see these neoplasms treated with anti-HER-2/neu modalities such as Herceptin and quite likely in combination with other chemotherapeutic agents especially for patients with advanced disease, and possibly for individuals with high-risk lesions in an adjuvant setting.

Overall recent progress in the past few years has led to the development of new agents based on MoABs that have been established in clinical oncology. Indeed, in some cases, neoplastic cells express antigenic targets at higher levels than normal cells. Today, there are two main types of MoABs that can be either conjugated to cytotoxic drugs or radio-active compounds or be non-conjugated (50). Among the last category, some are currently used in the treatment of patients, including two MoABs targeting receptors with tyrosine kinase activity (HER2: Herceptin (DCI: trastuzumab), EGFR: Erbitux (DCI: cetuximab). A third MoAb commonly employed in anti-neoplastic treatment targetting CD20 is a transmembrane marker of B lymphoma (MabThera (DCI: rituximab)). Both Herceptin and MabThera have been associated with improved survival in patients with breast carcinoma and lymphoma, respectively. New promising agents are under observation, such as anti-EGFR in colon and head and neck carcinoma or new compounds such as anti-VEGF. These examples outline the importance of recent progress in selectively targeting tumor antigens or TAAs and the potential impact of these approaches in clinical anti-neoplastic immunotherapy.

In the middle of November 2001, the group of Dr. Scheinberg from the Memorial Sloan-Kettering Cancer Center (New York) reported the employment of a combination of a single atom of actinium-225, a radioactive atom that emits α particles that can kill a target cell, enclosed in a molecular cage (molecular seized generators or nanogenerators), coupled to internalizing, engineered MoAB, programmed to target specific neoplastic cells. *In vitro* these molecular generators specifically destroyed leukemia, lymphoma, breast, ovarian, neuroblastoma, and prostate cancer cells at picocurie (becquerel) levels (51). Injections of single doses of these molecular-sized generators at nanocurie (kilobecquerel) levels into mice bearing solid prostate carcinoma or metastatic human lymphoma induced significant neoplastic regression and prolonged survival of the experimental mice. Untreated mice lived about 28 days, while treated animals lived 173 days and some as long as ten months. The authors expect that employing this treatment, oncologists will be able to catch metastatic neoplasms throughout the human body. The amount of radioactivity delivered to the patient would be small, because single atoms are employed.

Other alternative target molecules may be sought within the neoplasm's stromal tissue, which contains *de novo* formed capillaries, reactive cells of the connective tissues and proteins of the extracellular matrix (52). Rettig and co-investigators reported the expression of a cell surface glycoprotein of 165 kD molecular weight on vascular endothelial cells. The antigen, designated FB5/endosialin was identified immunocytochemically in 67% of 128 malignant tumors (including 18 of 25 PNETs, 41 of 61 sarcomas, and 26 of 37 carcinomas). The complex angiogenic process is also regulated by several cell surface and extracellular adhesion molecules (53). A fibronectin isoform (B-FN) with an oncofetal extra domain B (ED-B) has been discovered in the stroma of fetal and neoplastically transformed tissues. This extra domain B is also expressed during angiogenesis, but is undetectable in normal blood vessels (54,55). The development of human antibody fragments which react with the ED-B of fibronectin by a new, so-called phage antibody technology was recently reported (53). This oncofetal extra domain B is highly conserved in several mammalian species.

6. HUMAN ANTIBODIES

Immunotherapeutic experience accumulated from trials and basic research with mouse anti-human MoABs, have demonstrated that the development and use of human MoABs may have a lot of advantages. Most importantly, only human MoABs can identify TAAs that are immunogenic only to man. Human MoABs may add a new dimension in the identification of newly expressed TAAs that are not readily immunogenic to other (animal) immune systems. Human MoABs have several features that make them very important immunotherapeutic agents against human cancer:

1) reduced or even absent cross-reactivity with normal human tissues;
2) detection of polymorphic antigenic epitopes;
3) well developed interaction with the host's immune effector cells;

4) less allergic and anaphylactic reactions in patients; and
5) decreased host's anti-globulin responses avoiding neutralization of the particular human MoAB employed, decreased development of immunocomplexes causing immunosuppression, and decreased production of immunocomplexes causing kidney deposits.

In some circumstances, human MoABs can exhibit a higher degree of discrimination than xenoantibodies (as in tissue typing, where human antibodies have a wider range of specificities against allotypic determinants of Human Leukocyte Antigens [HLA]). Three gangliosides (the monosialoganglioside GM2 and the disialogangliosides GD2 and GD3) are associated with human melanoma, but they have also been detected on cells of neuroectodermal origin - OFA I-1 is identical to ganglioside GM2 and OFA I-2 to ganglioside GD2 (56). The human MoAB HJM1 produced by EBV transformation or fusion with the human cell line LICR2, followed by a second fusion with mouse myeloma cells (57). This human MoAB identifies a shared ganglioside epitope, common to GD3, GD2, GD1b and GM3. This type of discrimination was not observed in mouse MoABs raised against GD2 or GD3 (58,59). In fact, the two human MoABs that were developed, 2-139-1 and 14-31-10 (60,61), demonstrated more tumor tissue restriction than any of the mouse anti-human MoABs. Two other human MoABs 2-39M and 32-27M did not react with human cells if they were grown in human serum containing regular or synthetic medium, but did so if the culture medium was enriched with fetal bovine serum. Yamaguchi and co-workers (57) developed another human MoAB, designated 3-207 following immunization of a melanoma patient with a vaccine consisting of purified GM2 and *Bacillus Calmette-Guerin* (BCG). The human MoAB 3-207 also reacted strongly with GD2.

Unfortunately, the production and immunotherapeutical use of human x human hybrids has turned out to be more difficult than researchers expected (62). The lack of suitable human fusion myelomas and the use of interspecies hybrids (mouse x human) has resulted in the loss of human chromosomes, and a significant degree of instability (63). The fusion frequency and growth of human x human hybrids was consistently poorer than those of mouse x mouse hybrids.

7. ANTI-IDIOTYPIC ANTIBODIES

Since Jerne proposed the network theory of immune regulation in 1974 (64), the properties and abilities of anti-idiotypic antibodies have been observed widely (65). Immunoglobulins have an unique region in their variable part, called the idiotype. Anti-idiotypic antibodies (Ab2) have been shown to induce specific immunity against murine and human tumours (66). The anti-idiotype MoABs in some situations are neoplastic cell specific. Therefore, anti-idiotypic antibodies to anti-tumor associated antigen (TAA) antibodies represent potential candidates for active specific immunotherapy, since they may share structural similarities with TAA and also have a role in the regulation of the immune system (67). Ab2 immunizations may induce protective immunity against a neoplasm. In recent clinical trials, cancer patients were immunized with Ab2 (that mimic a gastrointestinal TAA) and they developed highly specific anti-neoplastic immune responses. Thus, Ab2 is a promising vaccine candidate in cancer immunotherapy. Future studies should be devoted to collect the necessary experimental data with Ab2, in order to devise an optimal immunization protocol. A second strategy is to immunize a tumor-bearing host with idiotypic Ig, obtained from the autologous neoplastic cells. Once in place, such an "active immune response" should suppress autologous tumor growth on a continuing basis. Future developments will include the combination of anti-idiotype antibodies with other biologic therapies, as well as chemotherapy (68). Therapy of B cell nodular lymphomas employing anti-idiotype (69-71) or other MoABs (72) produced the first complete remission (longer than four years). Single doses of 1 mg to 1,500 mg and total doses of 6,000 mg were used with minimal toxicity, although rarely, brief thrombocytopenia, neutropenia, hypotension and renal dysfunction were observed. The presence of blocking antigen (idiotype) had to be overcome before adequate tumor cell targeting could be accomplished. Lymphoma cells escaped *via* IP alterations, using the presence of idiotype and genotype heterogeneity, as well as the mechanism of spontaneous antigen modulation.

Studies using mouse serum as a source of complement defined that the IgM antibodies are potent activators of the complement, while IgG2a and IgG2b antibodies are of medium or low potency (73). One factor that could determine whether or not a neoplastic cell responds to complement-fixing MoABs is the density of antigen expression on target tumour cells (74) The TAAs provide appropriate targets for active immunotherapy employing utilized anti-idiotypic MoABs to anti-TAA antibodies for a selective manipulation of the patients' immune response (67).

8. BISPECIFIC ANTIBODIES

Bispecific MoABs, with dual specificities against tumor-associated antigens on cancer cells and surface antigens on immune effector cells have also been developed (75, 76). These bispecific MoABs are effective in directing immune (cytotoxic) effector cells to lyse neoplastic target cells (77-76). Milstein and Cuello (80) developed the technique of "hybrid-hybridomas", producing hybrid MoABs, *i.e.* between mouse hybridomas of IgG2a isotype, directed to two antigens, to a melanoma and to a human or mouse T cell receptor antigen respectively. The hybrid molecule can be produced by chemical linkage between the two parental antibodies, or alternatively by the fusion of the two hybridomas selected. In the resulting quadroma cell, the hybridoma immunoglobulin chains recombine randomly to form the bispecific MoABs. Bispecific, heterocrosslinked (bifunctional) MoABs constructed against TAA and the Ti-CD3 receptor complex on CTLs have been observed to promote target neoplastic cell lysis by CTL in various *in vitro* and *in vivo* preclinical models, at nanomolar concentrations (81,82). These MoABs can be used for antigen specific retargeting of various cytotoxic cells, as well as in many other ways, and they could also increase the extremely low (about 1-2%) effectiveness of MoAB biodistribution on the tumor deposits. Production and therapeutic use of anti-glioma associated antigen x anti-CD3 bispecific antibody has also been described (83). Humanized bispecific antibodies reactive with the host's CTL and neoplastic cells overexpressing the HER-2 proto-oncogene (p185HER2) were developed by Shalaby and co-workers (84). The F(ab')2 fragment of a humanized mouse bispecific antibody reacted with the extracellular domain of p185HER2 (NR6/10 or breast cancer cells SK-BR-3) and was also directed against the human T lymphocyte marker CD3. Each fragment was separately secreted by *E. coli* and they were later joined by directed chemical coupling reaction *in vitro*, to produce the bispecific F(ab')2 fragment. The cytotoxic lysis of NR6/10 cells or the breast cancer cells SK-BR-3 by activated human cytotoxic T lymphocytes was observed in the presence of various doses of bispecific F(ab')2 fragment. As little as 10 ng/ml dose was sufficient to produce maximal enhancement of the cytotoxic activity of CTLs. In our opinion, the theoretical basis for the use of bispecific MoABs is very solid, but our present knowledge concerning the chemical conjugation of the various regions of the antibody must be enchansed in order to ensure as strong a binding against the antigenic epitope for the bispecific antibody as for the regular MoAB.

As was already mentioned above, a humanized anti-HER-2 MoAB 4D5 and a human-mouse chimeric anti-EGF receptor MoAB C225 are currently being observed in clinical trials for their anti-tumor activities. Exposure of OVCA420 human ovarian cancer cells to saturating concentrations of C225 (20 nM) for 7 days resulted in 40-50% growth inhibition, and exposure to 20 nM MoAb 4D5 also resulted in 30-40% growth inhibition. The growth inhibition of OVCA420 cells by MoAB C225 or 4D5 was associated with an increased G1 cell population; an increased level of a cyclin-dependent kinase

(CDK) inhibitor p27Kip1 with increased association of p27kip1 with CDK2, CDK4 and CDK6; and decreased activities of these CDKs. Combination treatment with concurrent exposure to MoABs C225 and 4D5 resulted in additive anti-proliferative effects on these cells, which was accompanied by enhanced G1 cell distribution, a greater increase in the levels of p27Kip1 and a greater decrease in the activities of CDK kinases (85-87).

9. RADIOLABELED ANTIBODIES

Radioactive labeling of MoABs can be relatively easily accomplished and the antibodies can specifically localize neoplastic cells, even in small deeply seated cancer xenografts in nude animals in sufficient quantity to produce gamma camera images (88-91). Experience with the use of radiolabeled MoABs in cancer patients is still very limited (92, 93). The early investigations were carried out with purified, heterogenous polyclonal animal sera. Unfortunately, highly effective labeling procedures for attaching labels to MoABs at high specific activities, with full retention of ability to bind to specific antigen are not yet available. The MoABs are enormously sensitive to both labeling conditions and architectonical modification caused by label incorporation (94, 95). The requirement for cancer cell penetration is the most critical when malignant cell radiosensitivity is considered. The cancer cell mass is less oxygenated, and this means it is less radiosensitive, especially its deeper layers of cells (96).

Goldenberg and co-investigators (97) first reported the detection of cancer with ^{131}I labeled polyclonal antibodies, directed against carcinoembryonic antigen (CEA). Mach and co-workers (98) also recommended that the anti-CEA antibodies can be used in imaging human alimentary tract tumors. They found that CEA antigen expression on tumor cells when compared to normal tissue ratios was generally 4 to 5 times greater, but only 0.1% of the injected radioactivity was detected in the tumors 6-8 days after administration of the radiolabeled MoAB. Mach and co-investigators (98-100) reported the first successful localization of neoplasmic mass in patients employing a MoAB. A number of studies followed, with radiolabeled MoABs in patients with colorectal carcinoma (101-109), breast carcinoma (101, 110) and melanoma (111-117). IFN-γ pretreatment for upregulation of expression of glycoprotein-72, a cancer-associated marker on a moderately differentiated human colon cancer cell line HT29 grown in athymic, nude mice (118) was employed prior to the use of a ^{131}I-radionuclide conjugated, CC49 MoAB. Administration of 300 microCi of 131I-CC49 to nude mice, bearing HT-29 tumors resulted in transient suppression of neoplastic progression. A sustained HT-29 neoplasm growth suppression was achieved in mice which received 300 microCi of ^{131}I-CC49 and IFN-γ. In fact, this combination of cytokine/radioimmunoconjugate eradicated any evidence of cancer in 30% of the animals. The efficacy and toxicity of bismuth (^{212}Bi) α-particle mediated radioimmunotherapy was evaluated in nude mice bearing a murine lymphoma, transfected with the human interleukin-2 α receptor (CD25, IL-2Rα or human Tac) (119). The therapeutic agent was a humanized mouse anti-Tac MoAB conjugated to ^{212}Bi. The bifunctional chelate cyclohexyl-diethylenetriamine-pentaacetic acid was employed to couple ^{212}Bi to the humanized anti-Tac MoAB. Intraperitoneally administrated doses of 150 or 200 microCi prevented tumor occurrence in 75% of the animals. Intravenous treatment 3 days following S.C. tumor cell inoculation prevented the neoplasm development only in 30% of the nude mice.

The use of radiolabeled antibodies in radioimmunoguided surgery (RIGS) has also been described. Ten colorectal cancer patients were given ^{131}I-labeled human-mouse chimeric Fab monoclonal antibody A7 (ch-Fab-A7) intravenously (i.v.) 2 to 7 days prior to RIGS. The RIGS was carried out using a portable gamma detecting probe (GDP). Tumor localization was identified by GDP intraoperatively in 4 of the ten patients, while liver metastasis and lymph node metastasis were identified in 2 patients and 1 patient, respectively. The GDP revealed tumor/surrounding tissue radio(gamma)count ratios of 1.5 or greater in 8 of the ten resected neoplasms (120).

The observations in these various human cancers demonstrated the following:

1) ^{111}In could be used to detect malignant tumors outside of the liver (103,110,114,115);
2) Single positron emission computerized tomography could be used to obtain dosimetry and tomographic images of neoplasms (121);
3) F(ab')2 fragments of the MoAB are useful for 6 to 24 hours during tumor imaging, and thus they are probably superior to whole immunoglobulin molecules (109, 117, 122);
4) pharmaco-kinetic observations on patients with melanoma and use of ^{111}In labeled MoAB directed against the p97 melanoma antigen, suggested that the radiolabeled MoAB remains primarily in circulation with a half-life of more than 30 hours (116). In fact, only a small part of the injected radiolabeled MoAB remains at the tumor site, while most of it is cleared by the cardiovascular system.

10. CONSTRUCTION OF IMMUNOTOXINS

Toxins are highly immunogenic foreign proteins. In the absence of immunosuppression, neutralizing antibody secretion develops within ten days following exposure to a toxin (124). When toxins are chemically linked to antibodies they are termed immunotoxins (125). To construct effective immunotoxins, the toxic protein must be modified so that interactions with cellular receptors are abolished (126). The ideal MoABs would be most concentrated at the tumor site, where they would initiate an anti-tumor effect by the cytotoxic agents conjugated to them, which would be highly concentrated within the tumor, but remain at a very low systemic concentration. Modified toxin entry is mediated by antibody binding. The activity of an immunotoxin is initially assessed by measuring its ability to kill target cells. Animal (preclinical) studies have shown that immunotoxins and recombinant toxins quickly attacked the targeted neoplastic cells and that the tumor masses consequently regressed within a few days. The toxin acts within the protoplasm of the target cells, therefore cell surface receptors and other proteins on the surface which naturally enter the cell *via* endocytosis are the best antigenic targets for immunotoxins. On the contrary, proteins that are fixed on the cell surface cannot be used as MoAB targets.

The protein toxins ricin (125, 128-133) and diphteria toxin (DT) (128) are the subjects of great interest because of the extreme potency of active A-chain fragments. Ricin is a 62 kD molecular weight, disulfide-linked heterodimer, purified from castor beans and for many years has been used for immunotoxin production (134). The toxic A-chain, a 30 kD enzyme, that inactivates the 60S subunit of ribosomes (causing the blockage of protein synthesis), normally requires a second peptide (B-chain) in order to bind to tumor target cells. When the A-chain is attached alone to a MoAB, the toxic fragment is inactivated. The delivery of ricin via antibodies required the removal of the B chain, blockage of the galactose binding site or attachment of a galactose-rich carbohydrate antigen to the B chain. More recently a recombinant form of the A chain has been produced in *E. coli*. Intact diphteria toxin conjugated to polyclonal antibodies is capable of inhibiting tumour growth (135, 136). However, the intact diphteria toxin retained such a significant toxicity to nontarget (normal) cells and tissues that made its future use impossible. Despite its systemic toxicity, the diphteria toxin B chain allows two cytotoxically important activities: it binds cell surface receptors and facilitates the passage of the A chain through the cell membrane's lipid bilayer into the cytoplasm. Chang and Neville (137) substituted a new binding moiety for the B chain. Human placental lactogen was linked to the A chain, employing a disulfide bond. Since the toxin A chain cannot bind cells, it had low toxicity on normal, nontarget human cells.

Similarly, chemotherapeutic agents, biological response modifiers (BRMs), and other cell cytotoxins and liposomes are being studied for development of effective MoAB conjugates (138). Many factors, such as technical difficulties in the stability, biological activity and specificity still remain, thereby limiting the use of these antigen directed "bullets" of cytotoxic MoABs *in vivo* and delaying their employment in humans. Perhaps the most promising further step is the development and use of genetic engineering to custom design new immunotoxins for cancer therapy (139). This advanced technology will perhaps eliminate normal cell directed cytotoxicity, and ensure cytosol entry activities. Imaginative experimental designs at the genetic level, will hopefully increase the cytotoxic efficacy against targeted cancer cells and naturally decrease the unwanted side effects of the newly developed immunotoxins.

In preclinical xenograft studies of BMS-182248-1 (BR96-doxorubicin immunoconjugate), a chimeric human/mouse monoclonal antibody directed against the Lewis-Y antigen, linked to approximately eight doxorubicin molecules, demonstrated significant anti-tumor activity, including cures. Patients with measurable metastatic breast cancer and immunohistochemical evidence of Lewis-Y expression on their cells received either BR96-doxorubicin conjugate 700 mg/m2 IV over 24 hours or doxorubicin 60 mg/m2 every 3 weeks. There was one partial response (7%) in 14 patients receiving the BR96-doxorubicin conjugate and one complete response and three partial responses (44%) in nine assessable patients receiving doxorubicin. No patient experienced a clinically significant hypersensitivity reaction. The toxicities were significantly different between the two treatment groups, with the BR96-doxorubicin conjugate group having limited hematologic toxicity, whereas gastrointestinal toxicities, including marked serum amylase and lipase elevations, nausea, and vomiting with gastritis, were prominent, probably due to the targeting of the antibody to these tissues (140). Although the BR96-doxorubicin immunoconjugate has limited clinical anti-tumor activity in metastatic breast cancer, similar approaches in the future may lead to the development of targeted, individualized treatment options for cancer patients.

11. MONOCLONAL ANTIBODIES: CARRIERS OF DRUGS, TOXINS AND CYTOTOXIC CELLS

Great effort has been expanded in the development of MoABs that deliver potent cytotoxic or biologically active agents to neoplastic cell targets (141-143). The success of such a strategy requires that the attachment alters neither the efficacy, nor the specificity of the MoAB, nor the biological effect of the attached agent. Moreover, the conjugate must also be stable *in vivo* and inactive and non-toxic until it reaches its target. The generation of a local field effect to kill nearby neoplastic cells that have evaded targeting would be optimal. Of the various isotopes, the β particle emitting ^{131}I has been most widely used (144,145), but ^{47}Sc and ^{90}Y, both also β-emitters have also been observed. α-emitters are also under investigation, including ^{212}Bi and ^{211}At, which emit much shorter ranged and vastly more energetic α-particles; ^{125}I and ^{77}Br which generate short ranged auger electrons, which are under constant observation and are potentially fissionable.

Oldham (146) reported the use of a cocktail (up to six) of mouse, anti-human MoABs conjugated to Adriamycin (n=24) or Mitomycin (n=19). The MoABs were selected based on the requirement that total binding to cancer cells to be greater than 80%. The immunoconjugates were administered intravenously. Adriamycin immunoconjugates were well tolerated in 22/24 patients with disseminated refractory malignancies, with 17/24 having significant foreign protein (mouse) related side effects. A total of up to 1g Adriamycin and 5g MoAB cocktail was administered to each advanced cancer patient. No response was detected after employment of Mitomycin C. The development of an IgM antibody specific against the mouse MoABs had the sensitivity to predict the clinical effectiveness of the immunotherapy. It is our opinion that the MoAB dosage employed was too high, but the major technical hurdle remained the lack of an effective drug conjugation method. Boanmycin (bleomycin A6, BM), an anticancer antibiotic, was conjugated to three MoABs R19, H111 and CCT2 (147). The immunoconjugates that were produced demonstrated selective, neoplastic cell specific cytotoxicity exclusively affecting only the targeted cells, including coecum cancer Hce-86943 cells, liver cancer BEL-7402 cells and leukemic CEM cells. The immunoconjugates were highly effective against related human xenografts in nude animals, *e.g.* the inhibition rate of R19-BM immunoconjugate against human coecum cancer cells reached 90%. The results are impressive. The problems that remain concern the extremely heterogenous IP of *in vitro* cultured tumor cells. Recently, doxorubicin (DOX) was conjugated to MoAB 425 developed against the human epidermal growth factor receptor (EGFR) (148). *In vitro* cytotoxicity assay showed that this immunoconjugate suppressed the growth of primary and secondary M24 metastatic melanoma tumors in mice with severe combined immunodeficiency and prolonged the life span of these animals. Free DOX treatment proved to be ineffective. So, it seems that the direct, antigen level targeting of cancer cells with mouse MoAB or antibody cocktails may open a new era of chemotherapeutical applications for a number of anti-cancer drugs.

The antimetastatic potential of anti-P-glycoprotein (P-gp) antibodies against multidrug-resistant (MDR) human small cell lung cancer (SCLC) cells expressing P-gp has been assessed. Human SCLC cells H69 (P-gp negative) and its etoposide-resistant variant H69/YP (P-gp positive) were used. H69 and H69/VP cells injected i.v. metastasized to the liver, kidneys and systemic lymph nodes of NK cell-depleted severe combined immunodeficient (SCID) mice. H69/VP cells, but not H69 cells, were resistant to treatments with vindesine. Treatment with mouse-human chimeric anti-P-gp antibody (MH162) and its mouse counterpart (MRK-16) reduced metastasis of H69/VP cells in various organs and prolonged the survival of tumor-bearing mice, although they were less effective if injected at late times (after 28 days). Treatment with another mouse anti-Pgp antibody, MRK-17, was effective only against liver metastasis. MH162 and MRK-16 efficiently induced antibody-dependent cellular cytotoxicity (ADCC) by peritoneal macrophages against H69/VP cells *in vitro*, but MRK-17 was less effective, in accordance with their *in vivo* antimetastatic potential. Gene transfection of macrophage colony-stimulating factor (M-CSF) into H69/VP cells to augment macrophage-mediated ADCC resulted in inhibition of metastasis to the liver and lymph nodes, but not kidneys. Combined treatment with a low dose of MRK-16 completely cured metastasis of M-CSF transfectant, but not of the mock transfectant. These findings suggest that while anti-P-gp antibodies had antimetastatic potential against SCLC cells expressing P-gp, combined treatment with M-CSF gene transduction to augment the therapeutic efficacy of anti-P-gp antibodies may be beneficial for eradicating metastatic MDR SCLC in humans (149, 150).

12. CLINICAL TRIALS WITH MONOCLONAL ANTIBODIES AND THEIR MINIMAL TOXICITY

Novel techniques of linking MoABs to radionuclides, drugs, toxins, enzymes, hormones, growth factors, colony stimulating factors, liposomes and effector cells have offered promising approaches for cancer cell detection and cell surface TAA directed, tumor-specific immunotherapy (141, 151)

Despite the many questions never answered concerning the mechanism of MoAB therapy in preclinical animal models, a small number of clinical trials in humans have been reported and with encouraging results. A key factor limiting the therapeutical effectiveness of antibodies, whether polyclonal or monoclonal, is their inadequate and non-uniform localization in solid neoplasms (151, 152). In order to improve the delivery of antibody molecules to the site of the tumour, researchers must first understand the physiological parameters of blood circulation, vascular permeability index in normal and neoplastic tissues and the effective diffusion coefficients of normal vascular and interstitial tissues (153-157). There have been no *in vivo* measurements of interstitial diffusion of antibodies to date (151, 154-156). Therefore a study was designed to measure the interstitial transport of polyclonal rabbit and sheep antibodies in normal (mature granulation) and neoplastic (VX2 carcinoma) tissues grown in the rabbit ear chamber (151). These trials were largely phase I and pilot studies:

1) They observed the adverse reactions to mouse immunoglobulin (foreign protein) and any technical problems associated with the administration of the MoAB and the binding of the MoAB to tumor sites;
2) Early clinical trials have been conducted against leukemias and lymphomas, the largest data bank has been obtained in trials that used the T65 (Leu-1, T101) antigen as a target. T65 is expressed on normal T cells, T cell leukemias and lymphomas, chronic lymphocytic leukemia (CLL), and cutaneous T cell lymphoma (CTCL) cells (158-162). Individual doses ranged from 1 mg to 500 mg and toxicity was minimal with low doses or with slow infusion of high doses. However, side effects of dyspnea and chest tightness occurred transiently in patients given infusions at rates *Fas*ter than about 0.5 mg/minute, possibly secondary to leukoagglutination of target cells.

The toxicity syndrome associated with the employment of mouse anti-human MoABs has always been minimal: fever, rigors, hypotension, dyspnea, pruritus, rash, headache, chills, nausea and vomiting, diarrhea, arthralgias and myalgias, and the more serious symptoms of tachycardia and bronchospasm have all been observed, but frequently corresponded to rapid infusion rates, high doses, or the presence of very large amount of circulating target antigen. Certainly, multiple injections of mouse anti-human IgG in cancer patients will be expected to stimulate an anti-mouse immune response. Tumour regressions occurred in about 50% of the patients with CTCL, but virtually all were only transient. Shawler and co-workers (162) observed T cell lymphoma patients, treated with a mouse anti-human MoAB T101. Repeated administration provoked a host anti-mouse IgG and IgM response, and the authors also noted a regular increase of the anti-idiotypic portion of the anti-mouse IgG after each infusion of MoAB T101. Schroff and colleagues (163) studied host anti-mouse immunoglobulin responses in patients with chronic lymphocytic leukemia, with cutaneous T cell lymphoma and malignant melanoma. Chronic lymphocytic leukemia (CLL) patients did not develop an anti-mouse globulin immune response.

Melanoma and T cell lymphoma patients, however, demonstrated a strong anti-mouse Ig immune response within two weeks following the administration of the MoABs.

Anti-idiotypic MoAB therapy of leukemias and lymphomas has been reported (164). Simple administration of MoABs reactive with these determinants has only a limited effect on the tumour load, due largely to the multiplicity of strategies by which a neoplastic cell can avoid such attack.

The use of MoABs against the CALLA antigen during a MoAB therapy of acute lymphoblastic leukemia (165) and MoABs against the myeloid differentiation antigens in the immunotherapy of acute myelogenous leukemia (166) have similarly demonstrated minimal toxicity and incomplete, transient tumor responses.

Among the solid neoplasms, colorectal carcinoma and melanoma have been observed in clinical trials, with mouse anti-human MoABs mostly in phase I studies. Sears and co-workers (167, 168) evaluated 20 patients each in phase I and phase II studies, using MoAB 17-1A. Again, toxicity was minimal, even at single doses of 1,000 mg. Eighty percent of the anti-mouse Ig also contained reactivity directed at the idiotypic determinants on the MoAB. Clinical responses were reported only in five of 40 patients and did not appear to be related to the dose of MoAB.

Pancreatic cancers rank amongst the most deadly malignant diseases with a 5 year survival percentage less than 2% and few available therapeutic approaches (169). Antigens such as CA 19-9, BW 494 and DU-PAN 2 have been reported to be associated with pancreatic cancer. However, MoABs produced have lacked the necessary specificity to warrant therapeutic utilization. The above mentioned method of antigen directed retargeting of cytotoxic cells will most likely be the solution and may result in effective tumor cell lysis.

Three research groups conducted melanoma therapy using MoABs directed at three different target antigenic epitopes (170-172). Two trials demonstrated no clinical response, but in the trial with an IgG3 isotype anti-GD3 ganglioside MoAB, which activated the complement system and mediates ADCC, three partial and two mixed responses occurred in a group of twelve patients (5/12).

13. CONCLUSIONS

Unfortunately, the employment of MoABs and antibody-immunoconjugates in the treatment of human cancer is still in its infancy.

1) The results of the early immunotherapies performed with the use of MoABs suggest that careful selection of both the antigen target and

the MoAB employed is critical to the success of MoAB mediated cancer therapy. It is most likely that most MoABs lyse the cancer cells by antibody-dependent cellular cytotoxicity (ADCC) mechanisms. MoABs:

a) enhance immune effector cell function;
b) substitute a special human immunoglobulin subclass (IgG1) for more efficient ADCC;
c) enhance complement mediated cytotoxicity;
d) penetrate through cancer masses;
e) provide more efficient delivery of drugs, toxins, radionuclides or biological response modifiers (BRMs) by ligating sequences directly into the immunoglobulin gene.

2) To the present day, the previously reported problems still remain the same:

a) antigenic modulation, immunophenotype heterogeneity and low immunogenicity of the tumour cells, as a result the neoplastic cells are capable of escaping the host's cellular immune defense reactions;
b) a MoAB directed toward the identical antigen in various human tumours may have different therapeutic efficacy;
c) possibility of hypersensitivity from part of the host's humoral immune reaction against foreign proteins (anti-idiotype IgG responses);
d) the immunoglobulin class or subclass of the MoABs to be used is significant in obtaining a desired clinical use (*i.e.* the mouse IgG2a subclass appears to be the major murine immunoglobulin class capable of activating human immunological effector cells to participate in directed antibody-dependent cytotoxic lysis of human neoplastic cells).

3) It is possible to genetically alter the structures of MoABs, changing their specificity (altering the variable region) or turning them into an antibody of another species (altering the constant region). It is possible to turn a mouse MoAB more human, to reduce its immunogenecity as a foreign protein administered during immunotherapy with mouse MoABs in the human body.

4) Future clinical trials should use mostly human-human MoABs produced by advanced hybridoma technology and the clinical application of conjugated MoABs in Phase II studies will provide additional information regarding their stability within the human body, their biodistribution and efficacy, and any adverse effects.

5) The regular use of MoABs in immunotherapy will suggest and develop the new therapeutic strategies and combinations of

immunomodalities for therapy, *i.e.* use of “MoAB cocktails” or multimodality treatments with MoABs in combination with chemotherapeutics, biological response modifiers, growth factors or colony stimulating factors.

6) The use of combinations of cocktails of non-conjugated MoABs (probably bispecific) and cytotoxic TILs may significantly increase the efficiency of neoplastic cell lysis.

7) The use of conjugated MoABs for antigen specific delivery and biodistribution of toxins, regulating cytokines and chemotherapeutical drugs will open a new era of diseased cellular metabolism directed, specific biological pharmacotherapy and will change all of our previous knowledge concerning therapeutical drug doses.

8) The recent clinical and commercial success of anticancer antibodies such as rituximab and trastuzumab has revived the initial great interest in antibody-based anti-neoplastic therapeutics for hematopoietic malignant neoplasms and solid tumors (173). Given the likelihood of lower toxic effects of antibodies that target neoplastic cells and have limited impact on nonmalignant bystander organs versus small molecules, the potential increased efficacy by conjugation to radioisotopes and other cellular toxins, and the ability to characterize the target with clinical laboratory diagnostics to improve the drug's clinical performance.

9) Current and future antibody therapeutics are likely to find substantial roles alone and in combination therapeutic strategies for treating patients in every stage of progression of their neoplastic disorder. It also is likely that conjugation strategies will add new radiolabeled and toxin-linked products to the market to complement the recent approvals of ibritumomab tiuxetan and gemtuzumab ozogamicin.

10) Recently the National Cancer Institute and the University of Alabama are developing radiolabeled humanized versions of the anticarcinoma MoAB CC49 devoid of the CH2 domain (designated HuCC49 delta CH2). The MoAB is undergoing phase I trials for the potential treatment of a variety of human neoplasms (174).

REFERENCES

1. Ehrlich P: Collected studies on immunity. Vol. II, John Wiley, New York, 1906.
2. Himmelweit B: The collected papers of Paul Ehrlich. Pergamon Press, Oxford, 1975.
3. Kufe DW, Nadler L, Sargent L, Shapiro H, Hand P, Austin F, Colcher D, Schlom J: Biological behavior of human breast carcinoma-associated antigens expressed during cellular proliferation. Cancer Res *43:* 851-857, 1983.

4. Hakomori S, Kanagi R: Glycosphingolipids as tumor-associated and differentiation markers. J Natl Cancer Inst *71:* 231-251, 1983.
5. Aziz KJ, Mamim PE, Hackett JL, Tsakeris TM: Perspectives in laboratory management. Tumor markers: the premarket review and approval process. Am Clin Lab *1:* 13-15, 1993.
6. Potter VR: On the road to the blocked ontogeny theory. Adv Oncol *4:* 1-8, 1988.
7. Cho-Chung YS: Site-selective 8-chloro-cyclic adenosine 3',5'-monophosphate as a biologic modulator of cancer: Restoration of normal control mechanisms. J Natl Cancer Inst *81:* 982-987, 1989.
8. Rittenhouse HG, Manderino GL, Hass GM: Mucin-type glycoproteins as tumor markers. Lab Med *16:* 556-560, 1985.
9. Feizi T: Demonstration by monoclonal antibodies that carbohydrate structures of glycoproteins and glycolipids are oncodevelopmental antigens. Nature (London) *314:* 53-57, 1985.
10. Reisfeld RA, Cheresh DA: Human tumor antigens. Adv Immunol *40:* 323-377, 1987.
11. Lloyd KO, Old LJ: Human monoclonal antibodies to glycolipids and other carbohydrate antigens: Dissection of the humoral immune response in cancer patients. Cancer Res *49:* 3445-3451, 1989.
12. Greiner JW, Ullmann CD, Nieroda C, Qi CF, Eggensperger D, Shimada S, Steinberg SM, Schlom J: Improved radioimmunotherapeutic efficacy of an anticarcinoma monoclonal antibody (131I-CC49) when given in combination with gamma-interferon. Cancer Res *53:* 600-608, 1993.
13. Huebner RJ, Todaro GJ: Oncogenes of RNA tumor viruses as determinants of cancer. Proc Natl Acad Sci USA *64:* 1087-1094, 1969.
14. Darlington DC: The plasmagene theory of the origin of cancer. Brit J Cancer *2:* 118-126, 1948.
15. Kelly K, Siebenlist U: The regulation and expression of c-myc in normal and malignant cells. Ann Rev Immunol *4:* 317-338, 1986.
16. Bishop JM, Baker B, Fujita D, McCombe P, Sheiness D, Smith K, Spector DH, Stehelin D, Varmus HE: Genesis of a virus-transforming gene. Natl Cancer Inst Monogr *48:* 219-223, 1978.
17. Bishop JM: Oncogenes. Scientific American *246:* 80-90, 1982.
18. Bishop JM: The molecular genetics of cancer. Science *235:* 303-311, 1987.
19. Karess RE, Hayward WS, Hanafusa H: Transforming protein encoded by the cellular information of recovered avian sarcoma viruses. Cold Spring Harbor Symp Quant Biol *44:* 765-771, 1979.
20. Klein G: The role of gene dosage and genetic transpositions in carcinogenesis. Nature (London) *294:* 313-318, 1981.
21. Eva A, Robbins KC, Andersen PR, Srinivasan A, Tronick SR, Reddy EP, Ellmore NW, Galen AT, Lautenberger JA, Papas TS, Westin EH, Wong-Staal F, Gallo RC, Aaronson SA: Cellular genes analogues to retroviral onc genes are transcribed in human tumour cells. Nature (London) *295:* 116-119, 1982.
22. Ingraham CA, Cox ME, Ward DC, Fults DW, Maness PF: c-src and other proto-oncogenes implicated in neuronal differentiation. Mol Chem Neuropathol *10:* 1-14, 1989.
23. Dalla-Favera R, Bregni M, Erikson J, Patterson D, Gallo RC, Croce CM: Human c-myc onc gene is located on the region of chromosome 8 that is translocated in Burkitt lymphoma cells. Proc Natl Acad Sci USA *79:* 7824-7827, 1982.

24. Dalla-Favera R, Martinotti S, Gallo RC, Erikson J, Croce CM: Translocation and rearrangements of the c-myc oncogene locus in human undifferentiated B-cell lymphomas. Science *219:* 963-967, 1983.
25. Mahaley MS, Gillespie GY: Immunologic considerations of patients with brain tumors. In: Oncology of the nervous system. Walker M.D. (ed.), Boston, The Hague, Dordrecht, Lancaster: Martinus Nijhoff Publishers, The Netherlands, 151-164, 1983.
26. Kohler G, Milstein C: Continuous culture of fused cells secreting antibody of predefined specificity. Nature (London) *256:* 495-497, 1975.
27. Songsivilai S, Lachmann PJ: Antibody engineering: Current status and future development. Asian Pacific J Allergy Immunol *8:* 53-60, 1990.
28. van der Bruggen P, Traversari C, Chomez P, Lurquin C, De Plaen E, Van den Eynde B, Knuth A, Boon T: A gene encoding an antigen recognized by cytolytic T lymphocytes on a human melanoma. Science *254:* 1643-1647, 1991.
29. Houghton AN, Scheinberg DA: Monoclonal antibodies: potential applications to the treatment of cancer. Semin Oncol *13:* 165-179, 1986.
30. Coiffier B: Monoclonal antibodies in the treatment of neoplastic hematologic diseases. Bull Cancer *87:* 839-845, 2000.
31. Penault-Llorca FM, Balaton AJ. Monoclonal antibodies in oncology: applications in diagnosis, prognosis and prediction of response to therapy on tissue specimens. Bull Cancer *87:* 794-803, 2000.
32. Gabrilove JL: Differentiation factors. Semin Oncol *13:* 228-233, 1986.
33. Talmadge JE, Herberman RB: The preclinical screening laboratory: evaluation of immunomodulatory and therapeutic properties of biological response modifiers. Cancer Treat Rep *70:* 171-182, 1986.
34. Skerra A, Pluckthun A: Assembly of a functional immunoglobulin Fv fragment in Escherichia coli. Science *240:* 1038-1041, 1988.
35. Better M, Chang CP, Robinson RR, Horwitz AH: *Escherichia coli* secretion of an active chimeric antibody fragment. Science *240:* 1041-1043, 1988.
36. Riechemann L, Foote J, Winter G: Expression of an antibody Fv fragment in myeloma cells. J Mol Biol *203:* 825-828, 1988.
37. Levy R, Miller RA, Stratte PT: Therapeutic trials of monoclonal antibody in leukemia and lymphoma: Biologic considerations. In: Boss B.D., Langman R., Trowbridge I., et al. (eds). Monoclonal Antibodies and Cancer, Academic Press, Orlando, Florida, USA, 5-16, 1983.
38. Herlyn D, Lubeck M, Sears H, Koprowski H: Specific detection of anti-idiotypic immune responses in cancer patients treated with murine monoclonal antibody. J Immunol Methods *85:* 27-38, 1985.
39. Maltzman W, Czyzyk L: UV irradiation stimulates levels of p53 cellular tumor antigen in nontransformed mouse cells. Mol Cell Biol *4:* 1689-1694, 1984.
40. Kastan MB, Onyekwere O, Sidransky D, Vogelstein B, Craig RW: Participation of p53 in the cellular response to DNA damage. Cancer Res *51:* 6304-6311, 1991.
41. Nagasawa H, Li CY, Maki CG, Imrich AC, Little JB: Relationship between radiation-induced G1 phase arrest and p53 in human tumor cells. Cancer Res *55:* 1842-1846, 1995.
42. Lee JM, Bernstein A: p53 mutations increase resistance to ionizing radiation. Proc Natl Acad Sci USA *90:* 5742-5746, 1993.
43. Lowe SW, Schmitt EM, Smith SW, Osborne BA, Jacks T: p53 is required for radiation-induced apoptosis in mouse thymocytes. Nature *362:* 847-849, 1993.

44. Fan S, El-Deiry WS, Bae I, Freeman J, Jondle D, Bhatia K, Fornace AJ JR, Magrath I, Kohn KW, O'Connor PM: p53 gene mutations are associated with decreased sensitivity of human lymphoma cells to DNA damaging agents. Cancer Res *54:* 5824-5830, 1994.
45. Fan S, Smith ML, Rivet DJ II, Duba D, Zhan Q, Kohn KW, Fornace AJ JR, O'Connor PM: Disruption of p53 sensitizes breast cancer MCF-7 cells to cisplatin and pentoxifylline. Cancer Res *55:* 1649-1654, 1995.
46. Finlay CA, Hinds PW, Tan T-H, Eliyahu D, Oren M, Levine AJ: Activating mutations for transformation by p53 produce a gene product that forms an hsc70-p53 complex with an altered half-life. Mol Cell Biol *8:* 531-539, 1988.
47. Bartek J, Bartkova J, Vojtesek B, Staskova Z, Lukas J, Rejthar A, Kovarik J, Midgley CA, Gannon JV, Lane DP: Aberrant expression of the p53 oncoprotein is a common feature of a wide spectrum of human malignancies. Oncogene *6:* 1699-1703, 1991.
48. Livingston RB, Esteva FJ: Chemotherapy and Herceptin for HER-2^+ metastatic breast cancer: the best drug? The Oncologist *6:* 315-316, 2001.
49. Ross JS, McKenna BJ: TheHER-2/neu oncogene in tumors of the gastrointestinal tract. Cancer Invest *19:* 554-568, 2001.
50. Penault-Llorca F, Etessami A, Bourhis J: Principal therapeutic uses of monoclonal antibodies in oncology. Cancer Radiother Suppl 1: 24s-28s, 2002.
51. McDevitt MR, Ma D, Lai LT, Simon J, Borchardt P, Frank RK, Wu K, Pellegrini V, Curcio MJ, Miederer M, Bander NH, Scheinberg DA: Tumor therapy with targeted atomic nanogenerators. Science *294:* 1537-1540, 2001.
52. Rettig WJ, Garin-Chesa P, Healey JH, Su SL, Jaffe EA, Old LJ: Identification of endosialin, a cell surface glycoprotein of vascular endothelial cells in human cancer. Proc Natl Acad Sci USA *89:* 10832-10836, 1992.
53. Neri D, Carnemolla B, Nissim A, Leprini A, Querze G, Balza E, Pini A, Tarli L, Halin C, Neri P, Zardi L, Winter G: Targeting by affinity-matured recombinant antibody fragments of an angiogenesis associated fibronectin isoform. Nature Biotech *15:* 1271-1275, 1997.
54. Carnemolla B, Balza E, Siri A, Zardi L, Nicotra MR, Bigotti A, Natali PG: A tumor-associated fibronectin isoform generated by alternative splicing of messenger RNA precursors. J Cell Biol *108:* 1139-1148, 1989.
55. Castellani P, Viale G, Dorcaratto A, Nicolo G, Kaczmarek J, Querze G, Zardi L: The fibronectin isoform containing the ED-B oncofetal domain: a marker of angiogenesis. Int J Cancer *59:* 612-618, 1994.
56. Wong JH, Irie RF, Morton DL: Human monoclonal antibodies: Prospects for the therapy of cancer. Sem Surgical Oncol *5:* 448-452, 1989.
57. Yamaguchi H, Furukawa K, Fortunato SR, Livingston PO, Lloyd KO, Oettgen HF, Old LJ: Cell-surface antigens of melanoma recognized by human monoclonal antibodies. Proc Natl Acad Sci USA *84:* 2416-2420, 1987.
58. Cheresh DA, Harper JR, Schulz G, Reisfeld RA: Localization of the gangliosides GD3 and GD2 in adhesion plaques and on the surface of human melanoma cells. Proc Natl Acad Sci USA *81:* 5767-5771, 1984.
59. Cheung N-KV, Saarinen UM, Neely JE, Landmeier B, Donovan D, Coccia PF: Monoclonal antibodies to a glycolipid antigen on human neuroblastoma cells. Cancer Res *45:* 2642-2649, 1985.
60. Imam A, Mitchell MS, Modlin RL, Taylor CR, Kempf RA, Kan-Mitchell J: Human monoclonal antibodies that distinguish cutaneous malignant melanomas from benign nevi in fixed tissue sections. J Invest Dermatol *86:* 145-148, 1986.

61. Kan-Mitchell J, Imam A, Kempf RA, Taylor CR, Mitchell MS: Human monoclonal antibodies directed against melanoma tumor-associated antigens. Cancer Res *46:* 2490-2496, 1986.
62. James K, Bell GT: Human monoclonal antibody production current status and future prospects. J Immunol Methods *100:* 5-40, 1987.
63. Thompson KM: Human monoclonal antibodies. Immunol Today *9:* 113-117, 1988.
64. Jerne NK: Towards a network theory of the immune system. Ann. Immunol. (Paris). *125:* 373-389, 1974.
65. Ludwig DS, Finkelstein RA, Karu AE, Dallas WS, Ashby ER, Schoolnik GK: Anti-idiotypic antibodies as probes of protein active sites: Application to cholera toxin subunit B. Proc Natl Acad Sci USA *84:* 3673-3677, 1987.
66. Herlyn D, Wettendorff M, Iliopoulos D, Koprowski H: Functional mimicry of tumor-associated antigens by antiidiotypic antibodies. Exper Clin Immunogenet *5:* 165-175, 1988.
67. O'Connell MJ, Chen ZJ, Yang H, Yamada M, Massaro M, Mittelman A, Ferrone S: Active specific immunotherapy with antiidiotypic antibodies in patients with solid tumors. Semin Surg Oncol *5:* 441-447, 1989.
68. Levy R, Miller RA: Therapy of lymphoma directed at idiotypes. Monogr J Natl Cancer Inst *10:* 61-68, 1990.
69. Miller RA, Maloney DG, Warnke R, Levy R: Treatment of B cell lymphoma with monoclonal anti-idiotype antibody. N Engl J Med *306:* 517-522, 1982.
70. Meeker TC, Lowder J, Maloney DG, Miller RA, Thielemans K, Warnke R, Levy R: A clinical trial of anti-idiotype therapy for B cell malignancy. Blood *65:* 1349-1363, 1985.
71. Raffeld M, Neckers L, Longo DL, Cossman J: Spontaneous alteration of idiotype in a monoclonal B cell lymphoma: Escape from the detection by anti-idiotype. New Engl J Med *312:* 1653-1658, 1985.
72. Nadler LM, Stashenko P, Hardy R, Kaplan WD, Button LN, Kufe DW, Antman KH, Schlossman SF: Serotherapy of a patient with a monoclonal antibody directed against a human lymphoma-associated antigen. Cancer Res *40:* 3147-3154, 1980.
73. Neuberger MS, Rejewsky K: Switch from hapten-specific immunoglobulin M to immunoglobulin D secretion in a hybrid mouse cell line. Proc Natl Acad Sci USA *78:* 1138-1142, 1981.
74. Capone PM, Papsidero LD, Chu TM: Relationship between antigen density and immunotherapeutic response elicited by monoclonal antibodies against solid tumors. J Natl Cancer Inst *72:* 673-677, 1984.
75. Liu MA, Kranz DM, Kurnick JT, Boyle LA, Levy R, Eisen HN: Heteroantibody duplexes target cells for lysis by cytotoxic T lymphocytes. Proc Natl Acad Sci USA *82:* 8648-8652, 1985.
76. Perez P, Hoffman RW, Shaw S, Bluestone JA, Segal DM: Specific targeting of cytotoxic T cells by anti-T3 linking to anti-target cell antibody. Nature (London) *316:* 354-356, 1985.
77. Nolan O, O'Kennedy R: Bifunctional antibodies: concept, production and applications. Biochim Biophys Acta *1040:* 1-11, 1990.
78. Nolan O, O'Kennedy R: Bifunctional antibodies and their potential clinical applications. Int J Clin Lab Med *22:* 21-27, 1992.
79. Fanger MW, Segal DM, Romet-Lemonne J: Bispecific antibodies and targeted cellular cytotoxicity. Immunol Today *12:* 51-54, 1991.
80. Milstein C, Cuello AC: Hybrid hybridomas and their use in immunohistochemistry. Nature *305:* 537-540, 1983.

81. Barr IG, MacDonald HR, Buchegger F, von Fliedner V: Lysis of tumour cells by the retargeting of murine cytotoxic T lymphocytes with bispecific antibodies. Int J Cancer *40:* 423-429, 1987.
82. Menard S, Canevari S, Colinaghi MI: Hybrid antibodies in cancer diagnosis and therapy. Int J Biol Markers *4:* 131-134, 1989.
83. Nitta T, Sato K, Yagita H, Okumura K, Ishii S: Preliminary trial of specific targeting therapy against malignant glioma. Lancet *335:* 368-374, 1990.
84. Shalaby MR, Shepard HM, Presta L, Rodrigues ML, Beverley PC, Feldmann M, Carter P: Development of humanized bispecific antibodies reactive with cytotoxic lymphocytes and tumor cells overexpressing the HER2 protooncogene. J Exp Med *175:* 217-225, 1992.
85. Ye D, Mendelsohn J, Fan Z: Augmentation of a humanized anti-HER2 mAb 4D5 induced growth inhibition by a human-mouse chimeric anti-EGF receptor mAb C225. Oncogene *18:* 731-738, 1999.
86. Mendelsohn J: Epidermal growth factor receptor inhibition by a monoclonal antibody as anticancer therapy. Clin Cancer Res *3:* 2703-2707, 1997.
87. Ciardiello F, Bianco R, Damiano V, De Lorenzo S, Pepe S, De Placido S, Fan Z, Mendelsohn J, Bianco AR, Tortora G: Antitumor activity of sequential treatment with topotecan and anti-epidermal growth factor receptor monoclonal antibody C225. Clin Cancer Res *5:* 909-916, 1999.
88. Hnatowich DJ, Layne WW, Childs RL, Lanteigne D, Davis MA, Griffin TW, Doherty PW: Radioactive labeling of antibody: a simple and efficient method. Science *220:* 613-615, 1983.
89. Shah SA, Gallagher BM, Sands H: Radioimmunodetection of small human tumor xenografts in spleen of athymic mice by monoclonal antibodies. Cancer Res *45:* 5824-5829, 1985.
90. Illidge TM, Brock S: Radioimmunotherapy of Cancer: Using Monoclonal Antibodies to Target Radiotherapy. Current Pharmaceut. Design *6:* 1399-1418, 2000.
91. Vuillez JP: Radioimmunotargeting: diagnosis and therapeutic use. Bull. Cancer *87:* 813-827, 2000.
92. Juric JG: Antibody Immunotherapy for Leukemia. Curr Oncol Rep *2:* 114-122, 2000.
93. Sands H, Loveless SE: Biodistribution and pharmacokinetics of recombinant, human interleukin-2 in mice. Int J Immunopharmacol *11:* 411-416, 1989.
94. Andres RY, Schubiger PA: Radiolabeling of antibodies. Methods and limitations. Nukl Med *25:* 162-166, 1986.
95. Ramjeesingh M, Zywulko M, Rothstein A, Whyte R, Shami EY: Antigen protection of monoclonal antibodies undergoing labelling. J Immunol Methods *133:* 159-167, 1990.
96. Britton KE, Mather SJ, Granowska M: Radiolabelled monoclonal antibodies in oncology. III. Radioimmunotherapy. Nucl Med Communicat *12:* 333-347, 1991.
97. Goldenberg DM, DeLand FH, Kimm EE: Use of radiolabeled antibodies to carcinoembryonic antigen for the detection and localization of diverse tumors by external photoscanning. N Engl J Med *298:* 1384-1388, 1978.
98. Mach JP, Carrel S, Forni M, Ritschard J, Donath A, Alberto P: Tumor localization of radiolabeled antibodies against carcinoembryonic antigen in patients with carcinoma. New Engl J Med *303:* 5-10, 1980.
99. Mach JP, Carrel S, Merenda C, Sordat B, Cerottini JC: In Vivo localization of radiolabelled antibodies to carcinoembryonic antigen in human colon carcinoma grafted into nude mice. Nature (London) *248:* 704-706, 1974.

100. Mach JP, Chatal JF, Lumbroso JD, Buchegger F, Forni M, Ritschard J, Berche C, Douillard JY, Carrel S, Herlyn M: Tumor localization in patients by radiolabeled monoclonal antibody against colon carcinoma. Cancer Res *43:* 5593-5600, 1983.
101. Epenetos AA, Mather S, Granowska M: Targeting of iodine-123-labelled tumor-associated monoclonal antibodies to ovarian, breast and gastrointestinal tumors. Lancet *2:* 999-1005, 1982.
102. Farrands PA, Perkins AC, Pimm MV, Embleton MJ, Hardy JD, Baldwin RW, Hardcastle JD: Radioimmunodetection of human colorectal cancers by an anti-tumor monoclonal antibody. Lancet *2:* 397-404, 1982.
103. Fairweather DS, Bradwell AR, Dykes PW, Vaughan AT, Watson-James SF, Chandler S: Improved tumor localization using indium-111 labelled antibodies. Br Med J *287:* 167-170, 1983.
104. Moldofsky PJ, Powe J, Mulhern CB Jr, Hammond N, Sears HF, Gatenby RA, Steplewski Z, Koprowski H: Metastatic colon carcinoma detected with radiolabeled F(ab')2 monoclonal antibody fragments. Radiology *149:* 549-555, 1983.
105. Chatal JF, Saccavini JC, Fumoleau P, Douillard JY, Curtet C, Kremer M, Le Mevel B, Koprowski H: Immunoscintigraphy of colon carcinoma. J Nucl Med *25:* 307-314, 1984.
106. Dillman R: Monoclonal antibodies in the treatment of cancer. CRC Crit Rev Oncol./Hematol *1:* 357-385, 1984.
107. Moldofsky PJ, Sears HF, Mulhern CB Jr, Hammond ND, Powe J, Gatenby RA, Steplewski Z, Koprowski H: Detection of metastatic tumor in normal sized retroperitoneal lymph nodes by monoclonal antibody imaging. N Engl J Med *311:* 106-107, 1984.
108. Delaloye B, Delaloye-Bischof A, Dudczak R, Koppenhagen K, Mata F, Penafiel A, Maul FD, Pasquier J: Clinical comparison of 99mTc-HMDP and 99mTc-MDP. A multicenter study. Eur J Nucl Med *11:* 182-185, 1985.
109. Delaloye B, Bischof-Delaloye A, Buchegger F, von Fliedner V, Grob JP, Volant JC, Pettavel J, Mach JP.: Detection of colorectal carcinoma by emission-computerized tomography after injection of 123-iodine labeled Fab or F(ab)2 fragments from monoclonal anti-carcinoembryonic antigen antibodies. J Clin Invest *77:* 301-311, 1986.
110. Rainsbury RM, Westwood JH, Coombes RC: Location of metastatic breast carcinoma by a monoclonal antibody chelate labelled with indium-111. Lancet *2:* 934-938, 1983.
111. Larson SM, Brown JP, Wright PW, Carrasquillo JA, Hellstrom I, Hellstrom KE: Imaging of melanoma with 131-I-labeled monoclonal antibodies. J Nucl Med *24:* 123-129, 1983.
112. Larson SM, Carrasquillo JA, Krohn KA, McGuffin RW, Williams DL, Hellstrom I, Lyster D: Diagnostic imaging of malignant melanoma with radiolabeled anti-tumor antibodies. JAMA *249:* 811-812, 1983.
113. Larson SM, Carrasquillo JA, Krohn KA, Brown JP, McGuffin RW, Ferens JM, Graham MM, Hill LD, Beaumier PL, Hellstrom KE: Localization of 131I-labeled p96-specific Fab fragments in human melanoma as a basis for radiotherapy. J Clin Invest *72:* 2101-2114, 1983.
114. Halpern SE, Dillman RO, Witztum KF: Radioimmunodetection of melanoma using 111In 96.5 monoclonal antibody: A preliminary report. Radiology *155:* 493-499, 1985.
115. Murray JL, Rosenblum MG, Sobol RE, Bartholomew RM, Plager CE, Haynie TP, Jahns MF, Glenn HJ, Lamki L, Benjamin RS: Radioimmunoimaging in malignant melanoma with 111In-labeled monoclonal antibody 96.5. Cancer Res *45:* 2376-2381, 1985.
116. Rosenblum MG, Murray JL, Haynie TP, Glenn HJ, Jahns MF, Benjamin RS, Frincke JM, Carlo DJ, Hersh EM: Pharmacokinetics of In-111-labeled anti-p97 monoclonal

antibody in patients with metastatic malignant melanoma. Cancer Res *45:* 2382-2386, 1985.

117. Buraggi GL, Callegaro L, Mariani G, Turrin A, Cascinelli N, Attili A, Bombardieri E, Terno G, Plassio G, Dovis M: Imaging with I-131-labeled monoclonal antibodies to a high molecular weight melanoma-associated antigen in patients with melanoma: efficiency of whole immunoglobulin and its F (ab')2 fragments. Cancer Res *45:* 3378-3387, 1985.
118. Greiner JW, Guadagni F, Roselli M, Ullmann CD, Nieroda C, Schlom J: Improved experimental radioimmunotharapy of colon xenografts by combining 131I-CC49 and interferon-gamma. Diseas Col Rect *37:* S100-S105, 1994.
119. Hartmann F, Horak EM, Garmestani K, Wu C, Brechbiel MW, Kozak RW, Tso J, Kosteiny SA, Gansow OA, Nelson DL: Radioimmunotherapy of nude mice bearing a human interleukin 2 receptor alpha-expressing lymphoma utilizing the alpha-emitting radionuclide-conjugated monoclonal antibody 212Bi-anti-Tac. Cancer Res *54:* 4362-4370, 1994.
120. Yamamoto K, Kitamura K, Nishida S, Ichikawa D, Okamoto K, Yamaguchi T, Takahashi T: Iodine-131 human-mouse chimeric Fab monoclonal antibody A7 guided surgery for colorectal cancer patients: a pilot study. Surg Today *29:* 190-193, 1999.
121. Berche C, Mach JP, Lumbroso JD, Langlais C, Aubry F, Buchegger F, Carrel S, Rougier P, Parmentier C, Tubiana M: Tomoscintigraphy for detecting gastrointestinal and medullary thyroid cancers: first clinical results using radiolabelled monoclonal antibodies against carcinoembryonic antigen. Brit Med J *285:* 1447-1451, 1982.
122. Mach JP, Buchegger F, Forni M, Ritschard J, Carrel S, Egley R, Donath A, Rohner A: Immunoscintigraphy of human carcinoma after injection of radiolabeled monoclonal anti-carcinoembryonic antigen antibodies. Curr Top Microbiol Immunol *104:* 49-55, 1983.
123. Casellas P, Brown JP, Gros O, Gros P, Hellstrom I, Jansen FK, Poncelet P, Roncucci R, Vidal H, Hellstrom KE: Human melanoma cells can be killed in vitro by an immunotoxin specific for melanoma-associated antigen p97. Int J Cancer *30:* 437-443, 1982.
124. Durrant LG, Byers VS, Scannon PJ, Rodvien R, Grant K, Robins RA, Marksman RA, Baldwin RW: Humoral immune responses to XMMCO-791-RTA immunotoxin in colorectal cancer patients. Clin Exp Immunol *75:* 258-264, 1989.
125. Blakey DC, Thorpe PE: Treatment of malignant disease and rheumatoid arthritis using ricin A-chain immunotoxins. Scand J Rheumatol Suppl *76:* 279-287, 1988.
126. Pastan I, Fitzgerald D: Recombinant toxins for cancer treatment. Science *254:* 1173-1177, 1991.
127. Trowbridge IS, Domingo D: Anti-transferrin monoclonal antibody and toxin-antibody conjugates affect growth of human tumor cells. Nature (London) *294:* 171-173, 1981.
128. Gilliland G, Steplewski Z, Collier J, Mitchell KF, Chang TH, Koprowski H: Antibody-directed cytotoxic agents: Use of monoclonal antibody to direct the action of toxin A chains to colorectal carcinoma cells. Proc Natl Acad Sci USA *77:* 4539-4543, 1980.
129. Houston LL, Nowinski RC: Cell-specific cytotoxicity expressed by a conjugate of ricin and murine monoclonal antibody directed against Thy 1.1 antigen. Cancer Res *41:* 3913-3917, 1981.
130. Blythman HE, Casellas P, Gros O, Gros P, Jansen FK, Paolucci F, Pau B, Vidal H: Immunotoxins: Hybrid molecules of monoclonal antibodies and a toxin subunit specifically kill tumor cells. Nature (London) *290:* 145-146, 1981.
131. Vitatta ES, Krolick KA, Uhr JW: Neoplastic B cells as targets for antibody-ricin A chain immunotoxins. Immunol Rev *62:* 159-183, 1982.

132. Thorpe PE, Ross WCJ: The preparation and cytotoxic properties of antibody-toxin conjugates. Immunol Rev *62:* 119-158, 1982.
133. Knowles PP, Thorpe PE: Purification of immunotoxins containing ricin A-chain and abrin A-chain using Blue Sepharose CL-6B. Anal Biochem *160:* 440-443, 1987.
134. Cumber AJ, Forrester JA, Foxwell BM, Ross WC, Thorpe PE: Preparation of antibody-toxin conjugates. Meth Enzymol *112:* 207-225, 1985.
135. Moolten FL, Capparell NJ, Zajdel SH: Antitumor effects of antibody-diphteria toxin conjugates. II. Immunotherapy with conjugates directed against tumor antigens induced by simian virus 40. J Natl Cancer Inst *55:* 473-477, 1975.
136. Bernhard MI, Foon KA, Oeltmann TN, Key ME, Hwang KM, Clarke GC, Christensen WL, Hoyer LC, Hanna MG Jr, Oldham RK: Guinea pig line 10 hepatocarcinoma model: Characterization of monoclonal antibody and In Vivo effect of unconjugated antibody and antibody conjugated to diphteria toxin A chain. Cancer Res *43:* 4420-4428, 1983.
137. Chang T, Dazord A, Neville DM, Jr: Artificial hybrid protein containing a toxic protein fragment and a cell membrane receptor-binding moiety in a disulfide complex. J Biol Chem *252:* 1515-1522, 1977.
138. Dumontet C: Immunotherapy and cancer: the role of monoclonal antibodies. J. Chir. (Paris) *126:* 682-686, 1989.
139. Youle RJ, Colomabtti M: Immunotoxins: monoclonal antibodies linked to toxic proteins for bone marrow transplantation and cancer therapy. In: Monoclonal Antibodies in Cancer: Advances in Diagnosis and Treatment, ed. J.A. Roth, Futura Publishing Company, Inc., Mount Kisco, New York, USA, 173-213, 1986.
140. Tolcher AW, Sugarman S, Gelmon KA, Cohen R, Saleh M, Isaacs C, Young L, Healey D, Onetto N, Slichenmyer W: Randomized phase II study of BR96-doxorubicin conjugate in patients with metastatic breast cancer. J Clin Oncol *17:* 478-484, 1999.
141. Moller G (ed.): Antibody carriers of drugs and toxins in tumor therapy. Immunol Rev *62:* 1-216, 1982.
142. Byers VS, Baldwin RW: Therapeutic strategies with monoclonal antibodies and immunoconjugates. Immunology *65:* 329-335, 1988.
143. Pietersz GA, Krauer K: Antibody-targeted drugs for the therapy of cancer. J. Drug Target. *2:* 183-215, 1994.
144. Redwood WR, Tom TD, Strand M: Specificity, efficacy, and toxicity of radioimmunotherapy in erythroleukemic mice. Cancer Res *44:* 5681-5687, 1984.
145. Ferens JM, Krohn KA, Beaumier PL: High level iodination of monoclonal antibody fragments for radiotherapy. J Nucl Med *25:* 367-370, 1984.
146. Oldham RK: Custom-tailored drug immunoconjugates in cancer therapy. Mole Biother *3:* 148-162, 1991.
147. Zhen Y, Peng Z, Deng Y, Xu H, Chen Y, Tian P, Li D, Jiang M: Antitumor activity of immunoconjugates composed of boanmycin and monoclonal antibody. Chinese Med Sciences J *9:* 75-80, 1994.
148. Sivam GP, Martin PJ, Reisfeld RA, Mueller BM: Therapeutic efficacy of a doxorubicin immunoconjugate in a preclinical model of spontaneous metastatic human melanoma. Cancer Res *55:* 2352-2356, 1995.
149. Yano S, Hanibuchi M, Nishioka Y, Nokihara H, Nishimura N, Tsuruo T, Sone S: Combined therapy with anti-P-glycoprotein antibody and macrophage colony-stimulating factor gene transduction for multiorgan metastases of multidrug-resistant human small cell lung cancer in NK cell-depleted SCID mice. Int J Cancer *82:* 105-111, 1999.

150. Sone S, Yano S, Hanibuchi M, Nokihara H, Nishimura N, Miki T, Nishioka Y, Shinohara T: Heterogeneity of multiorgan metastases of human lung cancer cells genetically engineered to produce cytokines and reversal using chimeric monoclonal antibodies in natural killer cell-depleted severe combined immunodeficient mice. Cancer Chemother Pharmacol *43:* S26-S31, 1999.
151. Clauss MA, Jain RK: Interstitial transport of rabbit and sheep antibodies in normal and neoplastic tissues. Cancer Res *50:* 3487-3492, 1990.
152. Dykes PW, Bradwell AR, Chapman CE, Vaughan ATM: Radioimmunotherapy of cancer: clinical studies and limiting factors. Cancer Treat Rev *14:* 87-106, 1987.
153. Swabb EA, Wei J, Gullino PM: Diffusion and convection in normal and neoplastic tissues. Cancer Res *34:* 2814-2822, 1974.
154. Jain RK: Transport of molecules across tumor vasculature. Cancer Metastasis Rev *6:* 559-594, 1987.
155. Jain RK: Transport of molecules in the tumor interstitium: a review. Cancer Res. *47:* 3039-3051, 1987.
156. Jain RK: Delivery of novel therapeutic agents in tumors: physiological barriers and strategies. J Natl Cancer Inst *81:* 570-576, 1989.
157. Chary SR, Jain RK: Direct measurements of interstitial convection and diffusion of albumin in normal and neoplastic tissues by fluorescence photobleaching. Proc Natl Acad Sci USA *86:* 5385-5389, 1989.
158. Levy RL, Miller RA: Biological and clinical implications of lymphocyte hybridomas: Tumor therapy with monoclonal antibodies. Ann Rev Med *34:* 107-116, 1983.
159. Dillman RO, Beauregard JC, Sobol RE, Royston I, Bartholomew RM, Hagan PS, Halpern SE: Lack of radioimmunodetection and complications associated with monoclonal anti-carcinoembryonic antigen: Antibody crossreactivity with an antigen on circulating cells. Cancer Res *44:* 2213-2218, 1984.
160. Dillman RO, Shawler DL, Dillman JB, Royston I: Therapy of chronic lymphocytic leukemia and cutaneous T cell lymphoma with T101 monoclonal antibody. J Clin Oncol *2:* 881-891, 1984.
161. Foon KA, Schroff RW, Bunn PA, Mayer D, Abrams PG, Fer M, Ochs J, Bottino GC, Sherwin SA, Carlo DJ: Effects of monoclonal antibody therapy in patients with chronic lymphocytic leukemia. Blood *64:* 1085-1093, 1984.
162. Shawler DL, Bartholomew RM, Smith LM, Dillman RO: Human immune response to multiple injections of murine monoclonal Ig. J Immunol *135:* 1530-1535, 1985.
163. Schroff RW, Foon KA, Beatty SM, Oldham RK, Morgan AC: Human antimurine immunoglobulin responses in patients receiving monoclonal antibody therapy. Cancer Res *45:* 879-885, 1985.
164. Stevenson FK, George AJ, Glennie MJ: Anti-idiotypic therapy of leukemias and lymphomas. Chem Immunol *48:* 126-166, 1990.
165. Ritz J, Pesando JM, Sallan SE, Clavell LA, Notis-McConarty J, Rosenthal P, Schlossman SF: Serotherapy of acute lymphoblastic leukemia with monoclonal antibody. Blood *58:* 141-152, 1981.
166. Ball ED, Bernier GM, Cornwell GG: Monoclonal antibodies to myeloid differentiation antigens: In Vivo studies of three patients with acute myelogenous leukemia. Blood *62:* 1203-1210, 1983.
167. Sears HF, Herlyn D, Steplewski Z, Koprowski H: Effects of monoclonal antibody immunotherapy on patients with gastrointestinal adenocarcinoma. J Biol Resp Modif *3:* 138-150, 1984.

168. Sears HF, Herlyn D, Steplewski Z, Koprowski H: Phase II clinical trial of a murine monoclonal antibody cytotoxic for gastrointestinal adenocarcinoma. Cancer Res *45:* 5910-5913, 1985.
169. Bellet D, Bidart JM, Rougier P, Bohuon C: Use of monoclonal antibodies in the treatment of cancer of the pancreas: towards new progress? Bull Cancer *77:* 283-288, 1990.
170. Oldham RK, Foon KA, Morgan AC, Woodhouse CS, Schroff RW, Abrams PG, Fer M, Schoenberger CS, Farrell M, Kimball E: Monoclonal antibody therapy of malignant melanoma: In Vivo localization in cutaneous metastasis after intravenous administration. J Clin Oncol *2:* 1235-1244, 1984.
171. Houghton AN, Gordon-Curdo C, Eisinger M: Differentiation antigens of melanoma and melanocytes. Int Rev Exp Pathol *28:* 217-248, 1986.
172. Goodman GE, Beaumier P, Hellstrom I, Fernyhough B, Hellstrom KE: Pilot trial of murine monoclonal antibodies in patients with advanced melanoma. J Clin Oncol *3:* 340-352, 1985.
173. Ross JS, Gray K, Gray GS, Worland PJ, Rolfe M: Anticancer antibodies. Am J Clin Pathol *119:* 472-485, 2003.
174. Sheikh N: Technology evaluation: CC49 humanized radioimmunoconjugates, National Cancer Institute. Curr Opin Mol Ther *5:* 428-432, 2003.

Chapter 9

CANCER-TESTIS ANTIGENS: PROMISING TARGETS FOR ANTIGEN DIRECTED ANTI-NEOPLASTIC IMMUNOTHERAPY

1. INTRODUCTION

Neoplasm associated markers (NAMs) are the biochemical and immunological counterparts of the morphology of neoplasms. The presence of NAM or NAA is also related to the tissue of origin and is not a random event. The number of human neoplastic cells related antigens recognized employing autologous $CD8^+$, cytolytic lymphocytes or antibodies has been expanded rapidly over the past decade (1). The employment of MoABs against oncofetal, NA, cell lineage specific, endothelial, and cell proliferation related antigens in the diagnosis and biological assessment of the prognosis in neoplastic diseases has gained increased importance. A direct correlation exists between the expression of certain molecules and the development of a malignant IP of neoplastic cells, allowing for the occurrence of neoangiogenesis and metastasis. Human antigens expressed exclusively in neoplastically transformed cells and male germ line cells, such as those of the cancer/testis antigen (CTA) families, a distinct class of differentiation antigens, have received much attention. Human melanoma cells express several of these CTAs which are recognized by autologous and specific, cytolytic T lymphocyte clones in association with HLA-class-I molecules and are potentially useful for anti-neoplastic immunotherapy as potential targets of human cancer vaccines and for T lymphocyte based or antigen specific and directed cellular immunotherapy because of their broad expression in various neoplasms (1, 2).

2. CANCER/TESTIS ANTIGENS

Several CTAs were recognized by autologous cytolytic T lymphocytes (CTL) on established human melanoma cell line MZ2-MEL (3). Other CTAs were detected by employing serological analysis of recombinant cDNA expression libraries (SEREX) using neoplastic mRNA and autologous patient serum (antibodies). The last method provides a powerful approach to identifying immunogenic NA antigens.

Melanoma-associated antigens (MAGE) are regarded as inducing strong tumor-specific immune response (4-6). During the last decade, 23 human and 12 mouse MAGE genes have been isolated and characterized in various mammalian neoplasms (7). The first member, MAGE-A1, of the MAGE gene family was discovered and characterized as genes encoding melanoma antigens (2). Numerous gene homologues to MAGE-A1 have been identified on the Xq28, Xp21,3 Xq26, and Xp11.23 and are classified as MAGE-A, -B, -C and -D (8-12). The MAGE-C1 gene appears to be located on band Xq26, whereas the MAGE-A and MAGE-B genes are located on Xq28 and Xp21, respectively. Like other MAGE genes, MAGE-C1 is expressed in a significant proportion of neoplastic cells of various histological types, whereas it is silent in normal tissues except testis (11). Previous observations have characterized the MAGE-A subfamily as: (a) they were not present in normal cells except for testis (primitive germ cells, spermatogonia) and placenta (13); and (b) some antigens encoded by the gene family are presented by the human leukocyte antigen (HLA) (14). These results suggest that immunoreaction against MAGE proteins could be expected to affect only neoplastically transformed cells (activation or derepression of normally silent CT genes) but not normal cells because testis and placenta lack HLA expression (15, 16). Hence, experimental vaccination with MAGE-A peptide provoked neoplasm regression in melanoma patients without significant side effects (1). Recently, it was reported that the MAGE-D genes are well conserved between man and mouse, suggesting that these genes have important functions. The COOH-terminal domain of MAGE-D3 is identified to be trophinin, a previously detected protein believed to be involved in embryo implantation (6). The genomic structure of the MAGE-D genes indicate that one of them corresponds to the founder member of the family and that all of the other MAGE genes are retrogenes derived from that common ancestral gene. It is important to emphasize that the MAGE-D genes are universally present in normal tissues and the scientific literature has not classified them as CTAs.

Another CT gene, named BAGE, codes for a putative protein of 43 aa and seems to belong to a family of several genes (3). This antigen was also recognized by autologous CTL consisting of BAGE-encoded peptide

AARAVFLAL bound to an HLA-Cw 1601 molecule. The gene BAGE is expressed in 22% of melanomas, 30% of infiltrating bladder carcinomas, 10% of mammary carcinomas, 8% of head and neck squamous cell carcinomas, and 6% of non-small cell lung carcinomas. Like the MAGE genes, it is silent in normal tissues with the exception of testis.

The members of the GAGE genes are located in the p11.2 to p11.4 region of chromosome X (17). The GAGE-1, GAGE-2, and GAGE-8 code for an immunogenic antigen, the peptide YRPRPRRY, which is presented on various human neoplastic cells by class I, HLA-Cw6 molecules. GAGE-1 and GAGE-2 genes are identified in 24% of melanomas, 25% of sarcomas, 19% of non-small cell lung cancers and head and neck tumors, and 12% of urinary bladder neoplasms (18). GAGE-3, GAGE-4, GAGE-5, GAGE-6, and GAGE-7B code for a different peptide, YYNPRPRRY, which is presented to neoplasm cells by HLA-A29 molecules. Cultured normal and tumor cells treated with demethylating agents (*i.e.* azadeoxycystidine), resulted in a transcriptional activation of the members of GAGE gene family. A demethylation process is possibly responsible for their re-expression in different types of neoplastically transformed cells (17). GAGE-7B is linked to the disease progression of prostate cancer (19).

The PAGE-1 gene is also identified on human X chromosome and demonstrates homology to the members of GAGE gene family (20). PAGE-1 is predominantly expressed in normal and neoplastic prostate cells as well as in normal testis and testicular cancer, fallopian tube and uterus and uterine cancer.

The GAGE/PAGE family includes three newly identified genes, called XAGE subfamily (21). The XAGE-1 gene consists of 4 exons and is located on chromosome Xp11.21-Xp11.22. The full length cDNA contains 611 bp and predicts a protein of 16,300 mol. weight, with a potential transmembrane domain at NH_2 terminus (22). XAGE-1 shares homology with GAGE/PAGE proteins in the COOH terminal. The XAGE-1b antigen, codes for a putative protein of 81 amino acids, harboring a functional bipartite nuclear localization signal and a C-terminal acidic transcription-activation-like domain. XAGE-1b has been shown to have a 50% homology with members of the GAGE family, at least on the nucleotide level (23). XAGE-1 and XAGE-2 genes are detected in Ewing's sarcoma, rhabdomyosarcoma, breast cancer and germ cell tumors.

The SCP-1 (synaptonemal complex protein 1) gene differs from the other members of CT genes by its localization on chromosome 1. It is responsible for coding the HOM-TES-14 antigen. SCP-1 is selectively expressed during the meiotic prophase of the cells of spermatogenesis and is also involved in the pairing of homologous chromosomes, the most significant step in generation of haploid cells. This gene is frequently detected in glial tumors, breast, renal and ovarian cancer (24).

Normal SSX expression has been located only in the testis and, at very low levels, the thyroid. Today these proteins are also considered to be members of the still growing family of CTAs. The SSX (synovial sarcoma on X chromosome) genes, also located on the X chromosome, and encode a family of highly homologous nuclear proteins (25). The SSX1 and SSX2 genes were initially identified as fusion partners of the SYT gene on chromosome 18 and one member of the SSX family (SSX1, SSX2, XXS4) in t(X;18)-positive synovial sarcomas (26-30). The SYT gene is a 5' transcript is mapped to chromosome 18 and encodes a protein, rich in glutamine, proline and glycine (31). In great majority of synovial sarcomas, SYT gene rearrangement produced an SYT-SSX fusion protein. The SYT protein has a unique QPGY domain and associates with native SNF/SWI (global transcriptional co-activator or chromatin remodelling factor) complexes, directly linked to cancerogenesis (32). Both SSX1 and SSX2 genes encode closely related proteins (81% identity) of 188 amino acids that are rich in charged amino acids (33). The N-terminal portion of each SSX protein exhibits homology to the Kruppel-associated box (KRAB), a transcriptional repressor domain previously identified only in Kruppel-type zinc finger protein. The SSX2 gene encodes the HOM-MEL-40 human NA antigen, present in 50% of melanomas, 30% of hepatocellular carcinomas, 25% of colon, and 20% of breast cancer cells (34). The SSX3 gene also maps within Xp11.2 to p11.1 (35). This third member, the SSX3 gene is homologous to SSX1 (90%) and to SSX2 (95%) at the cDNA level, but it is not implicated in the t(x;18)-positive synovial sarcomas and does not act as fusion partner to SYT. The highly homologous SSX4 and SSX5 genes, were identified during immunoscreening of testicular cDNA libraries with allogeneic serum from a melanoma patient. The observations confirmed that all five SSX genes were expressed in normal testis (36).

Recently, a new CT antigen, NY-ESO-1 (one of the ten gene or gene families encoding CTAs), has been identified on the basis of spontaneous antibody responses to NAMs (37). NY-ESO-1 encodes a putative protein of Mr 17.995, having no homology with any previously identified CTAs. The pattern of NY-ESO-1 expression indicates that it belongs to an expanding family of immunogenic testicular antigens that are expressed in a wide variety of human neoplasms, including melanoma, breast, urinary bladder and prostate cancers, and hepatocellular carcinoma (38). NY-ESO-1 appears to be one of the most immunogenic antigens known to date, with spontaneous immune responses observed in 50% of patients with NY-ESO-1 expressing neoplams, inducing both a humoral immune response and antigen specific, cellular, $CD8^+$ cytolytic T lymphocyte immunoreactivity (39-41).

Immunity to NY-ESO-1 is clearly antigen dependent, disappearing with neoplasm removal or regression (42, 43).

In 2001, a new member of the CT genes, named OY-TES-1, was isolated and defined to be the human homologue of proacrosin binding protein sp32 precursor, which was initially found in the mouse, guinea pig, and pig (44). OY-TES-1, containing ten exons, maps to chromosome 12p12-p13, with Southern blot analysis suggesting that there are two OY-TES-1-related genes in the human genome. OY-TES-1 mRNA was expressed only in the testis when normal tissues were observed for its presence. However, OY-TES-1 mRNA was detected in various neoplastically transformed tissues, including bladder, breast, lung, liver, and colon cancers. The research group of Ono and co-investigators (2001) demonstrated through a serological survey of 362 cancer patients the presence of anti-OY-TES-1 antibody in twenty-five of these patients. In healthy individuals (n=20), OY-TES-1 sera reactivity was not identified, which indicates that the OY-TES-1 antigen is immunogenic in the course of human neoplastic disease.

3. DETECTION OF CANCER/TESTIS ANTIGENS IN VARIOUS MALIGNANT NEOPLASMS AND THEIR THERAPEUTIC SIGNIFICANCE

Over the past few years, more than 150 articles have appeared in the medical literature concerning CTAs. It has become obvious that CTAs can be used for the diagnosis and microstaging of human neoplasms, as well as the population screening of smokers to show an increased risk for lung cancer in their own individual lifetimes and by extension, clearly showing the truly damaging effects of smoking. Thus far, the author has never observed others writing about specific antigen families all arrive at the same conclusion, except in the case of CTAs, in which all authors have concluded that these antigens could and should be used for antigen targeted and specific anti-neoplastic immunotherapy.

A study from this year presented observation of the expression of CTAs in cryo-preserved cutaneous T cell lymphoma (CTCL) tissues, including mycosis fungoides, pleomorphic cutaneous T-cell lymphoma, Sezary's syndrome (SS), and a non-malignant entity (small plaques parapsoriasis, SPP). The authors used a panel of eleven CT antigens (MAGE-1, MAGE-C1, MAGE-3, BAGE, GAGE, SSX-1, SSX-2, SSX4, SCP-1, NY-ESO-1 and TS85) (HOM-Tes-85), with mRNA expression for SCP-1 being identified in mycosis fungoides (MF) and pleomorphic CTCL patients but not in the small plaques parapsoriasis. SS patients demonstrated a more heterogeneous antigen expression pattern, including GAGE, MAGE-1, MAGE-3, MAGE-

C1, NY-ESO-1 and TS85. Haffner *et al.* emphasized that CTAs could provide defined targets for antigen-based vaccination and CTA-directed immunotherapy in a high percentage of cases with CTCL. The authors also suggested that SCP-1 antigen expression might serve as an additional diagnostic indicator in early and clinically indistinct lesions suspicious for cutaneous T-cell lymphoma (45).

Another recent article by Mashino and co-workers analyzed the expression of six CT genes including NY-ESO-1, LAGE-1, SCP-1, SSX-1, SSX-2, and SSX-4 in gastrointestinal and breast carcinomas using reverse transcription-polymerase chain reaction (46). Relatively high expression of SCP-1 and SSX-4 was described in gastric carcinoma, LAGE-1 and NY-ESO-1 in esophageal carcinoma, and SCP-1 in breast carcinoma. Frequent synchronous expression with MAGE was identified, including LAGE-1 in esophageal carcinoma, SSX-4 in gastric carcinoma, and SCP-1 in breast carcinoma. Immunocytochemical analysis of the neoplasms expressing both MAGE-4 and NY-ESO-1 genes demonstrated differences in the distribution between MAGE-4 and NY-ESO-1 in serial sections. The authors suggested that NY-ESO-1, LAGE-1, SCP-1 and SSX-4, in addition to MAGE, genes may be promising candidates for NAM directed immunotherapy. The group of Mashino also added that polyvalent cancer vaccines may indeed be useful for cases of gastrointestinal and breast carcinomaswith heterogeneous CTA gene expression profiles.

CTAs are also expressed in a high percentage in hepatocellular carcinomas. In 30 hepatocellular carcinomas, mRNA expression of SSX-1, SSX-2, SSX-4, SSX-5, SCP-1, and NY-ESO-1 has been described, with limited expressions of these genes being detected in few non-neoplastically transformed liver tissues (47).

The expression of CTAs has also been examined in human brain tumors (48). Meningiomas were found to express only HOM-TES-14/SCP-1. SSX-4 was found to be the only CT gene expressed in oligodendrogliomas and it was also expressed in oligoastrocytomas and astrocytomas. Astrocytoma cells proved to be the most positive for HOM-TES-14/SCP-1 and SSX-4, with expression of HOM-TES-85, SSX-2 and MAGE-3 also defined in these tumors. MAGE-3 was detected only in grade IV astrocytomas, while the expression of the other CT genes showed no clear correlation with histological grade. Sixty percent of astrocytomas analyzed were found to express at least one CT gene, 21% expressed two CT genes, and 8% coexpressed three CT genes. These authors, like many others in relation to other neoplasms, concluded that a majority of oligoastrocytomas and astrocytomas might be amenable to CTA directed, specific immunotherapeutic interventions.

Synovial sarcomas are high-grade, extremely malignant mesenchymal neoplasms with two, biphasic (BSS) and monophasic (MSS) variants that carry a pathognomonic cytogenetic alteration, t(X;18), involving the SYT gene on chromosome 18 and one of several SSX genes on chromosome X, usually SSX1 or SSX2, a translocation characteristically involving t(X; 18) to (p11.2; q11.2) between SYT and SXX [31]. A recent study analyzed the expression of three CTAs, NY-ESO-1, MAGE-A1 and MAGE-C1 (CT7), through the use of immunocytochemistry, employing three MoABs, ES121 (anti-NY-ESO-1), MA454 (anti-MAGE-A1) and CT7-33 (anti-CT7), in synovial sarcomas (49). The results suggested that NY-ESO-1 is highly expressed in a homogeneous pattern in synovial sarcomas of both morphologic variants and both translocation types, making these neoplasms an attractive target for NY-ESO-1 antigen-based, and directed immunotherapy.

In the search for biomarkers for early lung cancer detection and possible receptor targeted immunotherapy, the frequencies of the expressional activation of MAGE-A1, MAGE-A3, and MAGE-B2 genes in non-small cell lung cancers (NSCLCs) were observed by Jang and co-workers (50). Expression of these genes was evaluated by reverse transcription-PCR (RT-PCR) in 20 primary NSCLC samples, with corresponding normal lung tissues, and in 20 bronchial brush specimens from former smokers without lung cancer. mRNA *in situ* hybridization was done to confirm the gene expression pattern at the cellular level. Among the twenty primary NSCLC samples analyzed, 14 expressed MAGE-A1 and 17 expressed both MAGE-A3 and MAGE-B2 while a substantial number of normal lung tissues adjacent to NSCLC also had detectable levels of MAGE expression, specifically 65% for MAGE-A1, 75% for MAGE-A3, and 80% for MAGE-B2. The activation of MAGE-A1, MAGE-A3, and MAGE-B2 genes was found to be common not only in NSCLC, but also in bronchial epithelium with severe carcinogenic insult. I agree with the authors that these results suggest that MAGE genes may be activated during the earliest stages of lung carcinogenesis, but more significantly, I feel that these genes should not only be considered as specific targets for lung cancer therapy, but that lung cancer prevention could be carried out *via* the population screening of smokers.

Since it has been both demonstrated and is now common knowledge that the MAGE genes are expressed in a wide variety of human neoplasms but only in the mitotic spermatogonia (germ cells) and in the primary spermatocytes in the normal testis, Aubry and co-authors chose to examine the expression of MAGE-A4 in a panel of testicular germ cell tumors (13). Classical seminomas uniformly and specifically expressed MAGE-A4 while anaplastic seminomas and nonseminomatous germ cell tumors (NSGCTs) were negative for this antigen. MAGE-A4 was also present in the fetal precursors of the stem germ cells from 17 weeks of gestation onward, which is in agreement with the fact that carcinomas *in situ* (CIS) can arise from prespermatogonia in the fetus. These results reveal that the MAGE-A4 antigen can be considered a specific marker for normal premeiotic germ cells and germ cell neoplasms and can in fact be employed as immunocytochemical markers to characterize classical seminomas.

The MAGE-3 gene is present in a variety of neoplasms, including lung cancer, but not in normal tissues except for the testis and placenta (51). The scientific aim of a recent observation by Eifuku and co-workers was to clarify whether HLA-A2 restricted MAGE-3 peptide (FLWGPRALV) was a lung cancer associated antigen recognizable by cytotoxic, $CD8^+$ T lymphocytes (CTL). MAGE-3-derived peptide-specific CTL were induced from the peripheral blood mononuclear cells (PBMC) of HLA-A0201-positive healthy donors and the regional lymph node lymphocytes (RLNL) of HLA-A2-positive patients with lung cancer by multiple stimulations with peptide-pulsed HLA-A0201-positive antigen-presenting cells. The lymphocytes stimulated with MAGE-3 peptide produced an antigen directed, specific lysis of Epstein-Barr virus-transformed B cells (EBV-B) pulsing with MAGE-3 peptide. Stronger and specific activity for MAGE-3-presenting targets was found after the second antigenic stimulation, and the activity increased with every repeated stimulation. The peptide-specific activity was inhibited by the addition of MoABs against MHC class I and HLA-A2. Such CTL also recognized established tumor cell lines expressing both HLA-A2 and MAGE-3 in an MHC class I-restricted manner, but did not recognize tumor cell lines that did not express HLA-A2 or MAGE-3. These results suggest that the MAGE-3 peptide could be used as a potential target for antigen directed, specific immunotherapy for HLA-A2 patients with lung cancer (51).

As neoplastically transformed cells spread beyond their primary site, they undergo changes in their gene expression that may be detectable and useful for microstaging of neoplastic disease (52). The CT antigens like MAGE 1-3, NY-ESO-1, SSX 1-5, and others are potential NAMs for microstaging melanoma. One CTA, CTp11, was shown to be expressed by metastasizing melanoma cell lines but not by nonmetastasizing variants. It was found that CTp11 tended to be expressed by primary melanomas. There was a statistically significant difference in the distribution of the expression of CTp11 and NY-ESO-1 in melanoma from different stages of progression. The authors concluded that NY-ESO-1 may be a prognostically significant CTA, detecting of more advanced neoplastic disease and CTp11 of less advanced disease.

Today it is well known that HLA-A*0201 melanoma patients can frequently develop a CTL response to the CT antigen NY-ESO-1. In a study by Romero *et al.*, the relative antigenicity and *in vitro* immunogenic effectivity of natural and modified NY-ESO-1 peptide sequences was analyzed (53). The results revealed that, although suboptimal for binding to the HLA-A*0201 molecule, peptide NY-ESO-1 157-165 is, among natural sequences, very efficiently recognized by specific CTL clones derived from three melanoma patients. In contrast, peptides NY-ESO-1 157-167 and NY-ESO-1 155-163, which bind very strongly to HLA-A*0201, are recognized less efficiently. Substitution of peptide NY-ESO-1 157-165 COOH-terminal C with a number of other amino acids resulted in a significantly increased binding to HLA-A*0201 molecules as well as in an increased CTL recognition, although variable at the clonal level. Among natural peptides, NY-ESO-1 157-165 and NY-ESO-1 157-167 exhibited good *in vitro* immunogenicity, whereas peptide NY-ESO-1 155-163 was poorly immunogenic. The fine specificity of interaction between peptide NY-ESO-1 C165A, HLA-A*0201, and T lymphocyte receptor was also analyzed at the molecular level employing a series of variant peptides containing single alanine substitutions. I agree with the authors that the results submitted in this article will have significant implications for the further development of NY-ESO-1 peptide based vaccines and for the monitoring of either natural or vaccine-induced NY-ESO-1 specific CTL responses in neoplasm patients.

The identification and characterization of L552S, an over-expressed, alternatively spliced isoform of XAGE-1 was reported in lung adenocarcinoma (54). Genomic sequence analysis has revealed that L552S and XAGE-1 are alternatively spliced isoforms, and expression of both L552S and XAGE-1 isoforms are present in lung adenocarcinoma. L552S was shown to be expressed at levels greater than 10-fold in lung adenocarcinomas compared with the highest expression level found in all normal tissues tested. L552S was present in both early and late stages of lung adenocarcinoma development, but it was not detected in large cell carcinoma, small cell carcinoma, or atypical lung neuroendocrine carcinoid. Immunocytochemical analysis employing affinity purified L552S polyclonal antibodies demonstrated specific nuclear staining in lung adenocarcinoma samples. Furthermore, antibody responses to recombinant L552S protein were observed in lung pleural effusion fluids of lung cancer patients. These results strongly imply that the L552S protein is a strong immunogen and the authors suggested that it could be employed as a vaccine target for lung cancer (54).

In recent experiments of suppression subtractive hybridization, the comparison of mRNA expression profiles of common nevocellular nevi and melanoma metastases has been employed to identify new potential markers of melanoma progression (23). The metastatic melanoma tissues contained XAGE-1b. Expression of XAGE-1b in normal tissues was mainly restricted to testis, while placenta and brain were sporadically positive. In general, expression of XAGE-1b was much more prominent than expression of the longer XAGE-1 transcript, isolated from Ewing's sarcoma. In the different stages of melanocytic tumor progression, expression was exclusively seen in melanoma metastases, while all tested common and atypical nevi as well as primary melanomas were negative. Upregulation of receptor expression after treatment with demethylating agent 5-aza-2'-deoxycytidine was identified in only one of four human melanoma cell lines tested. After transfection of XAGE-1 gene into COS cells, the corresponding protein could direct the coupled fluorescent protein to the nucleus, showing a distinct speckled staining aspect. These experimental data imply the nuclear CT associated XAGE-1b to be a marker for late melanocytic tumor progression (23).

An extensive review of the expression of the SSX genes in a variety of human tumors has been provided by Tureci and co-workers (55). The authors assessed the expression of SSX-1, -2, -3, -4 and -5 using RT-PCR in a sample of 325 tumors of various histological origins. The expression of at least one of the SSX family members was found to be most frequent in cancers of the head and neck, followed by ovarian cancer, malignant melanoma, lymphoma, colorectal cancer and breast cancer while some leukemias, leiomyosarcomas, seminomas and thyroid cancers were found not to express any SSX gene. These findings further illustrate the amenability of CTAs for use in the targeting of various antigen specific immunotherapeutical approaches.

Extensive immunocytochemical observations performed by me have confirmed the expression of the MAGE family of peptides in numerous neoplastically transformed tissues including breast and lung carcinomas, malignant melanomas and childhood astrocytomas. Of particular significance is the absence of these markers on the surface of neoplastically non-transformed cells in normal tissues examined employinging the same protocol, further substantiating the prevailing conclusions in regards to the specificity of the CTAs to genetically dedifferentiated populations of neoplastic cells and the distinct and very enticing possibility for their targeting in various immunotherapeutical approaches.

4. CONCLUSIONS

- Cancer/testis-antigens (CTAs) are a novel and expanding family of a distinct class of early differentiation antigens and also potent immunogenic proteins detected by serological screening of recombinant cDNA expression libraries.

- In the last years, following the mapping of the MAGE genes, MAGE peptides are already being employed for CT antigen directed anti-neoplastic immunotherapy. MAGE proteins affect only neoplastically transformed cells (activation or derepression of normally silent CT genes) but not normal cells, because testis and placenta lack HLA expression. This expression pattern in various neoplasms might contribute to the genetic instability of neoplastically transformed cells.
- Specifically, the newly detected CT antigen NY-ESO-1, is regarded as one of the most immunogenic antigens (proteins) ever isolated, inducing spontaneous host immune responses in 50% of patients with NY-ESO-1-expressing neoplasms. Numerous articles clearly stated that NY-ESO-1 specific, cytolytic, $CD8^+$ T lymphocyte responses can be induced by intradermal immunization with NY-ESO-1 peptides, and that immunization with NY-ESO-1 may have the potential significance to alter the natural course of NY-ESO-1-expressing human neoplasms.
- Immunotherapeutical treatments employing NA peptides derived from CT antigens already have been initiated to analyze the induction of antigen specific CTL responses *in vivo*. Immunological and clinical parameters for the assessment of peptide-specific reactions have been defined, *i.e.*, delayed-type hypersensitivity (DTH), CTL, autoimmune, and neoplasm regression responses. Early results show that NA peptides alone induce specific DTH and CTL responses and neoplasm regression after intradermal administration. Adjuvants (*i.e.* GM-CSF) were added to enhance peptide-specific immune reactions by amplification of dermal antigen (peptides) presenting dendritic cells. Complete tumor regressions have been noticed also, but only in the context of measurable peptide-specific CTL.
- However, in numerous cases with systematic neoplastic disease progression after an initial anti-tumor response, either a loss of the respective NAA targeted by CTL or of the presenting MHC class I allele was detected, suggesting an immunization-induced immune escape. These immunoselective reactions can be prevented employing antigen modifying cytokines.
- I strongly recommend further clinical trials with antigenic constructs and antigen presenting cells to induce both humoral and cellular immune responses. These studies will determine whether the targeting of CTAs may represent a more effective way for antigen specific and directed, individualized immunotherapeutical treatment of human neoplasms.

REFERENCES

1. Boon T, Old LJ: Tumor antigens. Curr Opin Immunol *9:* 681-683, 1997.
2. Bruggen P van der, Traversari C, Chomez P, Lurquin C, DePlaen E, Eynde B van den, Knuth A, Boon T: A gene encoding an antigen recognized by cytolytic T lymphocytes on a human melanoma. Science *254:* 1643-1647, 1991.
3. Boel P, Wildmann C, Sensi ML, Brasseur R, Renauld JC, Coulie P, Boon T, van der Bruggen P: BAGE: a new gene encoding an antigen recognized on human melanomas by cytolytic T lymphocytes. Immunity *2:* 167-175, 1995.
4. Ma Z, Khatlani TS, Li L, Sasaki K, Okuda M, Inokuma H, Onishi T: Molecular cloning and expression analysis of feline melanoma antigen (MAGE) obtained from a lymphoma cell line. Vet Immunol Immunopathol *83(3-4):* 241-252, 2001.
5. Weiser TS, Ohnmacht GA, Guo ZS, Fischette MR, Chen GA, Hong JA, Nguyen DM, Schrump DS: Induction of MAGE-3 expression in lung and esophageal cancer cells. Ann Thorac Surg *71:* 295-301; discussion 301-302, 2001.
6. Chomez P, De Backer O, Bertrand M, De Plaen E, Boon T, Lucas S: An overview of the MAGE gene family with the identification of all human members of the family. Cancer Res *61:* 5544-5551, 2001.
7. Ohman Forslund K, Nordqvist K: The melanoma antigen genes--any clues to their functions in normal tissues? Exp Cell Res *265:* 185-194, 2001.
8. De Plaen E, Arden K, Traversari C, Gaforio JJ, Szikora JP, De Smet C, Brasseur F, van der Bruggen P, Lethe B, Lurquin C: Structure, chromosomal localization, and expression of 12 genes of the MAGE family. Immunogenetics *40:* 360-369, 1994.
9. Rogner UC, Wilke K, Steck E, Korn B, Poutska A: The melanoma antigen (MAGE) family is clustered in the chromosomal band Xq28. Genomics *29:* 725-731, 1995.
10. Lurquin C, De Smet C, Brasseur F, Muscatelli F, Martelange V, De Plaen E, Brasseur R, Monaco AP, Boon T: Two members of the human MAGEB gene family loacted in Xp21.3 are expressed in tumors of various histological origins. Genomics *46:* 397-408, 1997.
11. Lucas S, De Smet C, Arden KC, Viars CS, Lethe B, Lurquin C, Boon T: Identification of a new MAGE gene with tumor-specific expression by representational difference analysis. Cancer Res *58:* 743-752, 1998.
12. Pold M, Zhou J, Chen GL, Hall JM, Vescio RA, Berenson JR: Identification of a new, unorthodox member of the MAGE gene family. Genomics *59:* 161-167, 1999.
13. Aubry F, Satie AP, Rioux-Leclercq N, Rajpert-De Meyts E, Spagnoli GC, Chomez P, De Backer O, Jegou B, Samson M: MAGE-A4, a germ cell specific marker, is expressed differentially in testicular tumors. Cancer *92:* 2778-2785, 2001.
14. Gillespie AM, Coleman RE: The potential of melanoma antigen expression in cancer therapy. Cancer Treat Rev *25:* 219-227, 1999.
15. Chen Y-T, Old LJ: Cancer-testis antigens: targets for cancer immunotherapy. Cancer J from Scientific American *5:* 16-17, 1999.
16. Marchand M, van Baren N, Weynants P, Brichard V, Dreno B, Tessier MH, Rankin E, Parmiani G, Arienti F, Humblet Y, Bourlond A, Vanwijck R, Lienard D, Beauduin M, Dietrich PY, Russo V, Kerger J, Masucci G, Jager E, De Greve J, Atzpodien J, Brasseur F, Coulie PG, van der Bruggen P, Boon T: Tumor regressions observed in patients with metastatic melanoma treated with an antigenic peptide encoded by gene MAGE-3 and presented by HLA-A1. Int J Cancer *80:* 219-230, 1999.
17. De Backer O, Arden KC, Boretti M, Vantomme V, De Smet C, Czekay S, Viars CS, De Plaen E, Brasseur F, Chomez P, Van den Eynde B, Boon T, van der Bruggen P:

Characterization of the GAGE genes that are expressed in various human cancers and in normal testis. Cancer Res *59:* 3157-3165, 1999.

18. Van den Eynde B, Peeters O, De Backer O, Gaugler B, Lucas S, Boon T: A new family of genes coding for an antigen recognized by autologous cytolytic T lymphocytes. J Exp Med *182:* 689-698, 1995.
19. Chen ME, Lin SH, Chung LW, Sikes RA: Isolation and characterization of PAGE-1 and GAGE-7. New genes expressed in the LNCaP prostate cancer progression model that share homology with melanoma-associated antigens. J Biol Chem *273:* 17618-17625, 1998.
20. Brinkmann U, Vasmatzis G, Lee B, Yerushalmi N, Essand M, Pastan I: PAGE-1, an X chromosoma-linked GAGE-like gene that is expressed in normal and neoplastic prostata, testis and uterus. Proc Natl Acad Sci USA *95:* 10757-10762, 1998.
21. Brinkmann U, Vasmatzis G, Lee B, Pastan I: Novel genes in the PAGE and GAGE family of tumor antigens found by homology walking in the dbEST database. Cancer Res *59:* 1445-1448, 1999.
22. Liu XF, Helman LJ, Yeung C, Bera TK, Lee B, Pastan I: XAGE-1, a new gene that is frequently expressed in Ewing's sarcoma. Cancer Res *60:* 4752-4755, 2000.
23. Zendman AJ, van Kraats AA, den Hollander AI, Weidle UH, Ruiter DJ, van Muijen GN: Characterization of XAGE-1b, a short major transcript of cancer/testis-associated gene XAGE-1, induced in melanoma metastasis. Int J Cancer *97:* 195-204, 2002.
24. Tureci O, Sahin U, Zwick C, Koslowski M, Seitz G, Pfreundschuh M: Identification of a meiosis-specific protein as a member of the class of cancer/testis antigens. Proc Natl Acad Sci USA *95:* 5211-5216, 1998.
25. dos Santos NR, Torensma R, de Vries TJ, Schreurs MW, de Bruijn DR, Kater-Baats E, Ruiter DJ, Adema GJ, van Muijen GN, van Kessel AG: Heterogeneous expression of the SSX cancer/testis antigens in human melanoma lesions and cell lines. Cancer Res *60:* 1654-1662, 2000.
26. Fisher C: Synovial sarcoma. Ann Diagn Pathol *2:* 401-421, 1998.
27. Hiraga H, Nojima T, Abe S, Sawa H, Yamashiro K, Yamawaki S, Kaneda K, Nagashima K: Diagnosis of synovial sarcoma with reverse transcriptase-polymerase chain reaction: analyses of 84 soft tissue and bone tumors. Diagn Mol Pathol *7:* 102-110, 1998.
28. Tsuji S, Hisaoka M, Morimitsu Y, Hashimoto H, Shimajiri S, Komiya S, Ushijima M, Nakamura T: Detection of SYT-SSX fusion transcripts in synovial sarcoma by reverse transcription-polymerase chain reaction using archival paraffin-embedded tissues. Am J Pathol *153:* 1807-1812, 1998.
29. Argani P, Askin FB, Colombani P, Perlman EJ: Occult pulmonary synovial sarcoma confirmed by molecular techniques. Pediatr Dev Pathol *3:* 87-90, 2000.
30. Tamborini E, Agus V, Mezzelani A, Riva C, Sozzi G, Azzarelli A, Pierotti MA, Pilotti S: Identification of a novel spliced variant of the SYT gene expressed in normal tissues and in synovial sarcoma. Br J Cancer *84:* 1087-1094, 2001.
31. Clark J, Rocques PJ, Crew AJ, Gill S, Shipley J, Chan AM, Gusterson BA, Cooper CS: Identification of novel genes, SYT and SSX, involved in the t(x;18) (p11.2;q11.2) translocation found in human synovial sarcoma. Nat Genet *7:* 502-508, 1994.
32. Kato H, Tjernberg A, Zhang W, Krutchinsky AN, An W, Takeuchi T, Ohtsuki Y, Sugano S, de Bruijn DR, Chait BT, Roeder RG: SYT associates with human SNF/SWI complexes and the C-terminal region of its fusion partner SSX1 targets histones. J Biol Chem *277:* 5498-5505, 2002.
33. Crew AJ, Clark J, Fisher C, Gill S, Grimer R, Chand A, Shipley J, Gusterson BA, Cooper CS: Fusion of SYT to two genes, SSX1 and SSX2, encoding proteins with

homology to the Kruppel-associated box in human synovial sarcoma. EMBO J *14:* 2333-2340, 1995.
34. Tureci O, Sahin U, Schobert I, Koslowski M, Scmitt H, Schild HJ, Stenner F, Seitz G, Rammensee HG, Pfreundschuh M: The SSX-2 gene, which is involved in the t(x;18) translocation of synovial sarcomas, codes for the human tumor antigen HOM-MEL-40. Cancer Res *56:* 4766-4772, 1996.
35. De Leeuw B, Balemans M, Geurts van Kessel A: A novel Kruppel-associated box containing the SSX gene (SSX3) on the human X chromosome is not implicated in t(x;18)-positive synovial sarcomas. Cytogenet Cell Genet *73:* 179-183, 1996.
36. Gure AO, Tureci O, Sahin U, Tsang S, Scanlan MJ, Jager E, Knuth A, Pfreundschuh M, Old LJ, Chen YT: SSX: a multigene family with several members transcribed in normal testis and human cancer. Int J Cancer *72:* 965-971, 1997.
37. Lee L, Wang RF, Wang X, Mixon A, Johnson BE, Rosenberg SA, Schrump DS: NY-ESO-1 may be a potential target for lung cancer immunotherapy. Cancer J Sci Am *5:* 20-25, 1999.
38. Chen YT, Scanlan MJ, Sahin U, Tureci O, Gure AO, Tsang S, Williamson B, Stockert E, Pfreundschuh M, Old LJ: A testicular antigen aberrantly expressed in human cancers detected by autologous antibody screening. Proc Natl Acad Sci U S A *94:* 1914-1918, 1997.
39. Stockert E, Jäger E, Chen Y-T, Scanlan MJ, Gout I, Karbach J, Arand M, Knuth A, Old LJ: A survey of the humoral immune response of cancer patients to a panel of human tumor antigens. J Exp Med *187:* 1349-1354, 1998.
40. Jäger E, Nagata Y, Gnjatic S, Wada H, Stockert E, Karbach J, Dunbar PR, Lee SY, Jungbluth A, Jäger D, Arand M, Ritter G, Cerundolo V, Dupont B, Chen Y-T, Old LJ, Knuth A: Monitoring CD8 T cell responses to NY-ESO-1: correlation of humoral and cellular immune responses. Proc Natl Acad Sci USA *97:* 4760-4765, 2000.
41. Jäger D, Jäger E, Knuth A: Immune responses to tumour antigens: implications for antigen specific immunotherapy of cancer. J Clin Pathol *54:* 669-674, 2001.
42. Jäger E, Stockert E, Zidianakis Z, Chen Y-T, Karbach J, Jager D, Arand M, Ritter G, Old LJ, Knuth A: Humoral immune responses of cancer patients against "cancer-testis" antigen NY-ESO-1: correlation with clinical events. Int J Cancer *84:* 506-510, 1999.
43. Kurashige T, Noguchi Y, Saika T, Ono T, Nagata Y, Jungbluth A, Ritter G, Chen Y-T, Stockert E, Tsushima T, Kumon H, Old LJ, Nakayama E: NY-ESO-1 expression and immunogeneicity associated with transitional cell carcinoma: correlation with tumor grade. Cancer Res *61:* 4671-4674, 2001.
44. Ono T, Kurashige T, Harada N: Identification of proacrosin binding protein sp32 precursor as a human cancer/testis antigen. Proc Natl Acad Sci USA *98:* 3282-3287, 2001.
45. Haffner AC, Tassis A, Zepter K, Storz M, Tureci O, Burg G, Nestle FO: Expression of cancer/testis antigens in cutaneous T cell lymphomas. Int J Cancer *97:* 668-670, 2002.
46. Mashino K, Sadanaga N, Tanaka F, Yamaguchi H, Nagashima H, Inoue H, Sugimachi K, Mori M: Expression of multiple cancer-testis antigen genes in gastrointestinal and breast carcinomas. Br J Cancer *85:* 713-720, 2001.
47. Chen CH, Chen GJ, Lee HS, Huang GT, Yang PM, Tsai LJ, Chen DS, Sheu JC: Expressions of cancer-testis antigens in human hepatocellular carcinomas. Cancer Lett *164:* 189-195, 2001.
48. Sahin U, Koslowski M, Tureci O, Eberle T, Zwick C, Romeike B, Moringlane JR, Schwechheimer K, Feiden W, Pfreundschuh M: Expression of cancer testis genes in human brain tumors. Clin Cancer Res *6:* 3916-3922, 2000.

49. Jungbluth AA, Antonescu CR, Busam KJ, Iversen K, Kolb D, Coplan K, Chen Y-T, Stockert E, Ladanyi M, Old LJ: Monophasic and biphasic synovial sarcomas abundantly express cancer/testis antigen NY-ESO-1 but not MAGE-A1 or CT7. Int J Cancer *94:* 252-256, 2001.
50. Jang SJ, Soria JC, Wang L, Hassan KA, Morice RC, Walsh GL, Hong WK, Mao L: Activation of melanoma antigen tumor antigens occurs early in lung carcinogenesis. Cancer Res *61:* 7959-7963, 2001.
51. Eifuku R, Takenoyama M, Yoshino I, Imahayashi S, So T, Yasuda M, Sugaya M, Yasumoto K: Analysis of MAGE-3 derived synthetic peptide as a human lung cancer antigen recognized by cytotoxic T lymphocytes. Int J Clin Oncol *6:* 34-39, 2001
52. Goydos JS, Patel M, Shih W: NY-ESO-1 and CTp11 expression may correlate with stage of progression in melanoma. J Surg Res *98:* 76-80, 2001.
53. Romero P, Dutoit V, Rubio-Godoy V, Lienard D, Speiser D, Guillaume P, Servis K, Rimoldi D, Cerottini JC, Valmori D: $CD8^+$ T-cell response to NY-ESO-1: relative antigenicity and in vitro immunogenicity of natural and analogue sequences. Clin Cancer Res *7:* 766s-772s, 2001.
54. Wang T, Fan L, Watanabe Y, McNeill P, Fanger GR, Persing DH, Reed SG: L552S, an alternatively spliced isoform of XAGE-1, is over-expressed in lung adenocarcinoma. Oncogene *20:* 7699-7709, 2001.
55. Tureci O, Chen Y-T, Sahin U, Gure AO, Zwick C, Villena C, Tsang S, Seitz G, Old LJ, Pfreundschuh M: Expression of SSX genes in human tumors. Int J Cancer *77:* 19-23, 1998.

PROLOGUE

"It is not enough to have a good mind, the main thing is to use it well."
- Descartes (1596-1650)

After everything said in this book, what is the most advanced anti-neoplastic therapy at the beginning of the twenty-first century? While the answer to this question may not be clear, it is obvious that cancer immunotherapy is still not accepted as a common modality and is still in essence viewed as "experimental." Nonetheless, immunotherapy is step by step taking its rightful place as the fourth modality in the fight against cancer. Incredible advances have been in the field recently, indicating that the medical community is more and more accepting of the idea that immunotherapy can only strengthen and increase the options available in not only treating but also preventing the disease. The countless articles and publications in this field are indicative of the vitality of this field of study for years to come. A worldwide, unified concept and effort does, however, need to exist. Different nations respond to these "alternative" therapies in different ways, but such options should be made available to all. The worldwide medical community has, for example, decided to fight AIDS on a global level and as such, has had many successes in the battle. Such efforts need to be extended in the fight against cancer to allow for the flow of ideas and research and thereby, help as many people suffering from cancer as much as possible.

Current research has undertaken many endeavors to finding a cure and even preventing the disease. For example, newly invented glucuronide prodrugs have shown promising efficacy in anti-neoplastic therapy due to their increased specificity and reduced systemic toxicity (1). Prodrugs can be employed in prodrug monotherapy (PMT), which is based on the idea that neoplasms demonstrate elevated β-glucuronidase activity. β-Glucuronidase is capable of activating the low-toxic prodrugs into highly cytotoxic agents, specifically in the neoplastic cell site. The antigenic specificity of the prodrugs can be further improved by combined employment of MoABs directed against tumor-specific antigens, namely antibody-directed enzyme prodrug therapy (ADEPT). The potency of the prodrugs can be enormously enhanced with the incorporation of an appropriate radionuclide in the combined chemo- and radio-therapy of cancer (CCRTC) strategy. The prodrugs can also be utilized to modify liposomes for efficient delivery of anti-neoplastic drugs.

Hepatocellular carcinoma (HCC) is one of the most common human malignancies in the world, responsible for an estimated one million deaths annually (2). It has a poor prognosis due to its rapid infiltrating *in situ* growth and the complications of secondary liver cirrhosis. Surgical resection, liver transplantation and cryosurgery are considered the best curative options, achieving a high rate of complete response, especially in patients with early stage and small in size HCC, plus a good residual liver function. The so-called regional interventional therapies have led to a major breakthrough in the management of unresectable HCC, which include transarterial chemoembolization (TACE), percutaneous ethanol injection (PEI), radiofrequency ablation (RFA), microwave coagulation therapy (MCT), laser-induced thermotherapy (LITT), etc. As a result of the technical development of locoregional approaches for HCC during the last two decades, the range of combined interventional therapies has been continuously extended. Most combined multimodal interventional therapies reveal their enormous advantages as compared with any mono-therapeutic regimen alone, and their significance is growing rapidly in treating unresectable HCC.

RC represents about 3% of all adult tumors, with an estimate of 31,900 new cases diagnosed in 2003 in the United States. In the earliest stage of its progression, RC is potentially curable by surgery, but if the disease presents any signs of metastasis, the chances of survival are minimal, even though in some, rare cases long survival have been reported (3). In fact, the treatment of metastatic RC remains unsatisfactory. Systemic treatment with single agents and with polychemotherapy, with or without cytokine-based immunotherapy, has not been successful, obtaining very low clinical response rates without a significant benefit in prognosis and overall survival. New approaches such as employment of MoABs, a variety of vaccine development, gene therapy, angiogenesis inhibitors and allogeneic cell transplantation and their possible clinical applications are still under discussion.

Systemic therapy for metastatic colorectal cancer (CC) is developing rapidly after many years without significant change. The employment of standard adjuvant therapy for rectal cancer, on the other hand, remains much the same as it has been since the early 1990s. Recent newly developed chemotherapeutics added to the treatment of CC include capecitabine, irinotecan, and oxaliplatin (4). Use of these agents in metastatic CC and potential for use in localized rectal cancer are discussed in this recently published article. In addition to these "classical" chemotherapeutic agents, there are a number of exciting, individually designed immunotherapies to target specific molecular biologic features of malignant cellular IP. These agents are now being tested in CC, in addition to chemotherapy and to radiotherapy, with the hope that improvements in outcome will be made without substantial increases in treatment toxicity.

Over the past several decades, improvements in chemotherapeutic agents and supportive care have resulted in significant progress in treating patients with acute myeloid leukemia (AML). Research advances in understanding the biology of AML have resulted in the identification of new, molecular biological therapeutic targets. The success of all-trans-retinoic acid in acute promyelocytic leukemia and of imatinib mesylate in chronic myeloid leukemia have demonstrated that the targeted biological therapy may be more effective and less toxic when well defined targets are available. At the same time, understanding mechanisms of multiple drug resistance and the possibilities to overcome them has led to newly developed changes of the structure of some of the existing cytotoxic agents. Rational design and conduct of clinical trials is necessary to ensure that the full potential of these new agents is realized.

Malignant melanoma (MM) is a neoplasm with an increasing tendency of incidence in the US that is rising at a rate second only to lung cancer in women (5). Early stage MM is curable, but in advanced metastatic stage is almost uniformly fatal, even at the beginning of the twenty-first century. Despite the evaluation of a variety of chemotherapy and immunotherapy drugs, the median survival in metastatic MM remains in the range of 6 to 9 months. The close relationship of MM with the immune system has led to a recent resurgence in the observation of immunotherapy in the treatment of this disease.The two most widely employed immunotherapy drugs for MM are IFN-α and IL-2. The role of IFNα-2b in the adjuvant therapy of patients with localized MM at high risk for relapse has recently been established by the results of three large randomized trials conducted by the US Intergroup; all three trials demonstrated an improvement in relapse-free survival and two in overall survival. Administration of rIL-2 has an overall response rate of 15-20% in metastatic MM and is capable of producing complete and durable remissions in about 6% of patients treated. Based upon these data, the US FDA has recently approved the use of high-dose bolus administration of rIL-2 for the therapy of patients with metastatic MM. Results of combination chemotherapy and immunotherapy regimens containing rIL-2 and IFNα (biochemotherapy) are promising, but conclusions regarding an advantage for this therapy in terms of survival must await the completion of ongoing randomized trials. Melanoma cell antigens are among the most immunogenic TAAs. The development of therapeutic vaccines is an ongoing area of research, and clinical trials of several types of vaccines (whole cell, carbohydrate, peptide) are being conducted in patients with intermediate and late-stage MM. In the setting of adjuvant therapy, to date, no vaccine has demonstrated a survival benefit in comparison with either observation or IFNα. Vaccines are also being tested in patients with metastatic MM to determine their immune effects and to define their activity in combination with other immunotherapeutic agents such as IL-2 or IFNα. We suggest that only autologous and highly individualized vaccines will be effective in the treatment of metastatic MM.

The relative ineffectiveness of current therapies for high grade malignant gliomas has led to the need for novel therapeutics. Immunotherapies based on biologic response modifiers (BRMs) are among a variety of anti-neoplastic treatments currently in use or under experimental evaluation and have shown great promise, especially since several potent stimulators of the immune system have been cloned and are now available in purified form and unlimited quantity for clinical employment (6). Early attempts at glioma therapy based on the employment of BRMs, however, have failed to demonstrate significant effectiveness. Despite limited clinical success, we could suggest that an advanced development in understanding of the molecular biology and tumorimmunology of brain tumor cells will pave the way for more effective, individualized immunotherapeutical regiments.

As was mentioned above, adenoviruses have been critical in the development of the molecular approaches to brain tumors (7). They have been engineered to function as vectors for delivering therapeutic genes in gene therapy strategies, and as direct cytotoxic agents in oncolytic viral therapies. Clinical trials of adenovirus-mediated p53 gene and two conditionally replicative adenoviruses (CRAds) ONYX-015 and Delta 24 are suggested as new immunotherapeutic approches in brain tumors.

Gene therapy using viral vectors for the treatment of primary brain tumors has proven to be a promising novel biologic therapeautic modality (8). Much effort in the past has been placed on utilizing replication-defective viruses to this end but they have shown many disadvantages. Much recent attention has been focused on the potential of replication competent viruses to discriminatingly target, replicate within, and destroy tumor cells *via* oncolysis, leaving adjacent post-mitotic neurons unharmed. The engineered tumor cell selective herpes simplex-1 virus (HSV-1) mutants G207 and HSV1716 have already completed Phase I in the treatment of recurrent high-grade glioma. This recent article also provides information about the manipulation and development of other viruses for the treatment of malignant glioma, including Newcastle disease virus, reovirus, poliovirus, vaccinia virus, and adenoviruses, in particular the adenovirus mutant ONYX-015.

REFERENCES

1. Chen X, Wu B, Wang PG: Glucuronides in anti-cancer therapy. Curr Med Chem Anti-Canc Agents *3:* 139-150, 2003.
2. Qian J, Feng GS, Vogl TP: Combined interventional therapies of hepatocellular carcinoma. World J Gastroenterol *9:* 1885-1891, 2003.
3. Gattinoni L, Alu M, Ferrari L, Nova P, Del Vecchio M, Procopio G, Laudani A, Agostara B, Bajetta E: Renal cancer treatment: a review of the literature. Tumori *89:* 476-484, 2003.
4. O'Neil BH: Systemic therapy for colorectal cancer: focus on newer chemotherapy and novel agents. Semin Radiat Oncol *13:* 441-453, 2003.
5. Agarwala S: Improving survival in patients with high-risk and metastatic melanoma: immunotherapy leads the way. Am J Clin Dermatol *4:* 333-346, 2003.
6. Marras C, Mendola C, Legnani FG, DiMeco F: Immunotherapy and biological modifiers for the treatment of malignant brain tumors. Curr Opin Oncol *15(3):* 204-208, 2003.
7. Vecil GG, Lang FF: Clinical trials of adenoviruses in brain tumors: a review of Ad-p53 and oncolytic adenoviruses.J Neurooncol *65:* 237-246, 2003.
8. Shah AC, Benos D, Gillespie GY, Markert JM: Oncolytic viruses: clinical applications as vectors for the treatment of malignant gliomas. J Neurooncol *65:* 203-226, 2003.

III. APPENDIX

Chapter 10

MATERIALS AND METHODS

1. TISSUES AND TISSUE HANDLING

During our systematic immunocytochemical study we employed formalin fixed, paraffin-wax embedded tissue sections, as well as frozen sections of 82 human primary childhood brain tumors. The brain tumor tissues were obtained following therapeutic surgery from the Department of Pathology at the University of Southern California. A number of brain tumor tissues were obtained at the time of neurosurgical resection and frozen in liquid nitrogen-cooled isopentane within 30 to 60 minutes after removal, followed by embedding in OCT compound (Tissue-Tek, 4583; Miles Scientific, Naperville, IL). Prior to sectioning all blocks were stored at -85 °C. Frozen sections (5-7 µm) were cut using a B1-H1 cryostat (model OTF/AS/EC/Mr-317 304H; Bright Instrument Company, Huntington, England), at -14 to -20 °C inner chamber temperature and transferred to chemically precleaned glass slides. After 24 hours drying at room 5temperature (20 to 23 °C), the tissue sections were fixed in cold (-20 °C) acetone (Fisher Scientific, Tustin, CA) for 5 to 10 minutes.

The diagnoses of the specific subtypes were established according to WHO guidelines for the classification of glioma by a clinical neurohistopathologist (1-6). WHO Histopathological Typing of Tumors by the CNS has shown progress for both the current members and possible new special types of brain tumors that may occur, especially for the meningiomas and neuro-epithelial/neuroglial type (7). The routine technique using light microscope examination was the most useful one for daily diagnosis for many years. Some immunohistochemistry techniques are needed for difficult cases, *e.g.*, GFAP, NE 14, NSE, S100, and MBP. Diagnostic problems could be caused by tissue- or cell-sampling errors, which are influenced by the tumor location itself. Thus, neurosurgeons encounter problems with biopsy intraoperatives or with the mishandling of tissues. Grading of CNS tumors must be put according to the Clinical interest for further management of the patient. CNS grading ranges from grade I (benign looking) to IV (malignant). Morphological grading is based on Kernohan and Adson (8) or Kernohan and Sayre (9).

Classification and grading of astrocytic tumors has been the subject of several controversies and no universally accepted classification system is yet available. Nevertheless, acceptance of a common system is important for assessing prognosis as well as easy comparative evaluation and interpretation of the results of multi-center therapeutic trials. Besides WHO guidelines (10), the other systems of classification are the Kernohan, Daumas-Duport (SAM-A) (11), and TESTAST-268 (12). Karak and co-workers set out to evaluate each system in order to establish which one is the most useful (13). They reported the results of a single center study on comparative survival evaluation along with assessment of inter-classification concordance in 102 cases of supratentorial astrocytic tumors in adults. Hematoxylin and eosin (H&E) stained slides of these 102 cases were reviewed independently by two pathologists and each case classified or graded according to the four different classification systems. The histological grading was then correlated with the survival curves as estimated by the Kaplan-Meier method. The most important observation was that similar survival curves were obtained for any one grade of tumor by all the four classification systems. Fifty three of the 102 cases (51.9%) showed absolute grading concordance using all 4 classifications with maximum concordant cases belonging to grades 2 and 4. Intra-classification grade-wise survival analysis revealed a statistically significant difference between grade 2 and grades 3 or 4, but no difference between grades 3 and 4 in any of the classification systems. It is apparent from the results of this study that if specified criteria related to any of the classification systems are rigorously adhered to, it will produce comparable results. The group thus recommended that one system be adhered to because it would allow for objectivity and reproducibility. Their final recommendation was for the Daumas-Duport (SAM-A) system since it appears to be the simplest, most objectivized for practical application and highly reproducible with relative ease.

Molecular aspects underlying the differences of the subtypes of astrocytic glioma can also be used to classify these tumors. Godard and co-workers demonstrated that human gliomas can be differentiated according to their gene expression (14). The researchers found that low-grade astrocytoma have the most specific and similar expression profiles while primary glioblastoma exhibit much larger variation between tumors. Secondary glioblastoma display features of both other groups. We identified several sets of genes with relatively highly correlated expression within groups that: (a). can be associated with specific biological functions; and (b). effectively differentiate tumor class. Such classification techniques allow for the development of targeted treatment strategies adapted to individual patients and allow for patient stratification (15). Moreover, using a genetic classification approach, classification success rates of up to 89% accuracy have been obtained (16).

Immunohistochemical markers are also proving to be a useful tool for characterizing medulloblastomas. Medulloblastomas occurring in children represent a histological spectrum of varying anaplasia and nodularity. In order to determine whether immunohistochemical markers might be useful parameters in subclassifying these tumors, Son and co-workers studied 17 pediatric medulloblastomas, including nine diffuse/non-anaplastic, four diffuse/anaplastic, three nodular/non-anaplastic and one nodular/anaplastic subtypes (17). The expression of neural cell adhesion molecule (NCAM), nerve growth factor receptor (NGFR), neurofilament (NF), synaptophysin (SYN), glial fibrillary acidic protein (GFAP), S100, Bcl-2, and Ki-67 was investigated by using the immunohistochemistry against specific antibodies. The study showed that NGFR, NF, GFAP and S100 were not detected in anaplastic subtypes of medulloblastomas (0/5), while non-anaplastic subtypes were mainly expressed within the nodules. All 17 tumors were reactive for NCAM, SYN and Bcl-2. In addition, Ki-67 labeling indices for anaplastic subtypes (39.0 +/- 7.42%) were significantly higher than that of non-anaplastic medulloblastomas (11.4 +/- 8.04%; $p < 0.0001$). These results suggest, according to the authors, that immunohistochemical markers are a useful adjunct in characterizing subtypes of pediatric medulloblastomas. In fact, scrutiny of the cytological variation found among medulloblastomas has recently led to the concept of the anaplastic medulloblastoma, which overlaps the large-cell variant and appears to share its poor prognosis. In contrast, the medulloblastomas with extensive nodularity, a distinctive nodular/desmoplastic variant occurring in infants, has a better outcome than most medulloblastomas in these young patients, proving that cytological and immunocytochemical data are indeed very relevant (18).

Perry conducted a similar study as Son and co-workers and found that medulloblastoma grading based on anaplasia demonstrated a statistically stronger association with patient outcome than clinical staging. Therefore, histologic grading of medulloblastomas seems warranted as a routine diagnostic aid (19). Classification of the MB histopathologically and according to profiles of molecular abnormalities will help both to rationalize approaches to therapy, increasing the cure rate and reducing long-term side-effects, and to suggest novel treatments (20).

Assessment of angiogenic potential by measuring the blood capillary density in histological sections assumes that a 4 mm thick section is representative of whole human solid neoplasm's vascularity. The number of blood capillaries was nonetheless counted at 1.0 mm^2 microscopic field (brain tumor tissue at 200x magnification), choosing microvessel hot spots on immunocytochemical slides stained by MoABs against endothelial cell markers (21).

2. LIBRARIES OF ANTIBODIES

Advances in immunohistochemistry have resulted in the characterization of hundreds upon hundreds of highly specific poly- and monoclonal antibodies recognizing only desired target antigens, which are useful in the IP assessment of normal and neoplastic cells. Monoclonal and polyclonal antibodies directed against the following antigenic epitopes were employed in our systematic immunocytochemical study:

1) neuroectodermal antigens, including synaptophysin, neuron specific nuclear protein (NSNP), chromogranin A;
2) hematopoietic markers;
3) lymphocyte differentiation CD antigens;
4) major histocompatibility complex (MHC) class I and class II molecules;
5) cytoskeletal proteins: glial fibrillary acidic protein (GFAP), high, medium and low molecular weight neurofilaments (NFs), cytokeratins, desmin, vimentin, tubulin-a and tubulin-b;
6) extracellular matrix: anti-tenascin, anti-fibronectin-Ab3;
7) cell cycle regulatory cyclins A, B1, D1, D2 and E;
8) cell proliferation markers: anti-Ki-67 antigen, anti-proliferating cell nuclear antigen (PCNA);
9) detecting oncogene related products (p53, p21, retinoblastoma (RB) gene protein);
10) markers of endothelial cell activation and proliferation, associated with neoplasm related and induced neoangiogenesis (endoglin/CD105, CD31, CD34, CD62E/ELAM-1, CD106/VCAM-1; anti-$VEGF_{121}$);
11) against apoptosis related antigens (anti-*Fas* (CD95); anti-Bcl-2, anti-Bcl-XL and anti-Bax; anti-caspase 3, anti-caspase 6, anti-caspase 8 and anti-caspase 9);
12) matrix metalloproteinases:

a) MMP-2 (Gelatinase A) Ab-4 (A-Gel VC2) from NeoMarkers, Inc. recognizing the pro and activated forms of the 72kDa MMP-2 molecule.
b) MMP-3 (Stromelysin-1) Ab-2 (Clone SL-1 IID4) from Neomarkers, Inc. recognizing both the glycosylated and unglycosylated pro and activated forms of human MMP-3.
c) MMP-9 (Gelatinase B) Ab-8 (Clone VIIC2) from NeoMarkers, Inc. recognizing the 95kDa MMP-9 molecule.
d) MMP-10 (Stromelysin-2) Ab-2 (Clone IVC5) from NeoMarkers, Inc. recognizing the pro and activated forms of MMP-10.
e) MMP-13 (Collagenase-3) Ab-2 (Clone ID3) from NeoMarkers, Inc. recognizing the glycosylated and unglycosylated pro and activated forms of MMP-13.

13) Antibodies against HOX gene products:

The anti-HOX-B3 gene product polyclonal antibody was generated against a specific peptide sequence (QSLGNAAPHAKSKEL) unique to the various epitopes of the target gene product by the Berkeley antibody company (BabCO – Catalog #PRB-215C). No antigen retrieval was required prior to application of this antibody.

The anti-HOX-B4 gene product polyclonal antibody was generated against a specific peptide sequence (ESSFQPEAGFGRRA) unique to the various epitopes of the target gene product by the Berkeley antibody company (BabCO – Catalog #PRB-220C). No antigen retrieval was required prior to application of this antibody.

The anti-HOX-C6 gene product polyclonal antibody was generated against a specific peptide sequence (EQGRTAPQDQKASIQ) unique to the various epitopes of the target gene product by the Berkeley antibody company (BabCO – Catalog #PRB-202C). No antigen retrieval was required prior to application of this antibody.

14) MoAB against Cancer-Testis Antigen (CTA): anti-MAGE-1.

15) MoAb against survivin: anti-survivin.

16) MoAB against epidermal growth factor (EGF):

Anti-c-erbB2 mouse monoclonal antibody was obtained from Labvision (Fremont, CA 94539, USA; Cat. #MS-441-R7) in a Ready-to-use for Immunohistochemical Staining form. Ab-12 is directed against the cytoplasmic domain of the human c-erbB-2 protein. Isotype: IgG1. Clone: CB11.

Anti-c-erbB3 mouse monoclonal antibody was obtained from Labvision (Fremont, CA 94539, USA; Cat. #MS-725-R7) in a Ready-to-use for Immunohistochemical Staining form. Ab-8's epitope is the extracellular domain. Isotype: IgG1/k. Clone: SGP1.

Anti-c-erbB4 mouse monoclonal antibody was obtained from Labvision (Fremont, CA 94539, USA; Cat. #MS-637-R7) in a Ready-to-use for Immunohistochemical Staining form. Ab-4's epitope is aa 1249-1264. Isotype: IgG2b. Clone: HFR-1.

A well-chosen library of monoclonal antibodies (MoABs) was employed to observe the desired antigens reflecting the actual maturation level and the direction of differentiation pathways of thymocytes and T lymphocytes and to identify the IP of stromal RE, DCs and accessory cells of the thymic microenvironment and the IP of the tumor infiltrating host's immune effector cells within the cellular microenvironment of brain tumors. In this way we were able to characterize the professional, specialized for antigen presentation intrathymic cell types and the IP changes during maturation of thymocytes into immunocompetent T lymphocytes.

The following antibody library was employed in our immunocytochemical observations on human and other vertebrate thymuses, brain tumors and bone marrows:

1) anti-endocrine and anti-epithelial: A_2B_5, Thy-1, and anti-epithelial membrane antigen (EMA);
2) hematopoietic: CLA (anti-HLe-1 directed against a 200-220 kD cell surface receptor), Leu-2/a (against $CD8^+$, cytotoxic/suppressor T-lymphocytes), Leu-3/a,3/b (for identification of $CD4^+$, helper lymphocytes), anti-CD34, Leu-7, Leu-11/b, Leu-14, Leu-19, anti-Interleukin-2 receptor (IL-2R), and Leu-M5 (for monocytes/macrophages);
3) Anti-MHC molecules: HLA-A,B,C, HLA-DR, and HLA-DP;
4) Anti-Thymic Stromal Cell antigens: TE-3, TE-4, TE-7, TE-8, TE 15, TE 16, and TE 19 (developed in the laboratory of Dr. Haynes and provided to us);
5) Intermediate Filaments: anti-GFAP, anti-vimentin, anti-desmin, anti-NF(H), anti-NF(M), anti-NF(L), S-100 protein, anti-MAP-1 (microtubules), anti-MAP-2, anti-MAP-5, anti-cytokeratins 13 and 18, desmosomal cytokeratin, AE1, AE2, AE3, AE5, and AE8;
6) Tumor markers: B18.7.7 and D14 [anti-carcinoembryonic antigen (CEA)];
7) Oncogenes and related protein products: anti-c-erbB-2, anti-c-erbB-3, anti-c-erbB-4, anti-c-myc, and anti-c-ras;
8) Cell cycle related cyclins, proliferation markers, and DNA repair related proteins: anti-p53, Ki-67, anti-$p34^{cdc2}$, anti-cyclin A, and anti-cyclin D; and
9) Growth factors: anti-TGF-β type II receptor (TGF-βIIR).

3. ANTIGEN RETRIEVAL TECHNIQUE

During our immunocytochemical observations of brain tumor IP, we employed the immunohistochemical method of “antigen liberation” or “antigen retrieval” (22-27). In the first step, antigen retrieval was sometimes achieved by single or combined enzymatic digestion (ficin, pepsin, and trypsin from Zymed Labs., South San Francisco, CA, USA) prior to the primary antigen-antibody reaction. Heat induced epitope retrieval (HIER) (28, 29), as modified by us, was also employed. Antigen retrieval required immersion of tissue sections, in a Target Retrieval Solution (DAKO Corp., Carpinteria, CA, USA) and heating in a water bath (95°C to 99°C). No single antigen retrieval solution works best with all antigenic epitopes. Our method with citrate-based (neutral pH) solution worked well with most antibodies applied in this study that required such pretreatment. Heat (microwave) targeted antigen retrieval resulted in an increase of antigen detection for a number of MoABs, and we also noticed a serious increase of immunoreactivity (*i.e.* staining intensity).

4. IMMUNOALKALINE PHOSPHATASE ANTIGEN DETECTION TECHNIQUE

We used the following immunoalkaline phosphatase cytochemical method, modified by us for antigen detection in formalin fixed, paraffin-wax embedded thymus tissues. The technique has been determined to be a highly sensitive, indirect, four to six step immunocytochemical method, which combines the biotin-streptavidin based ABC-method with enzyme-linked (alkaline phosphatase) immunohistochemistry. Briefly, following deparaffinization in three changes of Xylene substitute (Shandon-Lipshaw, Pittsburgh, PA, USA) for 20 to 30 minutes, rehydration was carried out employing descending dilutions of alcohol (100% to 50%) to TBS. An initial blocking step using 1% glacial acetic acid mixed with the working buffer for 10 minutes was necessary to eliminate the endogenous AP activity from the tissues. Use of levamisole solution is also described in our earlier observations. As we explained earlier in our papers, GAA inhibition was preferred because of the possible presence of levamisole-resistant AP iso-enzyme. The second blocking step was conducted with a purified mixture of proteins (Shandon-Lipshaw) from various species for 5-10 minutes to block cross-reactive antigenic epitopes. Excess serum was removed from the area surrounding the sections! The tissue sections were then incubated for 90-120 minutes with the particular primary antibody. Next, incubation with the secondary antibody, which was either a biotinylated, whole goat anti-rabbit or goat anti-mouse IgG molecule (IgG molecule diluted by ICN Biomedicals, Inc., Aurora, OH, USA) was carried out for 20 minutes. Streptavidin conjugation was accomplished by incubation with AP conjugated streptavidin for 20 minutes. Color visualization of the primary antigen-antibody (Ag-Ab) reaction was accomplished with an alkaline phosphatase (AP) kit I (Vector Laboratories, Burlingame, CA, USA) which contains AS-TR with Tris-HCl buffer at pH 8.2, added for 28-40 minutes to allow formation of a stable red precipitate. Sections were counterstained with a diluted solution of Gill's hematoxylin (Richard-Allan, Kalamazoo, MI, USA). The tissue slides were then dehydrated in ascending concentrations of alcohol (60% to 100%) to xylene substitute (Shandon-Lipshaw), in which they were kept overnight to ensure complete morphological clearing. The stained tissue sections were mounted using a solution specially designed for use following morphological clearing in xylene substitute (Shandon-Lipshaw).

5. IMMUNOPEROXIDASE ANTIGEN DETECTION TECHNIQUE

The following ABC method was employed for the immunocytochemical detection of antigens in thymus specimens. Briefly, following deparaffinization in three fresh changes of xylene substitute (Shandon-Lipshaw), a tissue rehydration employing descending grades of alcohols (100% to 40%) to PBS was performed. The tissue sections were never allowed to dry-out before being moved to the next solution! After 30 minutes incubation with 0.6% H_2O_2 in methanol to block the endogenous H_2O_2, a good rinse with running tap water and PBS for 2-8 minutes was performed. The second blocking step required 5 minutes incubation in non-specific protein mixture solution (Shandon-Lipshaw). The incubation time with the primary MoABs, as with any primary antibody, depended on the developers instructions, but it was usually between 60 to 120 minutes at room temperature. Next, we applied the secondary antibody for 20 minutes (on paraffin-wax sections, the whole anti-rabbit or anti-mouse IgG antibody was used as secondary antibody directed against the IgG of the original species of the primary antibody), followed by incubation for 30-40 minutes in 8 to 10 drops of the streptavidin/biotinylated horseradish peroxidase H complex (ABC). The binding of the biotinylated antibody to streptavidin/peroxidase complexes occurs with an extremely high affinity (10^{-19} M) (BioWhitaker, Inc., Walkersville, MD, USA). Color visualization of the primary Ag-Ab reaction was accomplished using diamino-benzidine (DAB) or amino-ethyl-carbazol (AEC). The brown color was enhanced with copper sulfate and diluted 1:100 Gill's hematoxylin (Richard-Allan) was used for gentle nuclear counterstaining. After the employment of DAB solution and the appropriate counterstain, the tissue slides were dehydrated, cleared, and mounted in a manner identical to that described above.

6. CONTROLS IN IMMUNOCYTOCHEMISTRY

To accurately assess the specificity of the T lymphocyte differentiation antigens and MHC class I and II antigens observed in this study, we investigated the immunoreactivity of several normal human tissues: adult and fetal thymic, tonsil, spleen, thyroid, lung, liver, kidney, heart, pancreas, ovary, prostate, small intestine, large intestine and brain tissue sections, all included in one multitissue block (DAKO Corporation, Lot: 5935B) (30, 31). Additional controls for all employed MoABs included:

1) omission of the primary MoAB;
2) using only the enzymatic developer solution to detect the presence of endogenous peroxidase or alkaline phosphatase activity; and
3) use of MOPC 21 mouse myeloma IgG_1 (ICN Biomedicals, Inc.) as a replacement for the primary MoAB to determine non-specific myeloma protein binding to the antigen epitopes of the screened tissues.

7. EVALUATION OF THE IMMUNOREACTIVITY (IMMUNOSTAINING)

Qualitative and quantitative evaluation of the percent of antigen positive cells and the intensity of immunostaining were conducted using a light microscope (Olympus America, Inc., Melville, NY, USA) counting 100-200 cells from each of five to eight distinct areas in non-necrotic thymic and positive and negative control tissues. Artifacts were avoided, while, on the other hand, morphologically characteristic areas were sought out.

Quantitative evaluation: **(++++)** over 90% of the total cell number are positive; **(+++)** 50% to 90% of the total cell number are positive; **(++)** 10% to 50% of the total cell number are positive; **(+)** 1% to 10% of the total cell number are positive; **(±)** under 1% of the total cell number are positive; **(-)** negative.

Qualitative evaluation: **(A)** very intense red/brown staining; **(B)** strong red/brown staining; **(C)** light red/brown staining; **(D)** negative staining.

8. TISSUE PROCESSING FOR TISSUE CULTURE EXPERIMENTS

The *in vitro* protocols described below were conducted using human postnatal thymic tissues obtained at the time of corrective cardiovascular surgery from 76 patients between the ages of 6 months and 10 years. The tissue samples were stored in RPMI 1640 (Gibco Labs., Grand Island, NY) medium under sterile conditions. Thymic tissues were turned into cell suspensions for *ex vivo* IP studies.

9. PREPARATION OF THYMIC, PERIPHERAL BLOOD AND BONE MARROW CELL SUSPENSIONS

To minimize cellular autolytic changes the fresh thymic tissue from the RPMI 1640 medium was placed in phosphate buffered saline (PBS, Gibco) within 10-30 minutes of being received and the experimental protocol was carried out. The thymus from each patient was pooled and dissociated after removal of the capsular and pericapsular connective tissue, fat and debris. The cleaned tissue pieces were minced with cross blades until no gross pieces remained. The PBS diluted cell suspension contained $2\text{-}5\times10^9$ cells/ml of undifferentiated cortical thymocytes, individual or clusters of RE cells, TNCs and other thymic cell complexes, Hassall's bodies (HBs), non-RE thymic stromal cells (DCs such as LCs and IDCs), and other accessory cells, such as macrophages and fibroblasts.

10. ISOLATION OF CORTICAL THYMOCYTES

The thymic cell suspensions contained a large number of cells that are easily, dissociable from the cortex by mechanical means. These immature cortical thymocytes, were in various stages of T lymphocyte differentiation. To use the conventional Ficoll-Hypaque (Ficoll-Paque, Pharmacia Fine Chemicals, Piscataway, NJ) density gradient isolation, the original thymocyte suspension was diluted at least 10 times. After isolation, a portion of the thymocytes, peripheral blood or bone marrow cells ($2\text{-}5\times10^5$ cells/ml) was suspended in RPMI 1640 medium and treated with Tris-buffered ammonium chloride (Sigma, St. Louis, MI) to lyse the contaminating erythrocytes. Following this procedure, they were cultured using 24 well, flat bottomed tissue culture plates (Falcon Labware, Division of Becton-Dickinson, Oxnard, CA) and RPMI 1640 medium, enriched with the desired combinations of growth factors, mitogens, colony stimulating factors and the appropriate cytokines, depending on the experimental aims. After 3 days in culture the thymocytes were transferred into T25 culture flasks (Falcon). When one µl rIL-2 (generously provided by Cetus Corporation, Emeryville, CA) was added to the culture, the proliferation rate remained high for the first 2-3 *in vitro* weeks. After this period the thymocytes were restimulated every second week, using the combined action of a mitogen (one µg/ml PHA) (Wellcome Diagnostics, Dartford, DA1 5AH, England) and a cellular microenvironment rich in cell secreted growth factors, and antigens provided by irradiated (3.6-4 Gy) peripheral mononuclear cells (PMNC) as "feeder cells", obtained from blood donor's "buffy coats" (target:feeder cell ratio 1:5). The thymocyte cultures were examined daily, the cell numbers counted and optimal *in vitro* conditions were maintained. The presence and number of proliferating thymoblasts (lymphoblasts) was carefully recorded. Prior to culture enzymatic thymic tissue digestion, we were unable to induce thymocyte proliferation by using any combination of mitogens, RE cell supernatant and/or rIL-2.

The biological activity of rIL-2 was measured prior to the *in vitro* experiments using a 24 hour bioassay on IL-2 dependent human splenic lymphocytes and one unit was defined as the amount inducing 50% of maximal proliferation. In thymocyte cultures and proliferation assays IL-2, IL-4 and IL-13 was used at a concentration of 5, 2, 1 and 0.5 units per $2x10^5$ cells.

11. ISOLATION OF THYMIC NURSE CELLS (TNCS)

11.1 Gravity sedimentation

Since TNCs, as lymphoepithelial cell complexes, are by many times larger than any other kind of thymic cells, employing a simple four step unit gravity sedimentation procedure *via* fetal calf or adult human serum (both fom Gibco Labs., Chagrin Falls, OH) the TNC were easily isolated from the thymic cell suspension.

11.2 Enzymatic tissue digestion

The thymic and another tissue homogenate was transferred (2-3 mm^2 tissue pieces) into a trypsinization bottle, containing 50 ml cold (4°C) PBS and agitated for 10 min. The supernatant contained a large number of cortical thymocytes and was discarded or used for thymocyte cultures. The sedimented fraction was resuspended in 50 ml prewarmed solution of 0.25% trypsin-ethylenediamine tetraacetic acid (EDTA) (Gibco), 0.5 mg/ml dispase/collagenase (Boehringer/Mannheim Biochemical, Indianapolis, IN) or 0.5 mg/ml collagenase IV (Millipore Corporation, Bedford, MA) and 2 ml DNAase (Boehringer/Mannheim). The enzymatic digestion was carried out at 37°C in an incubator for 20 min. This procedure was repeated 3-4 times until complete disintegration of the typical thymic tissue structure was accomplished. After each time, the collection of dissociated cells was possible. This cell suspension, containing RE cells, TNCs and HBs was pooled in 3 ml cold (4°C) D-MEM, supplemented with 5% heat inactivated human serum (AB blood group, Rh negative), 1% l-glutamine (Boehringer/Mannheim), 1% MEM (Gibco), 1% sodium pyruvate (Flow Labs., Mclean, VI), 1% hepes buffer (Biofluids Inc., Rockville, VA) and 0.05 mg/ml gentamycin sulfate (Sigma). The suspension was gently layered over 10 ml undiluted, heat inactivated human serum in a 40 ml conical, glass centrifuge tube for the TNC enriching procedure by sequential sedimentation at unit gravity for 15 min. The collected sediment was resuspended in 0.7 l complete D-MEM and overlayed over 5 ml human serum for second sedimentation in 12 ml plastic tube (Falcon) for an additional 10 min. This last step was repeated and resulted in an enriched to 80% presence of TNCs. The enclosed TNC-Thy were released by mild sonication or simple overnight incubation.

12. THYMIC STROMAL CELL (RE & DC) CULTURES

1. Whole organ cultures: 2-3 mm^2 pieces of thymic tissue were put *in toto* in organ culture, without previous enzymatic tissue digestion, using enriched D-MEM medium. The thymocytes disappeared from the culture within 3 weeks if the medium was not supplemented with rIL-2. The medium was replenished every 2-3 days depending on cell count and pH. The tissue cultures were grown in an incubator at 37°C with a gas mixture of 5% carbon dioxide, 5% oxygen and 90% nitrogen.
2. The loose thymic tissue fragments remaining after enzymatic digestion were also placed in tissue cultures using enriched D-MEM medium and the same *in vitro* conditions as mentioned above.
3. The isolated TNCs, following morphological examination and release of the enveloped TNC-Thy, were also cultured.
4. During enzymatic tissue digestion, other thymic cellular elements were also isolated. HBs, DCs and stromal accessory cells were cultured following microscopic evaluation. These cultures were used as controls

to detect optimal *in vitro* conditions to obtain pure RE cell monolayers. The thymic RE cell cultures were highly susceptible to overgrowth by always present thymic fibroblasts. Repeated treatment with 0.02% EDTA (Pharmacia) diluted in PBS resulted in pure RE monolayers, expressing cytoskeletal proteins of an epithelial nature.

Culture supernatants were collected daily for one week and used to culture autologous thymocytes immediately or stored at -70°C prior to use. The cultures were observed daily, with a tissue culture microscope (Leitz, FRG) and representative fields were photographed.

13. PROLIFERATION ASSAY (PA) FOR THYMOCYTES AND PERIPHERAL BLOOD HEMATOPOIETIC CELLS

As was mentioned above, 2-5 x 10^5/ml cortical thymocytes or enriched peripheral blood $CD34^+$ were cultured in RPMI 1640 medium, supplemented with various combinations of growth factors (including T lymphocyte growth factor), mitogens, a variety of colony stimulation factors and cytokines. In some experiments RE cell culture supernatant was added. For PA 96 well (Falcon), flat bottomed culture plates were used. 0.2 ml thymocyte triplicates from every experimental condition, were pulsed with 1μCi/well of 3H-TdR (New England Nuclear, Boston, MA). The assay required use of a tissue culture incubator at 37°C in humidified air, containing 5% CO_2, for 4 hours. Cell harvesting onto glass fiber filters was performed using MASH harvesting equipment. The amount of 3H-TdR incorporation within the thymocyte DNA was quantitated using by a β scintillation counter.

14. TRANSMISSION ELECTRONMICROSCOPY (TEM) OF CULTURED THYMIC MEDULLARY CELLS (RE, DC, INCLUDING LC, & IDC) AND MACROPHAGES

After a short time *in vitro* stay, cultured DCs, RE cells and macrophages are attached to the culture dish surface, forming a monolayer. They were removed using gentle digestion with collagenase IV (Millipore Corp., Bedford, MA) for 5 minutes at 37°C, followed by pelleting at 600-800 rpm for an additional 5 minutes. The cell pellets were fixed in 3% glutaraldehyde

(Fischer Scientific), diluted in 0.1 M Sorenson's buffer at pH 7.2 for one hour at room temperature (20-23°C), followed by three washes in phosphate buffer. Postfixation was performed with 1% solution of osmium tetroxide for 30 minutes, followed by washing in 0.9% NaCl for another 30 minutes. Staining "*en bloc*" with 0.5% uranyl magnesium acetate, diluted in 0.9% NaCl for one hour in dark, at 4°C, followed by washing in saline for an additional 30 minutes. After dehydration in ascending concentrations of ethanol, the tissue microcultures were embedded in Epox (E.F. Fullam, Schenectady, NY). Ultrathin sections were stained secondarily with uranyl acetate and lead citrate and were examined and photographed under a Siemens Elmiskop IA, transmission electronmicroscope at 80 kV. The cell types were identified by their TEM nuclear and cytoplasmic characteristics.

15. SCANNING ELECTRONMICROSCOPIC (SEM) PROCEDURE FOR TISSUE SAMPLES

Small portions (1-2 mm^2) of thymic tissue were fixed for 2-4 hours in 2.5-4% glutaraldehyde diluted in 0.1M phosphate buffer at room temperature, and at 4°C in a regular refrigerator. Postfixation was carried out employing 1% OsO_4 diluted in phosphate buffer (kept in dark). After three washes in phosphate buffer, dehydration was carried out in ascending dilutions of ethanol. After processing in the critical point dryer (CPD), sputter coating was performed. Scope and photography of the thymic tissues was carried out with JEOL-JSM-35C Scanning Electronmicroscope (SEM) at 380 to 60,000 times magnifications.

REFERENCES

1. Zulch KJ, Wechsler W: Pathology and classification of gliomas. Progr Neurol Surg, vol. 2, Karger, Basel-New York, pp 1-84, 1968.
2. Zulch KJ: Histologic classification of tumours of the central nervous system. International Histological Classification of Tumours, No. 21, World Health Organization, Geneva, 1979.
3. Zulch KJ: Principles of the new World Health Organization (WHO) classification of brain tumors. Neuroradiology *19:* 59-66, 1980.
4. Szymas J: Histologic classification of central nervous system tumors by the World Health Organization. Pol J Pathol *45:* 81-91, 1994.
5. Szymas J: Diagnostic immunohistochemistry of tumors of the central nervous system. Folia Neuropathol *32:* 209-214, 1994.
6. Kleihues P, Louis DN, Scheithauer BW, Rorke LB, Reifenberger G, Burger PC, Cavenee WK: The WHO classification of tumors of the nervous system. J Neuropathol Exp Neurol. 2002 *61:* 215-225; discussion 226-9, 2002.

7. Soetrisno E, Tjahjadi G: Pathological aspects of brain tumors. Gan To Kagaku Ryoho *27s(2):* 274-278, 2000.
8. Kernohan JW, Mabon RF, Svien HJ, et al: A simplified classification of gliomas. Proc Staff Meet Mayo Clin *24:* 71-75, 1949.
9. Kernohan JW, Sayre GP: Tumors of the central nervous system. In: Atlas of Tumor Pathology, Section 10, Vols 35 & 37. Washington DC. Armed Forces Institute of Pathology, 1952.
10. Kleihues P, Burger PC, Scheithauer BW: Histological typing of tumors of the central nervous system. New York: Springer-Verlag, 1993.
11. Daumas-Duport C, Scheithauer B, O'Fallon J, Kelly P: Grading of astrocytomas. A simple and reproducible method. Cancer *62:* 2152-2165, 1988.
12. Schmitt HP, Oberwittler C: Computer-aided classification of malignancy in astrocytomas: II. The value of categorically evaluated histologic and non-histologic features for a numerical classifier. Analytical Cell Pathol *4:* 409-419, 1992.
13. Karak AK, Singh R, Tandon PN, Sarkar C: A comparative survival evaluation and assessment of interclassification concordance in adult supratentorial astrocytic tumors. Pathol Oncol Res *6:* 46-52, 2000
14. Godard S, Getz G, Delorenzi M, Farmer P, Kobayashi H, Desbaillets I, Nozaki M, Diserens AC, Hamou MF, Dietrich PY, Regli L, Janzer RC, Bucher P, Stupp R, de Tribolet N, Domany E, Hegi ME: Classification of human astrocytic gliomas on the basis of gene expression: a correlated group of genes with angiogenic activity emerges as a strong predictor of subtypes. Cancer Res *63:* 6613-6625, 2003.
15. Shai R, Shi T, Kremen TJ, Horvath S, Liau LM, Cloughesy TF, Mischel PS, Nelson SF: Gene expression profiling identifies molecular subtypes of gliomas. Oncogene *22:* 4918-4923, 2003.
16. Steiner G, Shaw A, Choo-Smith LP, Abuid MH, Schackert G, Sobottka S, Steller W, Salzer R, Mantsch HH: Distinguishing and grading human gliomas by IR spectroscopy. Biopolymers *72:* 464-471, 2003.
17. Son EI, Kim IM, Kim DW, Yim MB, Kang YN, Lee SS, Kwon KY, Suh SI, Kwon TK, Lee JJ, Kim DS, Kim SP: Immunohistochemical analysis for histopathological subtypes in pediatric medulloblastomas. Pathol Int *53:* 67-73, 2003.
18. Ellison D: Classifying the medulloblastoma: insights from morphology and molecular genetics. Neuropathol Appl Neurobiol *28:* 257-82, 2002.
19. Perry A: Medulloblastomas with favorable versus unfavorable histology: how many small blue cell tumor types are there in the brain? Adv Anat Pathol *9:* 345-50, 2002.
20. Ellison D: Classifying the medulloblastoma: insights from morphology and molecular genetics. Neuropathol Appl Neurobiol *28:* 257-82, 2002.
21. Rojiani AM, Dorovini-Zis K: Glomeruloid vascular structures in glioblastoma multiforme: an immunohistochemical and ultrastructural study. J Neurosurg *85:* 1078-1084, 1996.
22. Shi S-R, Key ME, Kalra KL: Antigen retrieval in formalin fixed, paraffin-embedded tissues: an enhancement method for immunohistochemical staining based on microwave oven heating of tissue sections. J Histochem Cytochem *39:* 741-748, 1991.
23. Shi S-R, Cote C, Kalra KL, Taylor CR, Tandon AK: A technique for retrieving antigens in formalin-fixed, routinely acid-decalcified, celloidin-embedded human temporal bone sections for immunohistochemistry. J Histochem Cytochem *40:* 787-792, 1992.
24. Key ME, Shi S-R, Kalra KL: Antigen retrieval in formalin fixed tissues using microwave energy. US Patent # 5,244,787; 14 September 1993.
25. Beckstead JH: Improved antigen retrieval in formalin-fixed, paraffin-embedded tissues. Appl Immunohistochem *2:* 274-281, 1994.

26. Cuevas EC, Bateman AC, Wilkins BS, Johnson PA, Williams JH, Lee AH, Jones DB, Wright DH: Microwave antigen retrieval in immunocytochemistry: a study of 80 antibodies. J Clin Pathol *47:* 448-452, 1994.
27. Chen T, Zhang D, Shi S-R, Taylor CR: Antigen retrieval pretreatment: microwaving procedures and alternative heating methods. BioLink *4:* 2, 1995.
28. Shi SR, Cote RJ, and Taylor CR: Antigen retrieval techniques: current perspectives. J Histochem Cytochem *49:* 931-937, 2001.
29. Shi SR, Cote RJ, and Taylor CR: Antigen retrieval immunohistochemistry: past, present and future. J Histochem Cytochem *45:* 327-343, 1997.
30. Battifora H: The multitumor (sausage) tissue block: novel method for immunohistochemical antibody testing. Lab Invest *55:* 244-248, 1986.
31. Battifora H and Mehta P: The checkerboard tissue block. An improved multitissue control block. Lab Invest *63:* 722-724, 1990.

INDEX

5

A

B

C

D

E

F

G

H

I

K

L

M

N

O

P

R

S

T

V

W

MIX
Papier aus verantwortungsvollen Quellen
Paper from responsible sources
FSC® C105338

If you have any concerns about our products,
you can contact us on
ProductSafety@springernature.com

In case Publisher is established outside the EU,
the EU authorized representative is:
Springer Nature Customer Service Center GmbH
Europaplatz 3, 69115 Heidelberg, Germany

Printed by Libri Plureos GmbH
in Hamburg, Germany